Cell Signaling Reactions

Yasushi Sako • Masahiro Ueda
Editors

Cell Signaling Reactions

Single-Molecular Kinetic Analysis

Springer

Editor
Yasushi Sako
Cellular Informatics Laboratory
RIKEN Advanced Science Institute
Wako, Japan
sako@riken.jp

Masahiro Ueda
Graduate School of Frontier Biosciences
Osaka University, and JST, CREST
Suita, Japan
ueda@phys1.med.osaka-u.ac.jp

ISBN 978-90-481-9863-4 e-ISBN 978-90-481-9864-1
DOI 10.1007/978-90-481-9864-1
Springer Dordrecht Heidelberg London New York

Library of Congress Control Number: 2010937431

Printed on acid-free paper

Springer is part of Springer Science+Business Media (www.springer.com)

Preface

The biological cell, the minimal unit of life, is an extremely complicated reaction web. The human genome project has revealed that 20,000–30,000 genes are encoded in single human cells; these genes are thought to produce more than 100,000 protein species through alternative splicing and chemical modification. The major challenge of biology in the post-genomic era is to address the issue of how such a multi-element system, composed of huge numbers of protein species and other macro- and micro-molecules, brings emergence of the complex and flexible reaction dynamics that we call "life."

Biological macromolecules such as proteins are themselves complicated systems made up of a huge number of atoms. Proteins often show complex structural and functional dynamics. It has been demonstrated that single-molecule techniques are powerful tools in the study of proteins, because time series of the individual events carried out by a single molecule provide information that cannot be obtained with ensemble-molecule measurements and that is indispensable in analyses of the complex behaviors of biological macromolecules. Single-molecule measurements have recently been extended to the study of multi-molecular systems and even living cells. Because these single-molecule techniques are so effective in resolving the complex reactions of individual molecules, they are now expected to offer a powerful technology for the study of the complicated reaction web in living cells.

This book deals with single-molecule analyses of the kinetics and dynamics of cell signaling reactions. Several other books have already introduced the techniques and applications of single-molecule measurements of various biological events. However, as far as we know, this book is the first to concentrate on cell signaling. Analysis of the cell signaling that regulates the complex behaviors of cells should provide the keys required to understand the emergence of life. We intend this book to contain as many kinetic analyses of cell signaling as possible. Although the single-molecule kinetic analysis of cellular systems is a young field compared with the analysis of single-molecule movements in cells, this type of analysis is important because it directly relates to the molecular functions that control cellular behavior. Because there have been many successful single-molecule kinetic studies of purified proteins, future single-molecule kinetic analysis will be largely directed towards cellular systems.

In this book, we have included not only the results of single-molecule analyses of cell signaling in both living cells and in vitro systems, but also recent progress in the single-molecule technology required to study cell signaling and theories of single-molecule data processing. We would like to thank all the contributors to this volume for preparing these valuable manuscripts, despite busy schedules. We hope that the book is useful to a wide range of readers interested in cell signaling and single-molecule measurements. We would be delighted if this book advances our understanding of complex life systems.

Yasushi Sako
Masahiro Ueda

Contents

Contributors

Terrence M. Dobrowsky
Department of Chemical and Biomolecular Engineering,
The Johns Hopkins University

Kayo Hibino
Cellular Informatics Laboratory, RIKEN Advanced Science Institute

Minako Hirano
Graduate School of Frontier Biosciences, Osaka University

Michio Hiroshima
Cellular Informatics Laboratory, RIKEN Advanced Science Institute

Takehiko Ichikawa
Laboratory of Spatiotemporal Regulations,
National Institute for Basic Biology

Toru Ide
Graduate School of Frontier Biosciences, Osaka University

Carey K. Johnson
Department of Chemistry, University of Kansas

Tamiki Komatsuzaki
Molecule & Life Nonlinear Sciences Laboratory,
Research Institute for Electronic Science, Hokkaido University and
Core Research for Evolutional Science and
Technology (CREST), Japan Science and Technology Agency (JST)

Chun Biu Li
Molecule & Life Nonlinear Sciences Laboratory,
Research Institute for Electronic Science, Hokkaido University

Mangala R. Liyanage
Department of Chemistry, University of Kansas

Satomi Matsuoka
Graduate School of Frontier Biosciences, Osaka University, and JST, CREST

Yukihiro Miyanaga
Graduate School of Frontier Biosciences, Osaka University, and JST, CREST

Hiroyuki Noji
The Institute of Scientific and Industrial Research, Osaka University

Shigetoshi Oiki
Department of Molecular Physiology and Biophysics,
University of Fukui Faculty of Medical Sciences

Kenneth D. Osborn
Department of Chemistry, University of Kansas
Department of Math and Science, Fort Scott Community College

Yasushi Sako
Cellular Informatics Laboratory, RIKEN Advanced Science Institute

Ken Sato
Department of Life Sciences, Graduate School of Arts and Sciences,
University of Tokyo

Marcel J. M. Schaaf
Molecular Cell Biology, Institute of Biology, Leiden University

Thomas S. Schmidt
Physics of Life Processes, Institute of Physics, Leiden University

Tatsuo Shibata
Center for developmental Biology, RIKEN, and JST, CREST

Kazuhito V. Tabata
The Institute of Scientific and Industrial Research, Osaka University

Masahiro Ueda
Graduate School of Frontier Biosciences, Osaka University, and JST, CREST

Denis Wirtz
Department of Chemical and Biomolecular Engineering and
Physical Science Oncology Center, The Johns Hopkins University

Haw Yang
Department of Chemistry, Princeton University

Asma Zaidi
Department of Biochemistry, Kansas City University of Medicine
and Biosciences

Chapter 1
Single-Molecule Kinetic Analysis of Receptor Protein Tyrosine Kinases

Michio Hiroshima and Yasushi Sako

Abstract Signaling pathways mediated by receptor tyrosine kinases (RTKs) are among the most important pathways regulating various functions and behaviors in mammalian cells. Although many studies performed over several decades have revealed the molecular mechanisms underlying the cellular events regulated by these pathways, the overall structures of the pathways remain unclear, especially their quantitative properties. A technology has emerged that can potentially address these issues. Recent developments in optical microscopy and molecular biology allow us to visualize the behaviors of single RTK molecules and their association partners with fluorescent probes in living cells. Using the quantitative nature of these single-molecule measurements, we studied the signaling of epidermal growth factor (EGF) and nerve growth factor (NGF), both of which stimulate RTK systems. Single-molecule analyses revealed molecular dynamics and kinetics that cannot be demonstrated with conventional biochemical methods. These include the kinetic transitions of these receptors induced by ligand binding, signal amplification by the dynamic interactions between active and inactive receptors, downstream signaling with a memory effect exerted by the receptor molecule, and shifts in the motional modes of ligand-receptor complexes. These novel insights obtained from single-molecule studies suggest that detailed models of RTK signaling, which involve signal processing depend on protein dynamics.

Keywords Adaptor protein · Allosteric conformational change · Association kinetics · Association rate constant · Calcium signaling · Clustering · Cluster size distribution · Diffusion coefficient · Dimerization · Dissociation constant · Dissociation kinetics · Dorsal root ganglion: DRG · Epidermal growth factor: EGF · Epidermal growth factor receptor: EGFR · ErbB family · Fluctuation · Fluorescence resonance

---

M. Hiroshima (✉) and Y. Sako
Cellular Informatics Laboratory, Advance Science Institute,
RIKEN Hirosawa 2-1, Wako, Saitama 351-0198, Japan
e-mail: m_hiroshima@riken.jp; sako@riken.jp

Y. Sako and M. Ueda (eds.), *Cell Signaling Reactions: Single-Molecular Kinetic Analysis*,
DOI 10.1007/978-90-481-9864-1_1, © Springer Science+Business Media B.V. 2011

energy transfer: FRET · Green fluorescent protein: GFP · Growth cone · Growth-factor-receptor-bound protein 2: Grb2 · Hill factor · Immobile phase · Kinetic intermediate · Lateral diffusion · Memory · Mobile phase · Multiple exponential function · Multiple-state reaction · Negative concentration dependence · Nerve growth factor: NGF · Neurotrophic tyrosine kinase receptor 1: NTRK1 · Noise · Oblique illumination · Off-time · Oligomer · On-time · Plasma membrane · Phosphorylation · Phosphotyrosine · Predimer · RAF · Ras · Ras-MAPK system · Reaction rate constant · Receptor tyrosine kinases: RTKs · Response probability · Retrograde flow · RTK systems · Semi-intact cell · Signal amplification · Signal transduction · Single-molecule imaging · Src homology 2 (SH2) domain · Stretched exponential function · Sub-state · Super-resolution · Switch-like · Total internal reflection: TIR · Total internal reflection fluorescence: TIRF · TrkA · Ultrasensitive response · Velocity · Waiting time

1.1 RTK Systems

Receptor protein tyrosine kinases (RTKs) form a large superfamily of receptor molecules on the plasma membranes of eukaryotic cells [71]. A typical member of the RTKs is a single-membrane-spanning protein consisting of an extracellular ligand-binding domain, a short membrane-spanning α helix, and a cytoplasmic domain with tyrosine kinase activity. Upon its association with a ligand, the kinase activity of the RTK is stimulated and several tyrosine residues are phosphorylated in the cytoplasmic domain of the RTK. These tyrosine phosphorylations are critical for the signal transduction activity of RTKs because the phosphotyrosine residues provide scaffolds for various cytoplasmic proteins involved in signaling to downstream reactions. One of the major cell signaling networks downstream from RTKs is the Ras-MAPK system (Fig. 1.1a). This signaling system is responsible for decisions regarding cell fates, such as proliferation, differentiation, apoptosis, and even carcinogenesis. Intracellular calcium signaling, cell movement, and morphological changes in cells are also stimulated by these systems during the processes of cell fate decision. Therefore, the RTK-Ras-MAPK systems play critical roles in various cellular activities.

This chapter deals with the single-molecule analysis of subsystems of the RTK-Ras-MAPK systems, which we call "RTK systems" (Fig. 1.1b). The RTK systems consist of extracellular ligands, the plasma membrane receptor RTKs, and cytoplasmic proteins containing the Src homology 2 (SH2) and/or phosphotyrosine-binding (PTB) domains, which recognize the phosphotyrosines on the activated forms of RTKs. In this chapter, two types of RTKs are featured: the epidermal growth factor (EGF) receptor (EGFR) and the TrkA nerve growth factor (NGF) receptor. The activation of EGFR is responsible for proliferation, morphological changes, chemotactic movement, and carcinogenesis in almost all types of mammalian cells, except blood cells. Signals from NGF induce the differentiation, neurite elongation, and survival of peripheral nerve cells. NGF has two types of membrane receptors, TrkA and p75. Only TrkA belongs to the RTK superfamily. Single-molecule analysis of the ligand-RTK interaction, the dynamics and

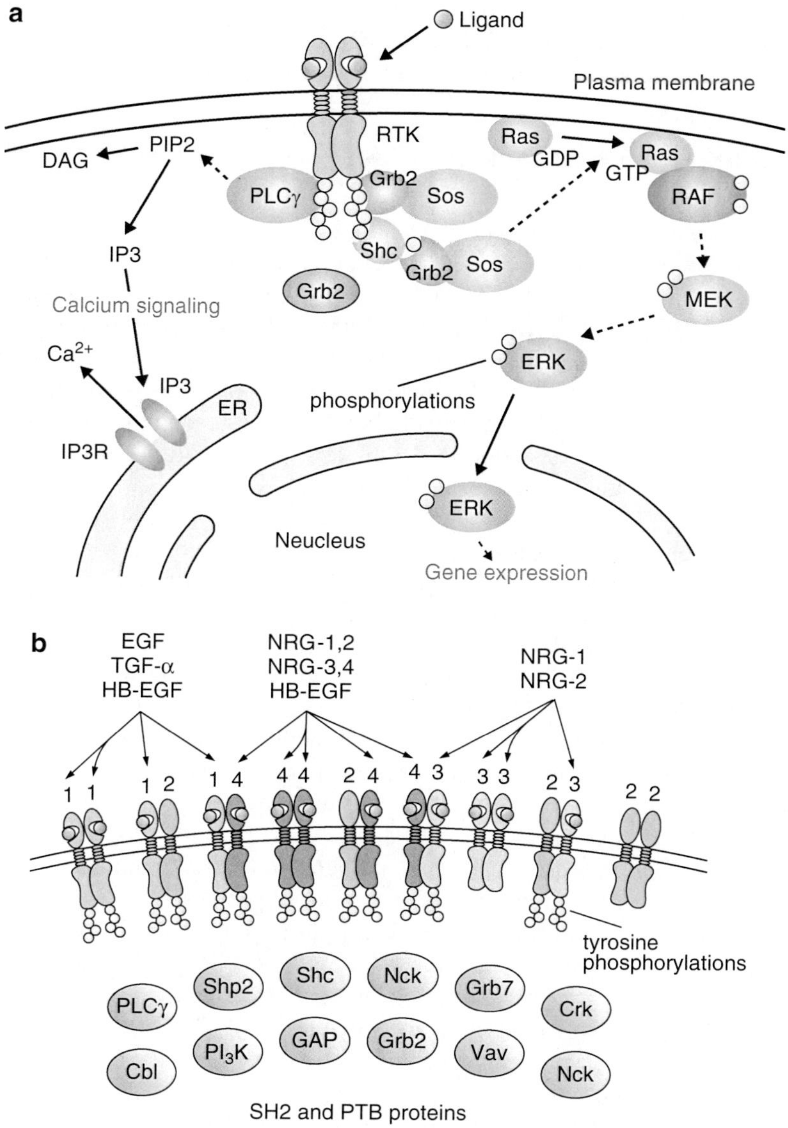

Fig. 1.1 RTK-Ras-MAPK systems and ErbB system. (**a**) Upon association of extracelluar ligands, receptor protein tyrosine kinases (RTKs) on the cell surface transduce signals downstream to a small GTPase, Ras that locates beneath the plasma membrane. Ras excites a cascade of three cytoplasmic kinases that called MAPK system to induce newly gene expressions. (**b**) RTK systems including the ErbB system are subsystems of the RTK-Ras-MAPK systems. The RTK systems are three-layer protein networks, containing an extracellular ligand, membrane receptors (RTK), and cytoplasmic proteins containing SH2 and/or PTB domains. In the ErbB system shown here, various extracellular ligands, including EGF and NRG, associate with ErbB1 to ErbB4 (1–4) to induce the phosphorylation of the cytoplasmic domains of the ErbBs, which are in turn recognized by various cytoplasmic proteins, including PLC_{γ} and Grb2. Grb2 is an adaptor protein responsible for Ras activation. Among the ErbB family members, ErbB2 (2) has no known ligand and ErbB3 (3) has no kinase activity. However, they are involved in cell signaling through heterodimer formation.

clustering of RTK molecules on the cell surface, the activation of RTK, the mutual recognition between activated RTK and cytoplasmic proteins, and the intracellular calcium response induced by RTK activation are the subjects of this chapter.

1.2 Single-Molecule Imaging of RTK Systems in Living Cells

Single-molecule imaging, one of the techniques most widely used in optical microscopy in recent years, can visualize the dynamic behavior of individual molecules and provide information lacking in the ensemble results obtained with conventional biochemical and biophysical methods. The superior feature of single-molecule imaging is its determination of the distributions and fluctuations in the dynamic or kinetic parameters of molecular interactions and movements. This feature of the technique allows detailed analysis of the reaction process, because it is independent of the dispersion in parameters caused by the nonsynchronized reaction starts when multiple molecules are measured.

Funatsu et al. [27] first demonstrated the single-molecule imaging of fluorophores in aqueous solution. They improved the contrast in fluorescence microscopy to detect single-molecules by limiting the excitation depth to a very narrow range (<200 nm) using the evanescent field produced by total internal reflection (TIR) illumination. The typical decay length of the evanescent field from the surface is less than a few 100 nm. In early works, a prism on a coverslip was used to produce illumination with a large incidental angle to generate TIR, and the specimen was observed from the opposite side of the illumination through another coverslip. However, because the samples and solutions were sandwiched between two coverslips, the observation of thick samples, the exchange of solutions, and sample manipulation were not easy. These difficulties in "prism-type" TIR microscopy were overcome with the use of an objective lens with a high numerical aperture (N.A. > 1.33) in an inverted microscope. The objective lens can produce an evanescent field by transmitting the incident light beyond the angle of TIR at the boundary between the coverslip and the solution. The concept of "objective-type" TIR microscopy (Fig. 1.2a) was proposed and demonstrated by Stout and Axelrod [77], and was applied to single-molecule imaging [84]. This method opened the way for single-molecule imaging in living cells, with the easy manipulation of experimental conditions.

In 2000, the first single-molecule imaging in living cells was reported independently by two groups [68, 72], one of which used objective-type TIR microscopy. Single-molecule imaging in living cells constituted a novel method in cell biology, which could be used to quantify biological phenomena in vivo at the molecular level. TIR fluorescence (TIRF) microscopy is now used for the observation of single molecules mainly on the basal (or ventral) surfaces of cells attached to a glass substrate. Cellular phenomena on the apical (or dorsal) surface, or in the cytoplasm, nucleus, or organelles, are observed as single molecules using oblique illumination (Fig. 1.2b) [47, 82, 83, 86]. The oblique illumination is achieved by changing the incident angle of the excitation laser beam slightly from the TIR critical angle, so that

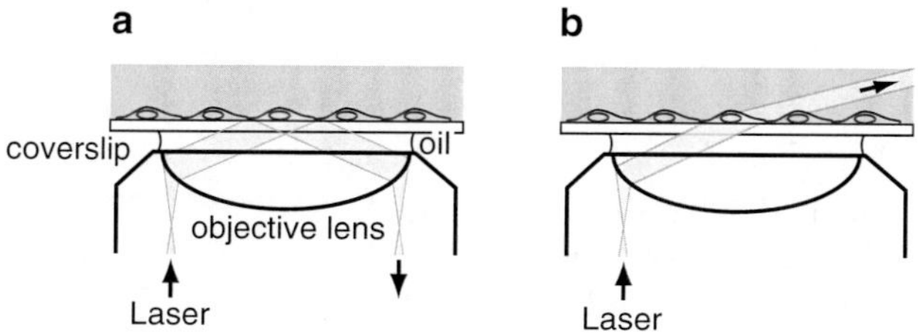

Fig. 1.2 Two illumination methods used for single-molecule microscopy in living cells. (**a**) Objective-type TIR illumination for imaging the basal cell surface. (**b**) Oblique illumination for imaging the apical cell surface.

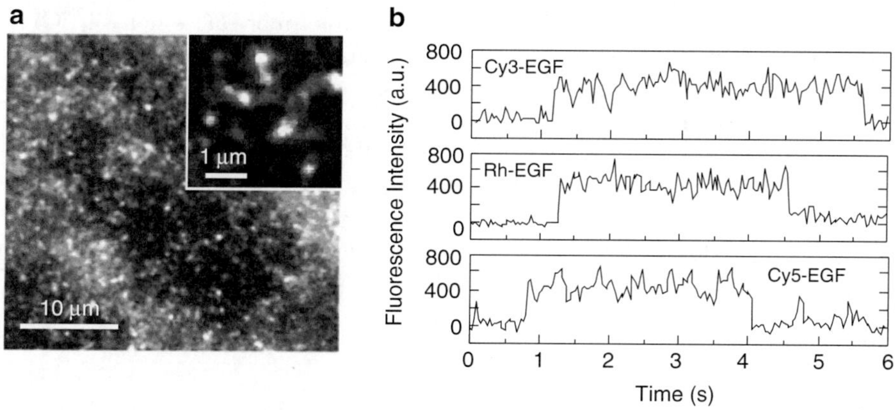

Fig. 1.3 Single-molecule imaging of fluorescently labeled EGF on living A431 cells. (**a**) A TIR fluorescent image of an A431 cell acquired in the presence of 1 nM Cy3-EGF in solution. Inset is a magnified view. (**b**) Typical traces of the fluorescence intensity of individual Cy3 (rhodamine [Rh] or Cy5)-EGF spots on the surfaces of living cells. Single-step increases and decreases in the fluorescence intensity indicate the association and photobleaching of single molecules, respectively.

the beam is transmitted through the cell at a low angle. Because fluorescent dyes outside the slice illumination are not excited by the oblique illumination, the background light is reduced, increasing the contrast and allowing single-molecule imaging. Oblique illumination microscopy was used for a ligand-binding assay of EGF and EGFR [82, 86], for which apical membrane imaging was suitable because the ligand does not easily access its receptors on the basal membrane when in tight contact with the substrate.

In early works [68], we observed the binding of single EGF molecules, conjugated with a fluorescent dye (Cy3, Cy5, or tetramethylrhodamine [Rh]), to the EGF receptors in the plasma membranes of living A431 cells (Fig. 1.3a). The derivation of the detected signals from single molecules was confirmed in two ways: by stepwise photobleaching and analysis of the quantal intensity distribution of the fluorescent spots. On the cell surface, Cy3-EGF emitted almost constant fluorescence, which was then photobleached in a single step before the dissociation or internalization of the complex from the cell surface (Fig. 1.3b). The intensity distribution of Cy3-EGF could be fitted to the sum of two Gaussian distributions. These two components were considered to arise from single and dual Cy3 molecules, respectively. Therefore, the monomeric and dimeric associations of EGF to EGFR could be quantified by integrating each Gaussian component.

Not only molecules labeled with chemical fluorophores like Cy3, but also proteins genetically labeled with fluorescent proteins (FPs) can be observed as single molecules. With progress in molecular biology, a target protein conjugated with an FP, e.g., green fluorescent protein (GFP), can be expressed in living cells. This technique allows one-to-one labeling, to visualize the behaviors of proteins of interest, and is currently used for various biological studies. EGFR-GFP was constructed and expressed in HEK293 and NIH3T3 cells by Carter and Sorkin [10] as the first FP chimera of an RTK. The construct reproduces normal EGFR functions of ligand binding, phosphorylation, and internalization. At present, a series of the FP-tagged RTKs have been introduced and used in many studies as useful probes for cellular imaging. Single-molecule imaging of FP chimeras in living cells was first successfully achieved with the Ras and Rho family of small GTPases [39]. Labeling with FPs can be used for the analysis of interactions between membrane proteins in the plasma membrane and cytoplasmic proteins [38]. In the case of RTKs, FP chimeras have mainly been used in single-particle tracking [53, 91, 92], to investigate the diffusion mechanism.

1.3 EGF and EGFR

EGF, a small 6-kDa protein, binds to its receptor (EGFR, also referred to as ErbB1), a member of the ErbB family of RTKs, consisting of ErbB1-B4. Since the first identification of EGF [16] and EGFR [9] by Cohen and coworkers, many ligands of EGFR have been identified besides EGF (Fig. 1.1). Like other RTKs, the EGFR molecule has three regions, extending from the N-terminus: the extracellular (ectodomain) region containing four subdomains (I-IV), the α-helical transmembrane (TM) region, and the cytosolic region, containing the juxtamembrane (JM), tyrosine kinase (TK), and C-terminal phosphorylation (CT) domains (Fig. 1.4a). Ullrich's group [64] and subsequent studies established that the binding of EGF to EGFR is an event that triggers the EGF signaling cascade and causes EGFR dimerization and the phosphorylation of tyrosine residues in its cytosolic region [70]. In this chapter, homodimers of liganded EGFR which are auto-phosphorylated and activate downstream signaling molecules, are called signaling dimers of EGFR. The ligand molecule, like EGF, associates only one of the EGFR molecules in the signaling dimers as shown later. Formation of signaling dimers is indispensable to start cellular responses against EGF or other EGFR ligands. It is now known that both the dimerization of EGFR molecules (homodimerization) and between EGFR and another ErbB family member (heterodimerization) can induce the neighboring

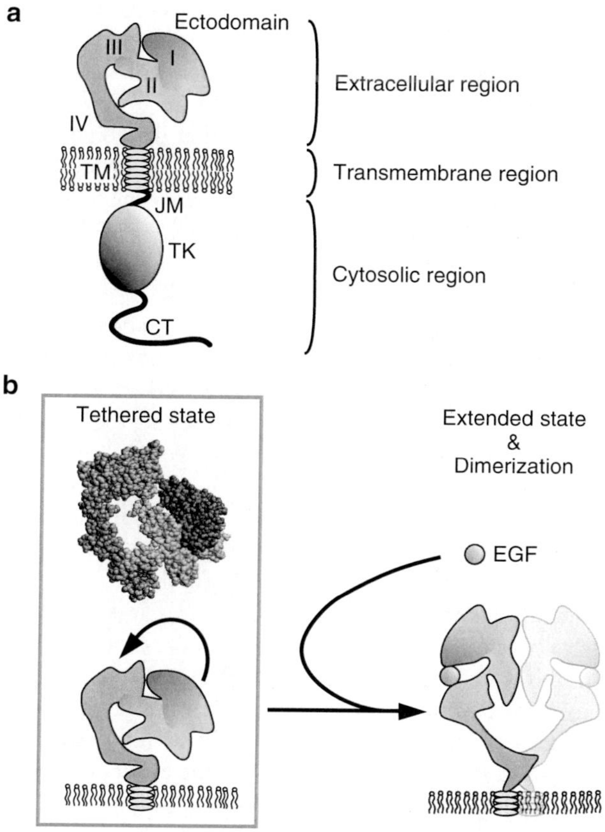

Fig. 1.4 Structure of the ErbB1 (EGFR) molecule. (**a**) ErbB1 consists of an extracellular (ectodomain), a transmembrane (TM), and a cytosolic region, reading from the N-terminus. Numerals I–IV refer to the subdomains of the extracellular region. The cytosolic region contains three domains: the juxtamembrane (JM), tyrosine kinase (TK), and C-terminal phosphorylation (CT) domains. (**b**) The tethered (*left*) and extended (*right*) states of the EGFR ectodomain. The X-ray crystallographic structure of the tethered state is shown in the *top* of the left column. The extended ectodomain dimerizes with its counterpart (semitransparent drawing) through interactions in subdomain II (back-to-back dimer).

cytoplasmic domains of ErbB family members to stimulate kinase activity [33]. However, structures of heterodimers of ErbBs have not been known yet.

Crystallographic studies [24, 29, 63] have revealed the structure of the extracellular region of the EGFR molecule (Fig. 1.4b). Without a ligand, the tethered conformation is adopted, in which subdomains II and IV of a single receptor molecule are in contact, and the ligand binding site containing the subdomains I and III opens wider than the size of the EGF molecule. When EGF binds to EGFR, the subdomains are rearranged and are configured in an "extended" conformation, in which the ligand can access both subdomains I and III simultaneously and the "dimerization loop" in subdomain II is exposed [24]. When ligands are bound, two different EGFR dimer structures occur [29]: a “back-to-back” configuration, in which two receptors are linked by the dimerization loops so that the associated ligands are located at opposite sites on the dimer, and a “head-to-head” configuration, in which subdomain I of each receptor interacts with subdomain III of its dimeric counterpart, so that the ligands are located at the center of the dimer. The back-to-back dimer has better conformational symmetry, a wider interface between the receptors, and a more conserved amino acid sequence at the dimer interface than the alternative head-to-head dimer. Therefore, the back-to-back dimer is favored as the biologically relevant conformation.

Scatchard analysis [19, 20] has shown that EGFR on the living cell surface exhibits two apparently different affinities for its ligands. The receptors with different affinities occur in different amounts and may induce different downstream signals. The high-affinity receptor constitutes only $<5\%$ of the total EGFR molecules but is thought to trigger all the early responses of the cell. The other low-affinity receptor comprises the major fraction and is thought to play roles in cellular hyperproliferation and apoptosis [7, 41]. However, when we assume a simple association between the receptors and ligands, the amount of high-affinity EGFR seems too low to induce the global cellular responses observed at low ligand concentrations. No direct correspondence has been shown between the different affinities for its ligands and the structures of EGFR, e.g., it is not clear that the high-affinity EGFRs form the back-to-back dimer upon its association with EGF. To understand the multifactorial mechanism of EGF-EGFR interactions, kinetic analysis is required to correlate the dynamic changes in affinity with the state of EGFR.

The ligand-induced signal transduction of the ErbB family becomes even more complex insofar as several studies have suggested the existence of oligomers larger than dimers [13–15, 55, 80, 90] and heterodimers between different ErbB species [85]. Each ErbB receptor activates specific intracellular signaling pathways. Therefore, the formation of these heterodimers produces various signal outputs, and cross-talk between the activated pathways occurs [31]. The influence and roles of oligomers and heterodimers of ErbBs in downstream signaling are not well understood, but appear to be concerned with signal amplification in the RTK system and the determination of cell fates. These issues have been the target of detailed investigations using single-molecule imaging because EGFR-ErbB interactions are key steps in signal transduction.

1.4 Extracellular and Intermembrane Events in the EGFR System

1.4.1 Association Between EGF and EGFR Induces the Formation of Signaling Dimers

Single-molecule imaging was applied to the kinetic analysis of EGF-EGFR binding and EGFR dimerization in living cells [82]. Rh-EGF (0.5–2 nM) was added to the culture medium of HeLa cells and its binding to the apical membrane was observed with the oblique illumination method at 10–50 s intervals for a period of 300 s (Fig. 1.5a). The time course of the number of EGF molecules bound to EGFR in the unit cell surface area was observed. The distributions of the fluorescence intensities at individual binding sites were also determined at every time point, to identify the fractions of monomeric and dimeric association sites of EGF (Fig. 1.5b).

In addition to this ensemble information from multiple single molecules, the waiting times for single binding events were measured directly for the first and second EGF molecules bound to individual binding sites (Fig. 1.5c and d, respectively). The reaction schemes for the first and second bindings were suggested from

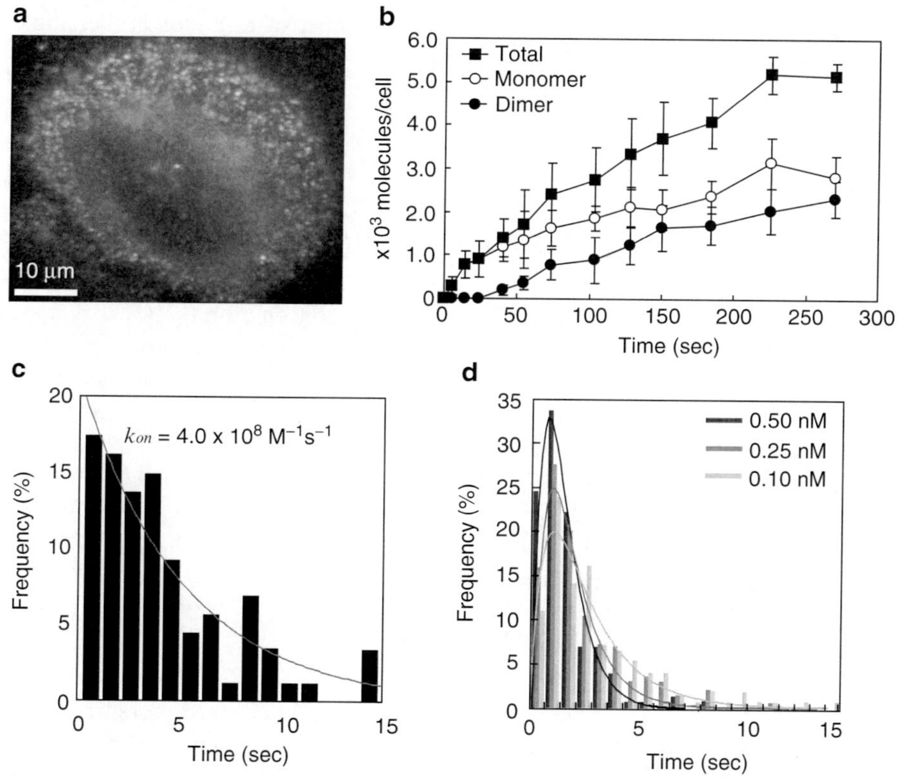

Fig. 1.5 Binding of Rh-EGF and the formation of the signaling dimer of EGFR. (**a**) An oblique illumination image of the apical surface of a HeLa cell 150 s after the addition of Rh-EGF (0.5 nM in solution). (**b**) Time course of EGF binding to the cell surface. The total number of EGF associations (*closed squares*) was subdivided into monomeric (*open circles*) and dimeric bindings (*closed circles*). (**c, d**) The waiting-time distributions for the first EGF binding (**c**) in the presence of 0.5 nM Rh-EGF and for the second binding (**d**) in the presence of 0.1 (*light gray*), 0.25 (*dark gray*), and 0.5 (*black*) nM Rh-EGF. The waiting-times for the first EGF binding are the durations from the application of Rh-EGF to the solution to the appearances of individual fluorescence spots on the cell surface. The waiting-times for the second EGF binding are the durations between the first and second bindings of Rh-EGF to the same association cites. The distributions are fitted to the functions described in the equations for Schemes 1 and 2 in the text.

the shape of the waiting-time distributions. The distribution for the first EGF binding was a single exponential, suggesting simple stochastic binding (Scheme 1). With the second binding, a peak was observed in the distribution, suggesting that the process had two sequential transitions (Scheme 2), so a kinetic intermediate during EGFR dimerization was proposed. The schemes are described below.

$$L + R_x \xrightarrow{k_{on}} LR_x \qquad \text{(Scheme 1)}$$

$$LR_x \xrightarrow{k_a} LR_2{}^* \xrightarrow{k_b} L_2 R_2 \qquad \text{(Scheme 2)}$$

Here, L and R indicate the ligand and receptor, respectively; k_{on}, k_a, and k_b are the reaction rate constants; and x (= 1 or 2) is the number of receptors at the single binding sites. $LR_2{}^*$ represents the kinetic intermediate. The waiting-time distributions are expressed by the following functions.

$$f(t) = C\exp(-k'_{on}t), \qquad \text{(first binding)}$$

$$g(t) = \frac{Ck_a k'_b}{k'_b - k_a}[\exp(-k_a t) - \exp(-k'_b t)], \qquad \text{(second binding)}$$

where $k'_{on} = k_{on}[L]$ and $k'_b = k_b[L]$ ($[L]$ is the EGF concentration).

k_{on}, k_a, and k_b are calculated by fitting the waiting-time distributions. The rate constant of the receptor-ligand association and the existence of a kinetic intermediate, which could not be identified by conventional methods, were revealed by single-molecule analysis. Using these values, the simplest and most plausible models of the binding of EGF to EGFR and the dimerization of EGFRs are proposed (Fig. 1.6a). The kinetic intermediate (L/R-R*) suggests that there is a conformational change in the EGFR dimer after the first EGF binds to it. By fitting of the time course of Rh-EGF binding (Fig. 1.5b) to the kinetic model (Fig. 1.6a), values for *k4*, *k6*, and the numbers of monomeric and preformed dimeric association sites of EGFR per cell before EGF addition were determined [82]. The model suggests the properties of the association and of the EGFR dimer in each state, as follows:

1. Approximately 1–2% of EGFR molecules form predimers. Therefore, if monomers and predimers, which have dimeric binding sites, are in equilibrium, the dissociation constant between the EGFR molecules would be quite large. However, because the association rate constant of EGF to the predimer is two orders of magnitude higher than that to the monomer, even at low concentrations of EGF, EGF preferentially binds to the predimers of EGFR. This facilitates the formation of signaling dimers because liganded EGFR molecules in predimers need not seek their association partner by moving around large areas on the plasma membrane.
2. The binding of the second EGF molecule to the EGFR dimer occurs immediately after the first EGF molecule binds to the predimer, because the rate

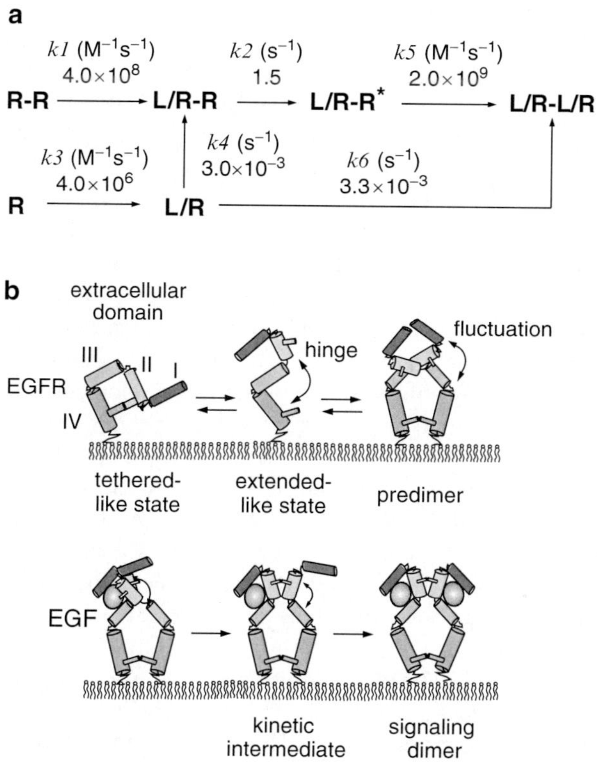

Fig. 1.6 A kinetic model of the formation of the EGFR signaling dimer. (**a**) The reaction network suggested by single-molecule analysis. L and R indicate EGF and EGFR, respectively. L/R-R* represents the kinetic intermediate newly identified by single-molecule analysis. The values for the reaction rate constants are shown. The values for k'_{on}, k_a, and k_b were determined from the waiting-time distributions (Fig. 1.5c and d). Others are the best-fit values suggested by modeling. (**b**) Structural explanation of the kinetics of signaling dimer formation. An extended EGFR molecule can switch between a tethered-like state and an extended-like state via fluctuations of the hinge connecting subdomains II and III. Predimer formation biases the structure toward the extended-like conformation, which has a higher association rate constant compared with that of the tethered monomer. EGF binding to one receptor in a predimer induces an allosteric conformational change in the other receptor, to form a kinetic intermediate with an even higher on-rate (Only the extracellular region of EGFR is shown in this diagram).

constant of the association between the second EGF and the dimer is one order of magnitude higher than that between the first EGF and the predimer. This means that the two EGF bindings to the EGFR dimer are cooperative and further facilitate the formation of the signaling dimer.

3. The formation of the signaling dimer from the EGFR predimer is much faster than its formation from the association of two EGF/EGFR complexes through their diffusion and collision on the cell surface. Therefore, in the early stage of signaling and/or at low concentrations of EGF, most of the signaling dimers are formed from predimers.

Therefore, predimer formation allows effective EGFR signaling by increasing association rate constants with EGF and avoiding time-consuming diffusion-mediated dimerization between EGFRs in the initial stages of cellular signaling.

Combined with the structural information about EGFR obtained from X-ray crystallography [12, 24, 29, 63], a conformational model of the formation of the signaling dimer is thought to explain the reaction kinetics as follows (Fig. 1.6b). EGFR molecules are in equilibrium between the tethered-like and extended-like conformations via fluctuations at the hinge region between subdomains II and III. Compared with the tethered-like conformation, the extended-like conformation has a higher association rate constant for EGF. In the monomeric form, the equilibrium is largely biased toward the tethered-like conformation, but in the predimer, the structure of each EGFR molecule is biased toward the extended-like conformation. The first binding event, in which EGF binds to one of the EGFR molecules in the predimer, induces an allosteric conformational change in the other EGFR molecule, elevating it to a further higher-association-rate state (kinetic intermediate). Finally, a signaling dimer is formed with the second EGF binding event. A recent study [95] that applied the same single-molecule analysis showed that kinetics of EGF association depends on the cell type and the expression level of EGFR, and a pharmacological inhibitor switched the reaction to the other pathway without a kinetic intermediate.

In the studies described above, only the association rate constants between EGF and EGFR were discussed and the origins of the binding sites with different affinities are still unclear. The dissociation rates of EGF are sufficiently slow ($<10^{-3}\ s^{-1}$), as suggested by biochemical analyses, to be ignored when considering only the early stage of EGF association. This situation makes the analysis of the association simple. However, the analysis of the dissociation kinetics, which is indispensable in determining the dissociation constant, is difficult. The dimeric structure of EGFR is also still unclear. Another image correlation spectroscopy study investigating the cluster size of EGFR [13, 15] indicated the existence of many more predimers and/or clusters before EGF stimulation. Predimers of EGFR with diverse structures, some of which show low association rate constants for EGF, possibly exist on the cell surface. Studies based on various single-molecule techniques should be performed to draw an accurate and complete picture of the initial reaction in cellular signaling.

The interaction between ligands bound to an EGFR dimer was measured by single-molecule fluorescence resonance energy transfer (FRET) [68]. After the addition of a mixture of Cy3-EGF and Cy5-EGF to the cells, Cy3 was excited and the fluorescence of both Cy3 and Cy5 were acquired at individual association sites using dual-view optics. Anticorrelated changes between the fluorescence intensities of Cy3 and Cy5 were observed, indicating that FRET from Cy3 to Cy5 occurred. Fluctuations in the FRET efficiency suggested conformational fluctuations in the dimers and/or clusters of EGFR. FRET can be detected when the distance between Cy3 (donor) and Cy5 (acceptor) is similar to or shorter than the Förster distance, approximately 6 nm. Therefore, the observed FRET indicated that the two EGF molecules are spaced several nanometers apart in a dimer or cluster of EGFR. However, the probability of detecting FRET in the dimers was only 5%, suggesting that the distance between Cy3 and Cy5 in the signaling dimer is greater than the Förster distance. This suggestion is consistent with the conformation of the back-to-back active dimer observed in crystallographic studies [29], in which the distance between the ligands is 11 nm, much longer than the detectable distance range of FRET. A recent single-molecule FRET study [90] calculated interligand distances of 5 and 8 nm and proposed two oligomeric conformations that differed from the back-to-back dimer to explain the former value (5 nm). One conformation is a head-to-head dimer, in which the distance between the EGF molecules is <5 nm, and the other is a tetramer, in which two adjacent back-to-back dimers are in side-by-side contact and the inter-EGF distances might be 4 nm. However, the distance of 8 nm could not be explained by known conformational models based on crystallographic evidence. The contribution of a kinetic and/or conformational intermediate was considered in that study, but the details of the correlations between the known structures and the observed kinetics are not well understood.

1.4.2 Amplification of the EGF Signal by Dynamic Clustering of EGFR Molecules

EGFR activation, which involves tyrosine phosphorylations in the cytoplasmic domain after the formation of signaling dimers, is a well known phenomenon, with the results averaged over numerous molecules in many biochemical studies [93, 94, 97]. However, few studies have analyzed the activation of EGFR from a quantitative perspective. We imaged Cy3-labeled mAb74, a monoclonal antibody that recognizes the conformational change in the cytoplasmic region of EGFR when it is activated, on the cytoplasmic sides of the plasma membranes of fixed A431 cells [68]. The binding of Cy3-mAb74 to the membrane was observed after Cy5-EGF stimulation, and the signals of the fluorescent spots of Cy5-EGF that colocalized with Cy3-mAb74 were about twice as intense as the other fluorescent spots. This result is consistent with the autophosphorylation of EGFR after its dimerization.

However, EGFR activation over the whole cell surface is a more complex process. EGF association and EGFR activation were quantified on the basal plasma membranes of the same A431 cells using Rh-EGF and Alexa-488-labeled antibody (mAb74) that recognizes activated EGFR molecules [40]. Pepsin-digested antigen-binding fragment (Fab') of the antibody, which contains single antigen association site, was used to allow one-to-one stoichiometry between EGF and antibody. In addition to intact cells, semi-intact cells in which the cytoplasm was replaced with an artificial solution, were used in the experiments. The experimental procedures differed for intact and semi-intact cells. Because the fixation of intact cells for immunostaining released significant amounts of Rh-EGF from EGFR, the ligand and antibody were independently quantified in the different populations of intact cell before and after fixation. Alternatively, EGF and the Fab' fragment of mAb74 were simultaneously quantified in the semi-intact cells [42]. In both experiments, the cells were incubated with 1 nM Rh-EGF for 1 min and washed, and the numbers of molecules at an individual fluorescent spot were quantified at various time points. In the experiment with semi-intact cells, an ATP regeneration system was added to the buffer to activate EGFR phosphorylation after the region inside the cells was equilibrated with the Fab' fragment.

In the initial phase of the reaction, the number of activated EGFR molecules in both preparations increased and exceeded those of bound Rh-EGF (Fig. 1.7a). The similar increments in the activated molecules in both preparations indicate that the same cellular processes occurred, regardless of the cell treatment. The excess amount of bound Fab' fragments relative to the bound EGF molecules indicates that EGF stimulation induced the activation of not only liganded EGFRs but also unliganded EGFRs. Therefore, the EGF signal was amplified. The activation phase of the intact cells was completed by the dephosphorylation and degradation of the EGFR molecules and the amount of activated EGFR molecules decreased thereafter. Conversely, activation continued to increase in the semi-intact cells because the solution inside the cells did not include the cytoplasmic components required to terminate the signal. Simultaneously, the size of the activated EGFR clusters continued to grow, even after the total number of activated EGFR spots reached a plateau.

The clustering of the ligands and activated receptors differed. As shown by the colocalized fluorescent spots of Rh-EGF and the Cy3-Fab' fragment, which are regarded as ligand-activated receptor sites, the percentage of EGFR oligomers larger than dimers increased, whereas the number of monomers decreased as the reaction proceeded. The total amount and the cluster size distribution of the bound Rh-EGF did not change with time and primarily contained both monomers and dimers, but not oligomers (Fig. 1.7b). These results suggest that the clusters of activated EGFRs consisted of both liganded and unliganded EGFRs, and that the latter had been activated by neighboring ligand-activated EGFR. In this amplification process, an interaction between the liganded and unliganded receptors occurred, with dynamic shuffling of the interaction pairs in the "hetero" clusters containing both liganded and unliganded EGFR molecules. The mechanism underlying EGFR heteroclustering is unknown, but suggests the interaction of an EGFR molecule with other EGFR molecules through extracellular [58] and/or cytoplasmic regions [96]. Such an interaction could be facilitated through an association with the actin cytoskeleton [87], and accumulation into the membrane domains [57]. From the points discussed above, a three-step model (Fig. 1.7c) is proposed:

Fig. 1.7 Quantification of EGFR activation. (**a**) Time course of EGFR activation in intact (*left*) and semi-intact (*right*) A431 cells after transient stimulation with EGF. Single-molecules of EGF and activated EGFR were detected using Rh-EGF and the fluorescently labeled Fab' fragment of an anti-activated-EGFR antibody (mAb74). The numbers on the line indicate the numbers of molecules per spot. (**b**) Cluster size distributions of EGF (*top row*) and activated EGFR (*bottom row*) in semi-intact cells. The numbers indicated by the *arrowheads* are the averaged cluster sizes.

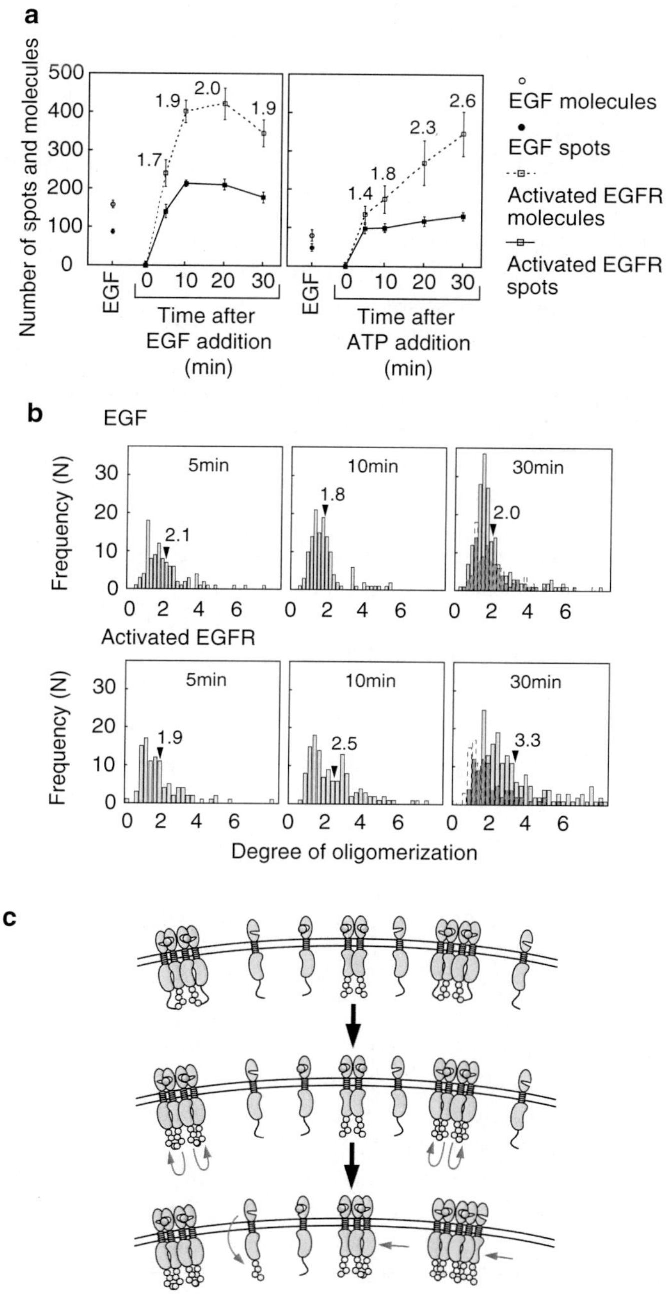

Fig. 1.7 (continued) Distributions at 5 min are superimposed to the distributions at 30 min using *dotted lines* and overlaps between 5- and 30-min distributions are shadowed. (**c**) Three-step model of the amplification of the EGF signal by the dynamic reorganization of the EGFR clusters. See text for details.

1. EGF preferentially binds to predimers of EGFR and stimulates the kinase activity that mutually phosphorylates the tyrosine residues in the cytoplasmic regions.
2. The ligand-activated EGFR activates the unliganded EGFR within a heterocluster by a transient interaction.
3. EGFR activation propagates on the cell surface through the dynamic interactions between monomers and clusters of EGFRs.

The correlation between signal amplification and EGFR expression levels was investigated because the probability of predimer and liganded dimer formation and heteroclustering should increase as the density of EGFR increases. HeLa cells, which express normal but much fewer EGFR molecules (5×10^4 EGFR/cell) [20] than A431 cells (3×10^6 EGFR/cell) [2], were examined for EGFR activation. Signal amplification was observed in semi-intact HeLa cells. Although the EGFR densities differ by a factor of 60 between the A431 and HeLa cells, the amplification rates were similar in both types of semi-intact cells. However, no amplification was observed in intact HeLa cells. This suggests that the activation processes are the same but the regulation by cytoplasmic factors caused the differences observed in the amplification in the two cell types. The normal expression of EGFR in intact HeLa cells might induce local amplification at individual EGFR clusters but it is insufficient to sustain the elevated activity in the face of the signal termination induced by cytoplasmic factors.

1.5 Cytoplasmic Events in the EGFR System

1.5.1 Interaction Between EGFR and Grb2

The EGF signal is transmitted from the plasma membrane to the nucleus through an interaction cascade of cellular proteins. Growth-factor-receptor-bound protein 2 (Grb2) is an element in the cascade, where it acts as an adaptor protein linking EGFR and the downstream signaling molecule, SOS. Grb2 has no enzymatic activity but recognizes the phosphotyrosine (pY) residues of EGFR and the Ras GTP exchange factor, SOS [11, 23, 28, 52, 65], through its unique SH2 domain and one of its two SH3 domains, respectively. pY1068 and pY1086 are the primary and secondary Grb2-binding sites of EGFR [1], and the former is phosphorylated to a greater extent than the latter in vivo when cells are stimulated with EGF [22].

The binding events between activated EGFR and Grb2 were measured by single-molecule analysis in vitro [59]. The plasma membrane fraction, including EGFR with or without EGF stimulation, was purified from A431 cells, attached to a coverslip, and incubated with Cy3-Grb2. Under TIRF microscopy, repetitive on and off Cy3 fluorescent signals were observed in the same fixed membrane fraction (Fig. 1.8a). "On-time" (the duration of the period between the association and dissociation of a single Cy3-Grb2 molecule) and "off-time" (the duration of the period between the dissociation

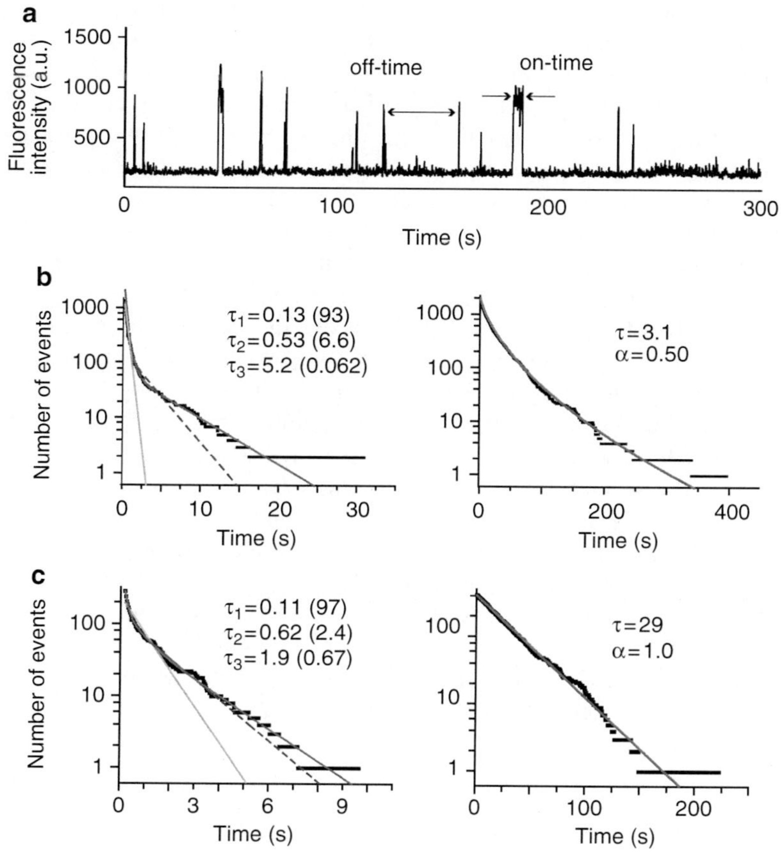

Fig. 1.8 Single-molecule analysis of the interactions between EGFR and Grb2. Fragments of the plasma membrane containing activated EGFR were attached to a coverslip and the association and dissociation of a single Grb2 molecule labeled with Cy3 were observed. (**a**) A typical trace of the reaction events at a single binding site for Grb2. Transient increases in the fluorescence intensity represent associations of Cy3-Grb2. The concentration of Cy3-Grb2 was 1 nM. (**b, c**) Cumulative histograms of on-times (*left*) and off-times (*right*) for wild-type (**b**) or Y1680F mutant (**c**) EGFR in the presence of 1 nM Cy3-Grb2. The *left-side* histograms (on-time distributions) are fitted to a single exponential function (*light gray lines*) and the sum of two (*dashed lines*) or three (*dark gray lines*) exponential functions. τ_1–τ_3 indicate the decay times for the three exponential fittings. The numbers in parentheses are the percentages of each fraction. The *right-side* histograms (off-time distributions) are fitted to a stretched exponential function (*dark gray lines*). τ and α indicate the time constant and exponent, respectively.

and association of the next Cy3-Grb2 molecule) were measured for individual interaction events (Fig. 1.8b). The dissociation and association kinetics were analyzed from the cumulative histograms for the on-time and off-time values, respectively.

The on-time distribution, which relates to the dissociation kinetics, was described with multiple exponential functions. The number of exponentials in an adequate fitting function refers to the number of sub-states in the reaction. The dissociation process was independent of the Grb2 concentration and exhibited at least three binding sub-states: the major component (89–98%) was fastest, with a rate constant of 8.1–7.5 s^{-1}; the rate constant for the second component (2–11%) was 1.6–2.6 s^{-1}; and the third minor component (<0.1%) was a slow but distinct event, with a reaction rate constant of 0.1–0.4 s^{-1} (Fig. 1.8b, left). None of the sub-states was attributed to a static subpopulation of the binding sites but multiple subpopulations appeared dynamically at every binding site.

The association process was more complex. Four or more components were required to describe the association kinetics with a multiple exponential function which could be applied globally to every Grb2 concentration examined (1–100 nM). The fraction of each component changed dynamically depending on the Grb2 concentration. Therefore, a clear unified view of the reaction was not obtained from the multiple exponential kinetic model. Another analysis of individual sites using exponential kinetics showed that the number of components differed between sites; one third of the sites exhibited single component and the others exhibited two components. The second-order association rate constants obtained for individual binding sites were distributed widely, without evident multiple peaks. These results suggest a model in which the system drifts through multiple unbound sub-states, exhibiting different association rate constants. A stretched exponential function, described as follows, can represent such a multiple-state reaction.

Stretched exponential kinetics (Fig. 1.8b, right) [26] was applied to explain the association between EGFR and Grb2. This kinetics is expressed using only two parameters: the time constant (τ) and the exponent (α).

$$\phi(t) = \phi_0 \exp\{-(t/\tau)^\alpha\}$$

The exponents in the stretched exponential kinetics had similar values (0.41–0.56) for different Grb2 concentrations, which means that the function provides a better representation than do multiple exponential kinetics. The stretched exponential kinetic model suggests the presence of many unbound sub-states. As the extreme cases, this kinetic model describes two scenarios: (1) multiple unbound sub-states exist together on the transition path to a unique unbound sub-state which has a path to the association state. In this scenario, α relates to the dimension of the transition network; (2) many unbound states exist, each of which transits to the association state with a different rate constant. In this case, α relates to the distribution of the rate constants. In the experiment, one third of the binding sites showed single exponential kinetics with a distinct association rate, so that the latter case is more suitable for the association between EGFR and Grb2. However, considering the kinetics of the other sites, which consist of multiple components, transitions are suggested between the unbound sub-states. Taken together, the association between EGFR and Grb2 seems to occur between the two extreme cases, i.e., many unbound sub-states with different association rate constants exist, with state transitions among them.

The reaction kinetics of a mutant EGFR (Y1068F) was examined (Fig. 1.8c). Y1068 has been reported to be the major binding site for Grb2 [1] and its phosphorylation induces a conformational change in the EGFR C-terminal tail [6]. In fact, this single-residue substitution resulted in a longer time constant, i.e., lower reaction frequency, in the stretched exponential kinetics at every Grb2 concentration compared with that of wild-type EGFR. At 1 nM Grb2, the association kinetics was reduced to a single exponential, meaning the disappearance of the multiple substates. However, at higher Grb2 concentrations (>10 nM), a multiplicity of associations was preserved. This means that the frequent interaction between EGFR and Grb2 induced a long tail in the association kinetics. The dissociation kinetics did not differ with the Y1068F mutant.

The interaction between EGFR and Grb2 showed a curious dependence on the Grb2 concentration. The apparent second-order association rate constant at individual sites (the inverse of the average off-time multiplied by the Grb2 concentration) increased by a factor of about three as the Grb2 concentration decreased by a factor of 10. This relationship (negative concentration dependence) between the Grb2 concentration and the association rate constant was maintained across a wide range (0.1–100 nM) of Grb2 concentrations in solution. This property of the association reaction suggests that the interaction between EGFR and Grb2 temporarily reduces the rate constant of the successive association events. To explain the mechanism, the proteins would be required to maintain a conformational memory after dissociation. The existence of memory in the sub-second to second regions in time was actually suggested by a non-Markovian function analysis of the single-molecule reaction trajectories. The average off-time observed for wild-type EGFR at 10 nM Grb2 was 9 s, meaning that a significant proportion of the off-time is within the seconds region of time. Therefore, the reaction was frequent enough to retain the memory induced by the molecular interaction. The reaction memory was not observed with the Y1068F mutant at 1 nM Grb2. This means that the conformational memory was diminished by the structural relaxation during the long off-time. As shown here, EGFR probably senses the concentration of its association partner, Grb2, as single molecules. The physical significance of this negative concentration dependence is thought to suppress the effects of changes in Grb2 expression in living cells. If this is the case, biased fluctuations in the receptor structure are exploited to regulate signal transduction.

1.5.2 Calcium Signaling

Calcium signaling is one of the primary pathways downstream from RTK. Phospholipase C_γ (PLC_γ) is activated by phosphorylated RTK and hydrolyzes a membrane lipid, phosphatidylinositol bisphosphate (PIP_2), into inositol triphosphate (IP_3) and diacylglycerol. IP_3 is released into the cytoplasm and binds to its receptor (IP_3R), which is a Ca^{2+} channel on the endoplasmic reticulum, to release Ca^{2+} thorough IP_3R into the cytoplasm. This cellular response has been investigated in many studies with various techniques [3, 54, 56, 61, 89]. The intracellular Ca^{2+} response is preceded by EGF-induced receptor dimerization. We tried to clarify the quantitative relationship between the EGF binding to EGFR and the Ca^{2+} response using single-molecule techniques. In A431 cells, single-molecule quantification of receptor-bound Cy3-EGF and measurement of the signal intensity of the Ca^{2+} indicator Fluo-3 in the cytoplasm showed that EGF dimers increased for 60 s after the application of Cy3-EGF to the extracellular medium and that the Ca^{2+} concentration subsequently increased [68]. This is consistent with the results of previous studies.

Uyemura et al. [86] advanced single-molecule research by demonstrating the minimal number of EGF molecules required to induce the cellular response in HeLa cells. The change in the cytoplasmic Ca^{2+} concentration, $[Ca^{2+}]_{in}$, indicated by the Fluo-4 fluorescence intensity, was measured in cells before and after transient Rh-EGF stimulation for 1 min (0.2–2.0 nM). The number of Rh-EGF molecules on the plasma membrane was then counted by scanning the focus of oblique illumination microscopy from the basal to apical membrane, with intervals of 0.3 μm. The number and cluster size distribution of the EGF molecules bound to the cells did not change during the observation, as shown in a previous experiment [40].

In the $[Ca^{2+}]_{in}$ plot constructed as a function of time, $[Ca^{2+}]_{in}$ increased steeply to a peak about 60 s after stimulation, and then decreased to a constant level similar to that observed before stimulation. The height of the $[Ca^{2+}]_{in}$ peak after stimulation tended to increase as the amount of bound EGF increased. The cell responsiveness to EGF was defined as positive when the peak $[Ca^{2+}]_{in}$ was >30 nM, which is the higher end of the $[Ca^{2+}]_{in}$ fluctuation in quiescent cells. The fractions of cells with positive responses showed a sigmoidal correlation with the number of bound EGF molecules on the cells (Fig. 1.9, *open circles*). This curve indicates an ultrasensitive response that can be fitted to Hill's equation. The apparent Hill factor was 3.9. About 300 EGF molecules were estimated to induce the calcium response in half the

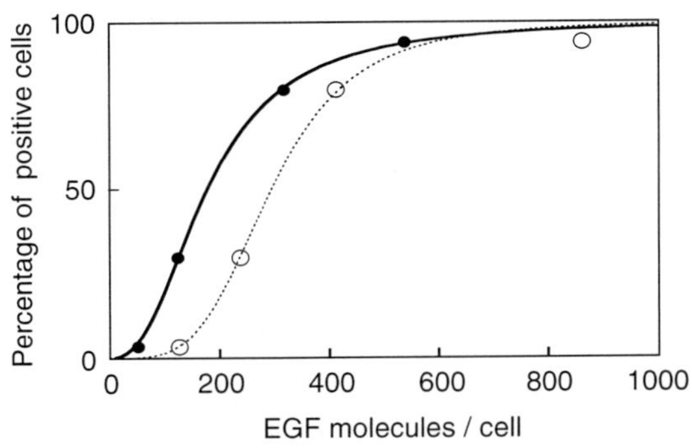

Fig. 1.9 Intracellular calcium response induced by EGF. The probability of the calcium response was plotted against the number of bound EGF molecules per single cell. According to the numbers of total (*open circle*) or dimeric-bound (*closed circle*) EGF molecules on the cell surface, the cells were classified into four groups and the response probability was calculated for each group. The lines are the results of fitting Hill's equation. The apparent Hill coefficients are 3.9 (total EGF) and 2.4 (dimeric-bound EGF).

cell population. Interestingly, this number of molecules is quite low compared with the number of EGFR molecules expressed on HeLa cells (50,000/cell) [2].

As described in the previous sections, the oligomerization of EGFR (including its dimerization) is critical for signal transduction. To investigate the role of EGFR oligomers in the calcium response, the oligomer size distribution was determined from the fluorescence intensity distributions of the EGF spots on cells. The results indicate that as the number of bound EGF molecules, i.e., the response probability, increased, the cells contained more EGF oligomers, especially dimers, but the number of monomers did not correlate clearly with the cellular responsiveness. The cell fraction with positive Ca^{2+} responsiveness changed the sigmoidal manner when plotted against the number of dimeric-bound Rh-EGF molecules instead of against the total EGF molecules (Fig. 1.9, *closed circles*). The number of EGF molecules required to induce the calcium response decreased to 180 molecules in half the cells, corresponding to 90 dimers, and the apparent Hill factor also shifted to 2.4. This decrease in the apparent Hill factor indicates that one of the mechanisms underlying this cellular ultrasensitivity is the dimerization of EGFR. However, the ultrasensitivity (Hill factor = 2.4) remained after the effect of EGFR dimerization was taken into account, indicating the presence of additional mechanisms. Several candidate mechanisms were considered: (1) the amplification of the EGF signal through receptor clustering, as shown in the previous section [40]; (2) the activation of PLC_{γ} by the multiple phosphorylation of its three tyrosines [44], when this phosphorylation occurs with multiple collisions, causing the ultrasensitive production of IP_3 [25]; and (3) the simultaneous opening of neighboring IP_3 receptors (IP_3R) under positive feedback control at submicromolar Ca^{2+} concentrations, although the small Ca^{2+} flux produced by a single IP_3R molecule cannot easily induce an increase in the whole-cell concentration of calcium.

The ultrasensitive and switch-like calcium response is thought to suppress the aberrant behavior accidentally induced by quite a small number of ligands or intrinsic noise in the cells. Whereas, the response can amplify a certain level of signal or noise that may occasionally exceed the threshold due to fluctuations. With such a mechanism, fluctuations in the signaling systems can drive various cell behaviors, even under the constant conditions.

1.6 NGF and NGF Receptors

Nerve growth factor (NGF) was first identified by Cohen and Levi-Montalcini [17]. Stimulation of the NGF on sensory and sympathetic neurons induces their axonal growth, with expanding lamellipodia, and promotes their survival. Growth cones at the distal edges of the lamellipodia guide the elongation of neurites. This guidance is a result of the NGF-induced signaling pathway in which the cytoskeleton is modulated to regulate the growth-cone behaviors at the downstream stage. TrkA, one of the NGF receptors (also called neurotrophic tyrosine kinase receptor 1, NTRK1), is a member of the RTKs and is involved in the NGF pathway. Like

EGFR, TrkA molecules in homodimers phosphorylate each other on their tyrosine residues upon association with NGF, and activate intracellular signaling molecules. The TrkA dimers are then endocytosed, trafficked to signaling endosomes [21, 32], and transported retrogradely to the soma [37]. Previous studies have demonstrated NGF properties in terms of its mobility, clustering [50], interaction kinetics with its receptor [43], internalization, and transport [8, 30, 49]. Fluorescence-labeled NGF has been used in these studies to image the somata of sensory neurons, but not single molecules.

1.6.1 Single-Molecule Behavior of NGF on the Growth Cone

The β subunit of NGF (βNGF) exists as a dimer of two identical 13-kDa polypeptides [51] and expresses the biological function of NGF [18]. We synthesized fluorescent βNGF by conjugating it with Cy3 (Cy3-NGF) and observed single molecules of Cy3-NGF on the surfaces of PC12 cells [67]. Tani et al. [81] improved the preparation of Cy3-NGF, and using the single-molecule imaging technique on the growth cones of dorsal root ganglion (DRG) neurons, observed the various behaviors of individual Cy3-NGF molecules. Conventional epifluorescence illumination, modified for single-molecule imaging, was used for the observations.

In the study of Tani et al., the fluorescent dots increased on the lamellipodia after the addition of Cy3-NGF to the solution, then reached a plateau after 10–20 min. The association rate constant calculated from the increase in Cy3-NGF molecules was $2.8\times10^{7}\ M^{-1}\ s^{-1}$ at 0.4 nM Cy3-NGF. The dissociation constant (K_d) determined from the number of bound molecules at the plateau stage of the association and the concentration in solution was 2.7×10^{-11} M. The affinity between NGF and the NGF receptor has been reported previously in chicken sensory ganglion neurons using ^{125}I-NGF. High- and low-affinity binding sites were identified and K_d for the high-affinity receptor was 2.3×10^{-11} M [79]. The structure of each binding site is unknown, but the high-affinity site at least is thought to contain TrkA. The value of K_d obtained with single-molecule measurements suggests that Cy3-NGF binds predominantly to the high-affinity receptor at a concentration of <0.4 nM Cy3-NGF in solution. Tani et al. calculated the value for K_d in the presence of a higher concentration (>5 nM) of Cy3-NGF by quantifying the total fluorescence intensities on the growth cones, instead of counting the indistinguishable individual molecules. The association was bimodal and two dissociation constants of $K_{d1} = 3.5\times10^{-11}$ M and $K_{d2} = 7.7\times10^{-9}$ M were calculated. K_{d1} was close to the value for the high-affinity receptor and K_{d2} was similar to that of the low-affinity receptor reported by Sutter et al. [79]. The numbers of low- and high-affinity receptors were 200 and 710 per single growth cone, respectively. The occupation of about 40 NGF binding sites (<5% of the total) induced lamellipodial expansion in the growth cone. Therefore, like the EGF-induced calcium response, the formation of several tens of RTK dimers induced a global cellular response.

Most Cy3-NGF on the lamellipodia of the growth cones bound as single molecule and was mobile. (Because βNGF is a dimeric molecule, it is possible that single Cy3-NGF molecules bound to the NGF receptor dimer). The molecules showed random diffusion, with a diffusion coefficient of 0.3 $\mu m^2 s^{-1}$ immediately after the association. This diffusion coefficient is rather large for an integral membrane protein. Thereafter, the motion of 22% of the molecules became unidirectional, with a transition rate of $6 \times 10^{-3} s^{-1}$. This transition requires kinase activity of TrkA (Tani, By addition of tyrosine Kinase inhibitor, the transition of TrKA was inhibited). After the transition, the molecules move rearward to the central region of the cell, with a velocity of 3.6 $\mu m\ min^{-1}$ and a small diffusion constant of 0.0096 $\mu m^2 s^{-1}$. This unidirectional movement constitutes transport on a retrograde flow of actin filaments in the lamellipodia, because Cy3-NGF moved in concert with the Alexa-647-phalloidin staining of actin filaments, and latrunculin B, an inhibitor of actin polymerization, blocked the movement. Therefore, about 20% of the complexes between NGF and its receptor were cross-linked to actin filaments after TrkA phosphorylation.

With this unidirectional retrograde movement, Cy3-NGF accumulated in the central region of the growth cone. When the Cy3-NGF molecules on the cell surface were removed by acid treatment, most of the diffusing Cy3-NGF dots on the lamellipodia disappeared, whereas the clusters of Cy3-NGF in the central region remained, indicating that the clusters of NGF-receptor complexes in the central region were not exposed to the extracellular medium, but were incorporated into vesicles in the cytoplasm. The endocytosis of the NGF-receptor complex was presumed to have occurred at the transition site between the lamellipodia and the central region. The internalization of the NGF-receptor complex could involve clathrin-dependent endocytosis [69] or pinocytosis, which depends on the chaperone protein, pincher [73].

1.6.2 Single-Molecule Behavior of NGF on PC12 Cells

NGF induces neuron-like morphological and functional changes in PC12, a cultured cell line derived from rat phenochromocytoma. We imaged single molecules of Cy3-NGF on the basal surfaces of PC12 cells [74]. Using objective-type TIR microscopy, both the intensity and positions of individual fluorescent spots of Cy3-NGF bound to the cell surface were measured in every video frame. Almost all the spots represented single NGF molecules. The trajectory of the movement of single NGF molecules, determined by connecting their positions in successive frames, suggested that the binding of Cy3-NGF and the diffusion of its bound receptors occurred in various regions of the plasma membrane. The lateral diffusion coefficient of the Cy3-NGF bound receptor was calculated by fitting the initial linear portion of the mean square displacement (MSD) for multiple single molecules to a linear regression line. The calculated value was 0.17 $\mu m^2 s^{-1}$, which is similar to that for the diffusive motion of Cy3-NGF on the growth cones, 0.3 $\mu m^2 s^{-1}$ [81].

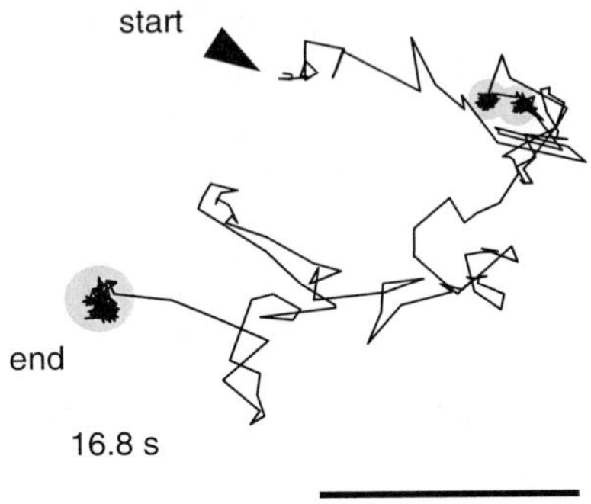

Fig. 1.10 A typical trajectory of the movement of the NGF-receptor complex on PC12 cells. Reversible transitions between mobile and immobile phases were observed in the movement of Cy3-NGF bound to NGF receptors. A trajectory over 16.8 s is shown. The immobile segments are *circled* in gray. Scale bar, 1 μm.

The single-molecule trajectories of the NGF-receptor complexes showed abrupt repeated switching between the "mobile phase" and the "immobile phase" (Fig. 1.10), which resembled the slowing or restricted motion of a membrane protein in fibroblasts [88]. In accordance with the strategy of Simson and Sheets [76], short segments in the trajectory were statistically classified into the two phases by detecting the period in which the receptor complexes were confined within a particular region for a longer period than the time that a free diffusant would stay within an equal area. The diffusion coefficients of the complexes in the mobile and immobile phases were 0.17 and 0.01 $\mu m^2\ s^{-1}$, respectively, calculated from the MSD during 134 and 97 segments of each period, respectively. The distributions of the durations of both phases fitted well to a single exponential decay function, with time constants of 1.0 s for the mobile phase and 1.5 s for the immobile phase. Therefore, the switching processes in both directions were inferred to contain single rate-limiting steps. In many trajectories, switching is observed multiple times. There was no significant difference between the durations of the consecutive mobile or immobile periods.

The transition from the mobile to the immobile phase was occasionally accompanied by a stepwise increase in fluorescence intensity, which reflected the colocalization of two Cy3-NGF molecules. This colocalization continued throughout the immobile phase. This coordinated phenomenon might be explained by one or both of the following mechanisms: (1) two randomly diffusing NGF-receptor complexes were simultaneously trapped and released at a membrane domain, and (2) immobilization of the complexes was induced and terminated by direct and/or indirect interactions between the complexes. The former mechanism is highly improbable because the synchronization of the behaviors of two independent complexes is very unlikely. The latter mechanism is more plausible. However, the association of two complexes resulting in the formation of a tetrameric molecule of TrkA alone would be insufficient to induce this immobilization, because the lateral diffusion coefficient of diffusants in lipid membranes is insensitive to the radius of the diffusant [48]. Therefore, an additional mechanism, such as an interaction with the cytoskeleton or the formation of membrane domains, is required to explain the immobilization observed with the receptor-receptor association.

A pharmacological study using k252a, an inhibitor of the TrkA tyrosine kinase, investigated the mechanism and physiological role of the immobilization of the ligand-receptor complex. k252a suppresses NGF-induced neurite formation in PC12 cells, but apparently does not affect the binding of Cy3-NGF to the cells [46]. After the PC12 cells were treated with 200 nM k252a for 30 min, the transition to the immobile phase of Cy3-NGF-bound receptors decreased significantly and 90% of the complexes were highly mobile, with no transition to the immobile phase. These results suggest that the immobilization of the NGF-TrkA complex is a physiological process related to receptor activation. The phosphorylation of TrkA induces the activation of the small GTPase, Ras, on the cytoplasmic side of the plasma membrane. Ras activation can be detected by the translocation of RAF, a cytoplasmic serine/threonine kinase, to the plasma membrane [39]. Interestingly, when Cy3-NGF and GFP-RAF were observed simultaneously as single molecules, NGF and RAF colocalized during the immobile phase of the NGF-receptor complexes. This result is consistent with the requirement for TrkA phosphorylation for this immobilization and suggests the physiological importance of the immobilization and clustering of the NGF-receptor complexes.

1.7 Conclusions and Perspectives

Single-molecule studies of the RTK systems have revealed novel aspects of RTK-mediated events that occur outside and inside cells. For example, in one of the first single-molecule imaging studies in living cells [68], it was estimated that the distance between ligand-binding sites in the EGFR signaling dimer is greater than 6 nm (Förster distance), based on the low FRET probability between Cy3-EGF and Cy5-EGF at those binding sites. This suggestion was confirmed in subsequent crystallographic studies, which demonstrated the back-to-back conformation of the signaling dimer, with a distance of 11 nm between the ligands. A most remarkable phenomenon in RTK signaling revealed by single-molecule imaging is that only several tens of RTK dimers can induce a global cellular response, such as calcium signaling (EGF) or neurite elongation (NGF). The high sensitivity of the RTK systems must be based on previously unknown molecular mechanisms of information processing, also identified with single-molecule measurements. In EGF signaling, the sensitivity for the extracellular ligand is improved when EGFR predimers and novel kinetic intermediates are used. The EGF signal is nonlinearly amplified based on the balance achieved between activation by the dynamic interaction of liganded and unliganded EGFR molecules and inactivation by cytoplasmic factors. There is a negative concentration dependence in the interaction between the EGFR cytoplasmic domain and Grb2. This negative concentration dependence suppresses the cell-to-cell deviation in signaling caused by fluctuations in the Grb2 concentration in multicellular systems. In single cells, the negative concentration dependence forces the Grb2 molecules to avoid rebinding to the same EGFR molecules. This mechanism may cause the propagation of signals through the cytoplasm and increase the information from single EGFR molecules, facilitating interactions with cytoplasmic proteins other than Grb2. In NGF signaling, the phosphorylation of TrkA changes the mobility of the NGF-receptor complex, in both neurons and PC12 cells. Activated (phosphorylated) TrkA tends to align on actin filaments (DRG) or be stationary (PC12) during signaling. This localized signaling could be important in increasing the efficiency of signal transduction. All these aspects suggest that both the intra- and intermolecular dynamics of the RTK systems, originally derived from thermal fluctuation and self assembly, are responsible for signal processing at the single-molecule level.

However, many issues still remain to be unraveled before we fully understand the RTK systems. One of these is the precise estimation of the RTK cluster size. Because RTKs require mutual interaction for their activation and signal amplification, the average cluster size, the distribution of the cluster sizes, and the spatial separation of the clusters, and how these factors relate to their movement are key factors in RTK signaling. Single-molecule imaging can be used to measure cluster sizes. However, if a cluster is located close to neighboring clusters within the optical resolution (~300 nm), because the RTK is highly expressed and/or has accumulated in a membrane microdomain, the cluster size will be overestimated because the fluorescence from unresolved clusters will overlap. FRET is quite sensitive at nanometer-range distances, so it is often used as a sensor of the interactions between protein pairs. However, FRET (even in single molecules) is not suitable for the detection of multiple-molecule interactions, such as the determination of cluster sizes. Recent advances in microscopy have allowed several "super-resolution" methods, such as stimulated emission depletion (STED) microscopy [36, 45], photoactivation localization microscopy (PALM) [4], stochastic optical reconstruction microscopy (STORM) [66], and structured illumination microscopy (SIM) [34, 35]. The spatial resolutions of these methods are 10 nm (PALM, STORM), 15 nm (STED), and 50 nm (SIM) in the best cases, so they can distinguish clusters located close together. At present, these super-resolution methods are used to acquire detailed qualitative images, but the methods should be improved for molecular quantification with high spatial resolution.

Other issues to be explored in the RTK system are the structures and functions of heterocomplexes of receptors. Every ErbB molecule, including EGFR, commonly activates several species of cytoplasmic signaling proteins, but at the same time, each activates some specific signaling proteins in particular pathways. The ErbB heterodimers drive signaling cascades related to the ErbB combination and can induce complicated cellular responses as the integrated results of the driven cascades. Two types of NGF receptor, TrkA and p75, are thought to form complexes that alter their affinities for NGF. Because TrkA and p75 emit different types of signals, NGF signaling can be complicated by the formation of heteroclusters of NGF receptors. However, structural heterogeneity of NGF-binding sites has not been studied on the cell surface. As described in this chapter, single-molecule studies have revealed the detailed kinetic properties of the interactions between

RTKs and their ligands [81, 82] or effector molecules [59]. Kinetic studies of various RTK molecules, including heterodimers and heterooligomers, are required to clarify the mechanisms of signaling. Recent technical progress in microscopy, with super-resolution using two differently colored fluorescent proteins [75, 78], will be a potent tool for quantitative studies of heterodimers.

Single-molecule measurements facilitate the analysis of the individual steps in signaling networks, i.e., protein interactions and enzymatic reactions, and have allowed us access to molecular-level signal processing. RTK systems, which form a complicated reaction web [62], could not be fully understood without correct information about single individual steps in these reactions. To complement the techniques of single-molecule analysis, systems biological analysis has been performed on the RTK-Ras-MAPK systems in many recent studies [5, 60]. Their purpose was to present reasonable computational models of the systems using information obtained from various experiments to understand and predict the system dynamics. By linking the mechanisms of signal transduction and processing revealed by single-molecule measurements with the computations of systems biology, a novel and precise understanding of RTK-mediated signal transduction will be achieved.

Acknowledgments We would like to thank Hiroaki Takagi, Tomomi Tani, Tatsuo Shibata, Masahiro Ueda, Toshio Yanagida, and members of our laboratory for their collaboration, continual encouragement, and helpful discussions.

References

1. Batzer AG, Rotin D, Urena JM, Skolnik EY, Schlessinger J (1994) Hierarchy of binding sites for Grb2 Shc on the epidermal growth factor receptor. Mol Cell Biol 14:5192–5201
2. Berkers JAM, Van Berger en Henegouwen PMP (1991) Three classes of epidermal growth factor receptors on HeLa cells. J Biol Chem 266:922–927
3. Berridge MJ, Heslop JP, Irvine RF, Brown KD (1984) Inositol trisphosphate formation calcium mobilization in Swiss 3T3 cells in response to platelet-derived growth factor. Biochem J 222:195–201
4. Betzig E, Patterson GH, Sougrat R, Lindwasser OW, Olenych S, Bonifacino JS, Davidson MW, Lippincott-Schwartz J, Hess HF (2006) Imaging intracellular fluorescent proteins at nanometer resolution. Science 313:1642–1645
5. Birtwistle MR, Hatakeyama M, Yumoto N, Ogunnaike BA, Hoek JB, Kholodenko BN (2007) Ligand-dependent responses of the ErbB signaling network: experimental and modeling analyses. Mol Sys Biol. doi:10.1038/msb4100188
6. Bishayee A, Beguinot L, Bishayee S (1999) Phosphorylation of tyrosine 992, 1068 and 1086 is required for conformational change of the human epidermal growth factor receptor C-terminal tail. Mol Biol Cell 10:525–536
7. Brabyn CJ, Kleine LP (1995) EGF causes hyperproliferation and apoptosis in T51B cells: involvement of high and low affinity EGFR binding sites. Cell Signal 7:139–150
8. Bronfman FC, Tcherpakov M, Jovin TM, Fainzilber M (2003) Ligandinduced internalization of the p75 neurotrophin receptor: a slow route to the signaling endosome. J Neurosci 23:3209–3220

9. Carpenter G, King L Jr, Cohen S (1978) Epidermal growth factor stimulates phosphorylation in membrane preparations in vitro. Nature 276:409–410
10. Carter RE, Sorkin A (1999) Endocytosis of functional epidermal growth factor receptor-green fluorescent protein chimera. J Biol Chem 52:35000–35007
11. Chardin P, Camonis JH, Gale NW, van Aelst L, Schlessinger J, Wigler MH, Bar-Sagi D (1993) Human Sos1: a guanine nucleotide exchange factor for Ras that binds to GRB2. Science 260:1338–1343
12. Cho HS, Leahy DJ (2002) Structure of the extracellular region of HER3 reveals an interdomain tether. Science 297:1330–1333
13. Clayton AHA, Orchard SG, Nice EC, Posner RG, Burgess AW (2008) Predominance of activated EGFR higher-order oligomers on the cell surface. Growth Factors 26:316–324
14. Clayton AHA, Tavarnesi ML, Johns TG (2007) Unligated epidermal growth factor receptor forms higher order oligomers within microclusters on A431 cells that are sensitive to tyrosine kinase inhibitor binding. Biochemistry 46:4589–4597
15. Clayton AHA, Walker F, Orchard SG, Henderson C, Fuchs D, Rothacker J, Nice EC, Burgess AW (2005) Ligand-induced dimer-tetramer transition during the activation of the cell surface epidermal growth factor receptor-a multidimensional microscopy analysis. J Biol Chem 280:30392–30399
16. Cohen S (1962) Isolation of a mouse submaxillary gland protein accelerating incisor eruption and eyelid opening in the newborn animal. J Biol Chem 237:1555–1562
17. Cohen S, Levi-Montalcini R (1957) Purification and properties of a nerve growth-promoting factor isolated from mouse sarcoma 180. Cancer Res 17:15–20
18. Connolly JL, Seeley PJ, Greene LA (1987) Rapid regulation of neuronal growth cone shape surface morphology by nerve growth factor. Neurochem Res 12:861–868
19. Defize LH, Boonstra J, Meisenhelder J, Kruijer W, Tertoolen LG, Tilly BC, Hunter T, van Bergen en Henegouwen PM, Moolenaar WH, Henegouwen PM Van Bergen en, Moolenaar WH, De Laat SW (1989) Signal transduction by epidermal growth factor occurs through the subclass of high affinity receptors. J Cell Biol 109:2495–2507
20. Defize LHK, Arndt-Jovin DJ, Jovin TM, Boonstra J, Meisenhelder J, Hunter T, de Hey HT, de Laart SW (1988) A431 cell variants lacking the blood group A antigen display increased high affinity epidermal growth factor-receptor number, protein-tyrosine kinase activity receptor turnover. J Cell Biol 107:939–949
21. Delcroix JD, Valletta JS, Wu C, Hunt SJ, Kowal AS, Mobley WC (2003) NGF signaling in sensory neurons: evidence that early endosomes carry NGF retrograde signals. Neuron 39:69–84
22. Downward J, Parker P, Waterfield MD (1984) Autophosphorylation sites on the epidermal growth factor receptor. Nature 311:483–485
23. Egan SE, Giddings BW, Brooks MW, Buday L, Sizeland AM, Weinberg RA (1993) Association of Sos Ras exchange protein with Grb2 is implicated in tyrosine kinase signal transduction and transformation. Nature 363:45–51
24. Ferguson KM, Berger MB, Mendrola JM, Cho HS, Leahy DJ, Lemmon MA (2003) EGF activates its receptor by removing interactions that autoinhibit ectodomain dimerization. Mol Cell 11:507–517
25. Ferrell JE Jr (1996) Tripping the switch fantastic: how a protein kinase cascade can convert graded inputs into switch-like outputs. Trends Biochem Sci 21:460–466
26. Flomenbom O, Velonia K, Loos D, Masuo S, Cotlet M, Engelborghs Y, Hofkens J, Rowan AE, Nolte RJ, Van der Auweraer M et al (2005) Stretched exponential decay and correlations in the catalytic activity of fluctuating single lipase molecules. Proc Natl Acad Sci USA 102:2368–2372
27. Funatsu T, Harada Y, Tokunaga M, Saito K, Yanagida T (1995) Imaging of single fluorescent molecules individual ATP turnovers by single myosin molecules in aqueous solution. Nature 374:555–559

28. Gale NW, Kaplan S, Lowenstein EJ, Schlessinger J, Bar-Sagi D (1993) Grb2 mediates the EGF-dependent activation of guanine nucleotide exchange on Ras. Nature 363:88–92
29. Garrett T, McKern N, Lou M, Elleman T, Adams T, Lovrecz G, Zhu H, Walker F, Frenkel M, Hoyne P et al (2002) Crystal structure of a truncated epidermal growth factor receptor extracellular domain bound to transforming growth factor. Cell 110:763–773
30. Gatzinsky KP, Haugland RP, Thrasivoulou C, Orike N, Budi-Santoso AW, Cowen T (2001) p75 and TrkA receptors are both required for uptake of NGF in adult sympathetic neurons: use of a novel fluorescent NGF conjugate. Brain Res 920:226–238
31. Graus-Porta D, Beerli RR, Daly JM, Hynes NE (1997) ErbB-2, the preferred heterodimerization partner of all ErbB receptors, is a mediator of lateral signaling. EMBO J 16:1647–1655
32. Grimes ML, Beattie E, Mobley WC (1997) A signaling organelle containing the nerve growth factor-activated receptor tyrosine kinase, TrkA. Proc Natl Acad Sci USA 94:9909–9914
33. Gschwind A, Fischer OM, Ullrich A (2004) The discovery of receptor tyrosine kinases: targets for cancer therapy. Nat Rev 4:361–370
34. Gustafsson MGL (2005) Nonlinear structured-illumination microscopy: Wide-field fluorescence imaging with theoretically unlimited resolution. Proc Natl Acad Sci USA 102:13081–13086
35. Heintzmann R, Jovin TM, Cremer C (2002) Saturated patterned excitation microscopy – a concept for optical resolution improvement. J Opt Soc Am A 19:1599–1609
36. Hell SW, Wichmann J (1994) Breaking the diffraction resolution limit by stimulated emission: stimulated-emission-depletion fluorescence microscopy. Opt Lett 19:780–782
37. Hendry IA, Stockel K, Thoenen H, Iversen LL (1974) The retrograde axonal transport of nerve growth factor. Brain Res 68:103–121
38. Hibino K, Shibata T, Yanagida T, Sako Y (2009) A RasGTP-induced conformational change in C-RAF essential for accurate molecular recognition. Biophys J 97:1277–1287
39. Hibino K, Watanabe T, Kozuka J, Iwane AH, Okada T, Kataoka T, Yanagida T, Sako Y (2003) Single- and multiple-molecule dynamics of the signaling from H-Ras to c-Raf1 visualized on the plasma membrane of living cells. Chem Phys Chem 4:748–753
40. Ichinose J, Murata M, Yanagida T, Sako Y (2004) EGF signaling amplification induced by dynamic clustering of EGFR. Biochem Biophys Res Commun 324:1143–1149
41. Kamer AR, Sacks PG, Vladutiu A, Liebow C (2004) EGF mediates multiple signals: dependence on the conditions. Int J Mol Med 13:143–147
42. Kano F, Sako Y, Tagaya M, Yanagida T, Murata M (2000) Reconstitution of brefeldin A-induced golgi tubulation fusion with the endoplasmic reticulum in semi-intact chinese hamster ovary cells. Mol Biol Cell 11:3073–3087
43. Kasaian MT, Jacobberger JW, Neet KE (1994) Flow cytometric analysis of fluorescein-labeled nerve growth factor binding to A875 human melanoma cells. Exp Cell Res 210:77–85
44. Kim HK, Kim JW, Zilberstein A, Margolis B, Kim JG, Schlessinger J, Rhee SG (1991) PDGF stimulation of inositol phospholipid hydrolysis requires PLC-gamma 1 phosphorylation on tyrosine residues 783 and 1254. Cell 65:435–441
45. Klar TA, Hell SW (1999) Subdiffraction resolution in far-field fluorescence microscopy. Opt Lett 24:954–956
46. Koizumi S, Contreras ML, Matsuda Y, Hama T, Lazarovici P, Guroff G (1988) K-252a: a specific inhibitor of the action of nerve growth factor on PC 12 cells. J Neurosci 8:715–721
47. Konopka CA, Bednarek SY (2008) Variable-angle epifluorescence microscopy: a new way to look at protein dynamics in the plant cell cortex. Plant J 53:186–196
48. Kucik DF, Elson EL, Sheetz MP (1999) Weak dependence of mobility of membrane protein aggregates on aggregate size supports a viscous model of retardation of diffusion. Biophys J 76:314–322
49. Lalli G, Schiavo G (2002) Analysis of retrograde transport in motor neurons reveals common endocytic carriers for tetanus toxin neurotrophin receptor p75NTR. J Cell Biol 156:233–239

50. Levi A, Shechter Y, Neufeld EJ, Schlessinger J (1980) Mobility, clustering, and transport of nerve growth factor in embryonal sensory cells in a sympathetic neuronal cell line. Proc Natl Acad Sci USA 77:3469–3473
51. Levi-Montalcini R (1982) Developmental neurobiology the natural history of nerve growth factor. Annu Rev Neurosci 5:341–362
52. Li N, Batzer A, Daly R, Yajnik V, Skolnik E, Chardin P, Bar-Sagi D, Margolis B, Schlessinger J (1993) The SH2 and SH3 domains of mammalian Grb2 couple the EGF receptor to the Ras activator mSos1. Nature 363:85–88
53. Lommerse PHM, Snaar-Jagalska BE, Spaink HP, Schmidt T (2005) Single-molecule diffusion measurements of H-Ras at the plasma membrane of live cells reveal microdomain localization upon activation. J Cell Sci 118:1799–1809
54. Majerus PW, Connolly TM, Deckmyn H, Ross TS, Bross TE, Ishii H, Bansal VS, Wilson DB (1986) The metabolism of phosphoinositide-derived messenger molecules. Science 234:1519–1526
55. Martin-Fernandez M, Clarke DT, Tobin MJ, Jones SV, Jones GR (2002) Preformed oligomeric epidermal growth factor receptors undergo an ectodomain structure change during signaling. Biophys J 82:2415–2427
56. Michikawa T, Miyawaki A, Furuichi T, Mikoshiba K (1996) Inositol 1,4,5-trisphosphate receptors calcium signaling. Crit Rev Neurobiol 10:39–55
57. Mineo C, Gill GN, Gerson R (1999) Regulated migration of epidermal growth factor receptor from caveolae. J Biol Chem 274:30636–30643
58. Moriki T, Maruyama H, Maruyama IN (2001) Activation of preformed EGF receptor dimers by ligand-induced rotation of the transmembrane domain. J Mol Biol 311:1011–1026
59. Morimatsu M, Takagi H, Ota KG, Iwamoto R, Yanagida T, Sako Y (2007) Multiple-state reactions between the epidermal growth factor receptor Grb2 as observed using single-molecule analysis. Proc Natl Acad Sci USA 104:18013–18018
60. Nagashima T, Shimodaira H, Ide K, Nakakuki T, Tani Y, Takahashi K, Yumoto N, Hatakeyama M (2007) Quantitative transcriptional control of ErbB receptor signaling undergoes graded to biphasic response for cell differentiation. J Biol Chem 282:4045–4056
61. Nishizuka Y (1984) Turnover of inositol phospholipids signal transduction. Science 225:1365–1370
62. Oda K, Matsuoka Y, Funahashi A, Kitano H (2005) A comprehensive pathway map of epidermal growth factor receptor signaling. Mol Syst Biol. doi:10.1038/msb4100014
63. Ogiso H, Ishitani R, Nureki O, Fuakai S, Yamanaka M, Kim J-H, Saito K, Sakamoto A, Inoue M, Shirouzu M, Yokoyama S (2002) Crystal structure of the complex of human epidermal growth factor receptor extracellular domains. Cell 110:775–787
64. Riedel H, Dull TJ, Schlessinger J, Ullrich A (1986) A chimeric receptor allows insulin to stimulate tyrosine kinase activity of epidermal growth factor receptor. Nature 324:68–70
65. Rozakis-Adcock M, Fernley R, Wade J, Pawson T, Bowtell D (1993) The SH2 and SH3 domains of mammalian Grb2 couple the EGF receptor to the Ras activator mSos1. Nature 363:83–85
66. Rust MJ, Bates M, Zhuang X (2006) Sub-diffraction-limit imaging by stochastic optical reconstruction microscopy (STORM). Nat Methods 3:793–795
67. Sako Y, Hibino K, Miyauchi T, Miyamoto Y, Ueda M, Yanagida T (2000) Single-molecule imaging of signaling molecules in living cells. Single Mol 1:159–163
68. Sako Y, Minoghchi S, Yanagida T (2000) Single-molecule imaging of EGFR signaling on the surface of living cells. Nat Cell Biol 2:168–172
69. Santini F, Keen JH (1996) Endocytosis of activated receptors clathrincoated pit formation: deciphering the chicken or egg relationship. J Cell Biol 132:1025–1036
70. Schlessinger J (1988) Signal transduction by allosteric receptor oligomerization. Trends Biochem Sci 13:443–447
71. Schlessinger J (2000) Cell signaling by receptor tyrosine kinases. Cell 103:211–225

72. Schütz GJ, Kada G, Pastushenko VP, Schindler H (2000) Properties of lipid microdomains in a muscle cell membrane visualized by single molecule microscopy. EMBO J 19:892–901
73. Shao Y, Akmentin W, Toledo-Aral JJ, Rosenbaum J, Valdez G, Cabot JB, Hilbush BS, Halegoua S (2002) Pincher, a pinocytic chaperone for nerve growth factor/TrkA signaling endosomes. J Cell Biol 157:679–691
74. Shibata SC, Hibino K, Mashimo T, Yanagida T, Sako Y (2006) Formation of signal transduction complexes during immobile phase of NGFR movements. Biochem Biophys Res Commun 342:316–322
75. Shroff H, Galbraith CG, Galbraith JA, White H, Gillette J, Olenych S, Davidson MW, Betzig E (2007) Dual-color superresolution imaging of genetically expressed probes within individual adhesion complexes. Proc Natl Acad Sci USA 104:20308–20313
76. Simson R, Sheets ED, Jacobson K (1995) Detection of temporary lateral confinement of membrane proteins using single-particle tracking analysis. Biophys J 69:989–993
77. Stout AL, Axelrod D (1989) Evanescent field excitation of fluorescence by epi-illumination microscopy. Appl Opt 28:5237–5242
78. Subach FV, Patterson GH, Manley S, Gillette JM, Lippincott-Schwartz J, Verkhusha VV (2009) Photoactivatable mCherry for high-resolution two-color fluorescence microscopy. Nat Methods 6:153–159
79. Sutter A, Riopelle RJ, Harris-Warrick RM, Shooter EM (1979) Nerve growth factor receptors. Characterization of two distinct classes of binding sites on chick embryo sensory ganglia cells. J Biol Chem 254:5972–5982
80. Szabó A, Horváth G, Szöllösi J, Nagy P (2008) Quantitative characterization of the large-scale association of ErbB1 and ErbB2 by flow cytometric homo-FRET measurements. Biophys J 95:2086–2096
81. Tani T, Miyamoto Y, Fujimori KE, Taguchi T, Yanagida T, Sako Y, Harada Y (2005) Trafficking of a ligand-receptor complex on the growth cones as an essential step for the uptake of nerve growth factor at the distal end of the axon: a single-molecule analysis. J Neurosci 25:2181–2191
82. Teramura Y, Ichinose J, Takagi H, Nishida K, Yanagida T, Sako Y (2006) Single-molecule analysis of epidermal growth factor binding on the surface of living cells. EMBO J 25:4215–4222
83. Tokunaga M, Imamoto N, Sakata-Sogawa K (2008) Highly inclined thin illumination enables clear single-molecule imaging in cells. Nat Methods 5:159–161
84. Tokunaga M, Kitamura K, Saito K, Iwane AH, Yanagida T (1997) Single molecule imaging of fluorophores enzymatic reactions achieved by objective-type total internal reflection fluorescence microscopy. Biochem Biophys Res Commun 235:47–53
85. Tzahar E, Waterman H, Chen X, Levkowitz G, Karunagaran D, Lavi S, Ratzkin BJ, Yarden Y (1996) A hierarchical network of interreceptor interactions determines signal transduction by Neu differentiation factor/neuregulin and epidermal growth factor. Mol Cell Biol 16:5276–5287
86. Uyemura T, Takagi H, Yanagida T, Sako Y (2005) Single-molecule analysis of epidermal growth factor signaling that leads to ultrasensitive calcium response. Biophys J 88:3720–3730
87. van Bergen en Henegouwen PM, den Hartigh JC, Romeyn P, Verkleij AJ, Boonstra J (1992) The epidermal growth factor receptor is associated with actin filaments. Exp Cell Res 199:90–97
88. Vrljic M, Nishimura SY, Brasselet S, Moerner WE, McConnell HM (2002) Translational diffusion of individual class II MHC membrane proteins in cells. Biophys J 83:2681–2692
89. Wahl MI, Nishibe S, Kim JW, Kim H, Rhee SG, Carpenter G (1990) Identification of two epidermal growth factor-sensitive tyrosine phosphorylation sites of phospholipase C-gamma in intact HSC-1 cells. J Biol Chem 265:3944–3948
90. Webb SED, Roberts SK, Needham SR, Tynan CJ, Rolfe DJ, Winn MD, Clarke DT, Barraclough R, Martin-Fernandez ML (2008) Single-molecule imaging and fluorescence

lifetime imaging microscopy show different structures for high- and low-affinity epidermal growth factor receptors in A431 cells. Biophys J 94:803–819
91. Xiao Z, Ma X, Jiang Y, Zhao Z, Lai B, Liao J, Yue J, Fang X (2008) Single-molecule study of lateral mobility of epidermal growth factor receptor 2/HER2 on activation. J Phys Chem B 112:4140–4145
92. Xiao Z, Zhang W, Yang Y, Xu L, Fang X (2008) Single-molecule diffusion study of activated EGFR implicates its endocytic pathway. Biochem Biophys Res Commun 369:730–734
93. Yarden Y, Schlessinger J (1987) Epidermal growth factor induces rapid, reversible aggregation of the purified epidermal growth factor receptor. Biochemistry 26:1443–1451
94. Yarden Y, Schlessinger J (1987) Self-phosphorylation of epidermal growth factor receptor: evidence for a model of intermolecular allosteric activation. Biochemistry 26:1434–1442
95. Yu C, Hale J, Ritchie K, Prasad NK, Irudayaraj J (2009) Receptor overexpression or inhibition alters cell surface dynamics of EGF-EGFR interaction: new insights from real-time single molecule analysis. Biochem Biophys Res Commun 378:376–382
96. Yu X, Sharma KD, Takahashi T, Iwamoto R, Mekada E (2002) Ligand-independent dimer formation of epidermal growth factor receptor (EGFR) is a step separable from ligand-induced EGFR signaling. Mol Biol Cell 13:2547–2557
97. Zhang X, Gureasko J, Shen K, Cole PA, Kuriyan J (2006) An allosteric mechanism for activation of the kinase domain of epidermal growth factor receptor. Cell 125:1137–1149

Chapter 2
Single-Molecule Kinetic Analysis of Stochastic Signal Transduction Mediated by G-Protein Coupled Chemoattractant Receptors

Yukihiro Miyanaga and Masahiro Ueda

Abstract Cellular chemotactic behaviors are typical examples of stochastic signal transduction in living cells that have been investigated in detail both experimentally and theoretically. In this chapter, we describe single-molecule kinetic analysis for stochastic signal transduction in chemotactic responses mediated by G protein-coupled chemoattractant receptors in order to give deeper understanding of the stochastic nature in chemotactic signaling processes. We also describe theoretical analysis of receptor-mediated chemotactic signaling, which reveals that noise generated in the transmembrane signaling by G protein-coupled chemoattractant receptors limits the precision of the gradient sensing. This suggests that receptor-G protein coupling and its modulation have an important role for improving the signal-to-noise ratio of chemotactic signals and thus cellular chemotaxis. Extending this beyond G protein signaling, combining single-molecule kinetic analysis with theoretical analysis offers a new tool in exploring the relationship between the kinetic properties of signaling molecules and their corresponding cellular responses in general.

Keywords *Dictyostelium* · Chemoattractant · cAMP · GPCR · cAR1 · PTEN · PI3K · Crac · TIRFM · Halo · Chemotaxis · Noise · Fluctuation · *Gain-fluctuation relation* · GFR · SNR · Kinetic heterogeneity · Polarity

---

Y. Miyanaga (✉) and M. Ueda
Laboratories for Nanobiology, Graduate School of Frontier Biosciences, Osaka University, Suita, Osaka 565-0871, Japan
and
Japan Science and Technology Agency, Core Research for Evolutional Science and Technology, Suita, Osaka 565-0871, Japan
e-mail: yukihiro@phys1.med.osaka-u.ac.jp; ueda@phys1.med.osaka-u.ac.jp

Y. Sako and M. Ueda (eds.), *Cell Signaling Reactions: Single-Molecular Kinetic Analysis*,
DOI 10.1007/978-90-481-9864-1_2,

2.1 Introduction

Intracellular signal transduction has a stochastic nature due to its underlying chemical and physical reactions. For example, when an extracellular ligand binds stochastically to a cell's membrane receptors, the number of occupied receptors fluctuates with time and space in cells. The average of this number corresponds to concentrations of the extracellular ligand, while fluctuations in this number are considered a counting error, or noise, a result of the receptor's stochastic reactions. Downstream of the receptor are many types of signaling molecules involved in the signal processing and transduction. These molecules also operate stochastically, making the signal transmission inevitably noisy. Revealing how cells identify environmental cues enmeshed inside a noisy signal is crucial to further our understanding of the computational principles involved in intracellular signal transduction. This is an important question, as the stochastic computation systems in living cells are either noise-robust or noise-utilizing information processors that require far less energy than current artificial systems, and therefore offers a new design paradigm.

For such studies, single-molecule imaging has been successfully applied to a variety of biomolecules both in vitro and in living cells, leading to advances in the understanding of these molecules' stochastic nature [9, 42, 51, 54, 61]. Examples include the random transitions between open and closed states of single ion channels, stochastic behaviors in catalytic reactions done by single enzyme molecules, and stepwise motions of molecular motors [20, 23, 40]. In living cells, vital unitary reactions for signal transductions involving the signaling molecules like complex formations, conformational changes, and diffusion have been visualized at the single-molecule level [14, 41, 42, 54]. The stochastic properties of these reactions, such as the kinetic rates of the association and dissociation of complexes and diffusion coefficients, have been determined experimentally in the context of the cellular microenvironments. Intracellular microenvironments such as ion concentrations, lipid compositions, and cytoskeletal organizations have the potentials to provide variations for the properties of individual signaling molecules. In fact, various kinds of signaling molecules have been found to exhibit heterogeneity in their behaviors in relation to the spatiotemporal changes of the intracellular microenvironment [13, 27, 54]. Such information is lost when making ensemble measurements of a large number of molecules. Thus, single-molecule imaging analysis provides a unique tool for revealing the kinetic details of signaling molecules in living cells.

Similarly, theoretical approaches have been applied to reveal how stochasticity in molecular reactions affects cellular behaviors. For example, Berg and Purcell revealed that molecule counting noise generated by receptors limits the precision of chemoreception on measurements of chemical concentrations in bacteria chemotaxis [2]. Bialek and Setayeshgar compared theoretical and experimental studies on the chemotaxis of *Escherichia coli*, showing that the bacterial chemotaxis is performed near the physical limits in chemoreception [3]. For amoeboid chemotaxis, Tranquillo et al. described a stochastic model in which fluctuations in ligand-receptor binding reactions affect motile behaviors and limit chemotactic accuracy [52]. However, these theoretical analyses have been performed despite knowing very little about the kinetic details of signaling molecules in living cells, and therefore further theoretical analyses are required to quantitatively describe the relationship between the inherent noise of the signaling molecules and the corresponding cellular response. Recently, Shibata and Fujimoto proposed the *gain-fluctuation relation* (GFR) for stochastic signal transduction, which can describe how stochastic noise is generated, amplified and propagated along a signaling cascade ([46], Chapter 13). Based on the kinetic properties of signaling molecules, the GFR can estimate the noise strength propagated along the signal transduction process.

In this chapter, we discuss the chemotactic signaling of eukaryote *Dictyostelium discoideum* cells as a typical example for stochastic signal transduction. Specific attention is given to single-molecule kinetic analysis and GFR-based theoretical analysis, which can quantitatively explain the stochastic cellular responses of this system by deriving the kinetic properties of the relevant signaling molecules.

2.2 Gradient Sensing and Directional Cell Migration in *Dictyostelium discoideum*

Dictyostelium discoideum is a well-established model organism for elucidating molecular mechanisms of amoeboid movements and regulation. This organism has many advantages for molecular and cellular biology studies including having its genome completely sequenced, well-established genetic engineering techniques, and advanced microscopic techniques [6, 11, 12, 32, 38, 62]. In addition, there are several more reasons why it is ideal for cellular motility research. For example, *Dictyostelium* cells are highly motile and exhibit fast amoeboid movements, with a velocity of 10–20 μm/min on glass substrates. The crawling movements on the surface take place spontaneously and randomly, even in the absence of extracellular cues [50]. Under a spatial heterogeneous environment of extracellular cues, cells exhibit tactic behaviors such as chemotaxis and electrotaxis [44, 48]. At the aggregation stage during their life cycle, cells exhibit chemotaxis in response to cyclic adenosine 3′5′-monophosphate (cAMP)(Fig. 2.1a), which induces multicellular organization [45]. About one hundred thousand of the cells move directionally toward the aggregation center by chemotaxis, and then form one multicellular aggregate. In this situation, more than 99% of cells exhibit synchronous chemotaxis, making it possible to prepare highly homogeneous cell populations. These properties have enabled investigators to study numerous essential intracellular activities including the roles of actin and microtubule cytoskeletons on cellular motility [11, 32, 62]. The molecular components required for chemotactic responses have been also identified extensively [12, 16, 33, 35, 43, 48]. Furthermore, imaging analysis of cytoskeletal proteins and their regulatory molecules has revealed the dynamic behaviors of these molecules during amoeboid movements in

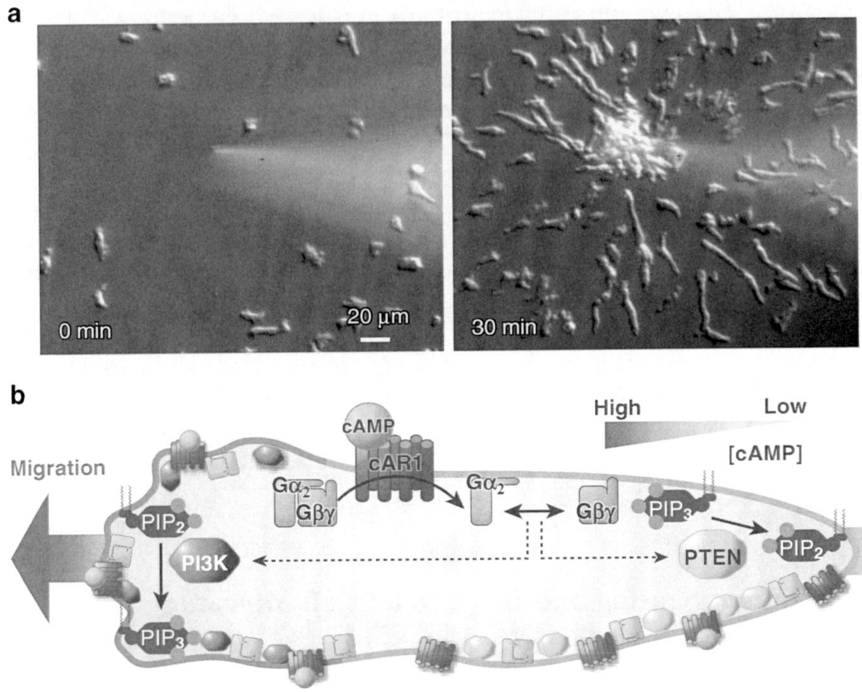

Fig. 2.1 Chemotactic signaling of *Dictyostelium* cells. (**a**) Chemotaxis of *Dictyostelium* cells toward the tip of a micropipette containing the chemoattractant cAMP. Images were captured at time 0 (*left panel*) and 30 min after (*right panel*) introduction of the micropipette. (**b**) cAMP signals are mediated by G protein-coupled cAMP receptors, especially cAR1. Receptor stimulation results in the activation and dissociation of heterotrimeric G protein, followed by the regulation of downstream signaling molecules such as PTEN and PI3K. PTEN catalyzes the dephosphorylation of PtdIns(3,4,5)P_3, while PI3K catalyzes the phosphorylation of PtdIns(4,5)P_2. Such antagonistic activities are essential for PtdIns(3,4,5)P_3 localization at the leading edge of chemotactic cells. Abbreviations: PTEN, phosphatase and tensin homologue deleted on chromosome 10; PI3K, phosphatidylinositol 3-kinase; PIP_2, phosphatidylinositol (4,5) bisphosphate or PtdIns(4,5)P_2; PIP_3, phosphatidylinositol (3,4,5) triphosphate or PtdIns(3,4,5)P_2.

response to chemotactic stimulations [5, 10, 12, 48]. In particular cases, signaling molecules have been observed at the single molecule level in living cells [27, 28, 31, 54, 57].

Chemotaxis is a fascinating phenomenon from the viewpoint of stochastic signal transduction in living cells. Cells show extreme sensitivity to subtle cAMP concentration gradients (Fig. 2.1b). The chemotactic ability of cells to cAMP gradients is studied quantitatively and signal inputs for chemotaxis are estimated [7, 26, 47, 55]. In a range of 0.1 nM to several μM cAMP, even a 2% gradient across the cell body can trigger a chemotactic response. When we consider a cell length of 10 μm with 80,000 receptors and $K_d = 100$ nM, for such a gradient (for example, ~3.3×10^{-3} nM/μm assuming 0.5 nM cAMP), the receptor occupancy is estimated to be ~400 molecules, while the difference in receptor occupancy between front and back halves of the cell is only on the order of 10 molecules. Such a small difference suggests that the stochastic properties of the receptor may affect gradient sensing ability. Assuming a Poisson process, fluctuations in receptor occupancy equal the square root of the averaged occupancy, which for 400 occupied receptors translates to ~20 molecules, which is larger than the front-back differences. This means that stochastic fluctuations can reverse the ligand-binding patterns along the length of a cell even when the extracellular chemoattractant gradient is stationary, which implies that signal inputs for chemotaxis are noisy. Thus, *Dictyostelium* cells can sense a faint signal within stochastic noise. Understanding how the signaling system is designed to reliably obtain information in such an environment is an important issue in chemotaxis studies.

Signaling components and their reaction networks for chemotaxis are largely shared among many eukaryotic cells (Fig. 2.1b) [12, 16, 33, 35, 43, 48]. In *Dictyostelium* cells, chemotaxis in response to cAMP is mediated by G protein-coupled cAMP receptors (cARs). Among them, cAR1 functions dominantly at the aggregation stage. cAMP binding to cAR1 activates intracellular heterotrimeric G protein, which facilitates the G protein to separate into its two subunits, $G\alpha_2$ and $G\beta\gamma$. These subunits transduce signals that regulate downstream molecules including Ras, PI3K, PTEN, Crac, guanylyl cyclase, $PLC\gamma$, and PLA_2. One key reaction in this signaling system is the localization of phosphatidylinositol 3,4,5-trisphosphate ($PtdIns(3, 4, 5)P_3$) on the membrane side facing a higher concentration of cAMP. $PtdIns(3, 4, 5)P_3$ localization is regulated by PI3K and PTEN, which catalyze $PtdIns(3, 4, 5)P_3$ production and degradation, respectively. Because PI3K and PTEN are reciprocally localized on the membrane sides facing higher and lower cAMP concentration, respectively, $PtdIns(3, 4, 5)P_3$ accumulates at the leading edge of chemotaxing cells, acting as a cue for pseudopod formation to bias cell movement. Thus, a chemotactic signaling system can convert small differences in extracellular signals into localized signals that facilitate pseudopod formation and maintenance. $PtdIns(3, 4, 5)P_3$ localization takes place in an all-or-none manner with respect to the direction of the chemoattractant gradient [18], suggesting that noisy chemotaxis signals are somehow processed to consistently reflect the direction of the chemoattractant gradient. To understand how signal inputs for chemotaxis are received, processed and transduced by stochastically-operating signaling molecules, it is important to uncover the stochastic nature of the signaling molecules at work.

2.3 Single-Molecule Imaging Analysis of Chemotactic Signaling System in Living Cells

We have applied total internal reflection fluorescence microscopy (TIRFM) for single molecule imaging in a living cell (Fig. 2.2a). Originally, TIRFM was used for imaging multi-molecules and organelles in living cells [1], and has been further refined for single-molecule imaging of biomolecules [9, 42, 51, 54]. The theoretical aspects of TIRFM are discussed elsewhere [1, 8, 59]. Also, experimental apparatus and sample preparation for *Dictyostelium* cells are described previously in detail [28, 31].

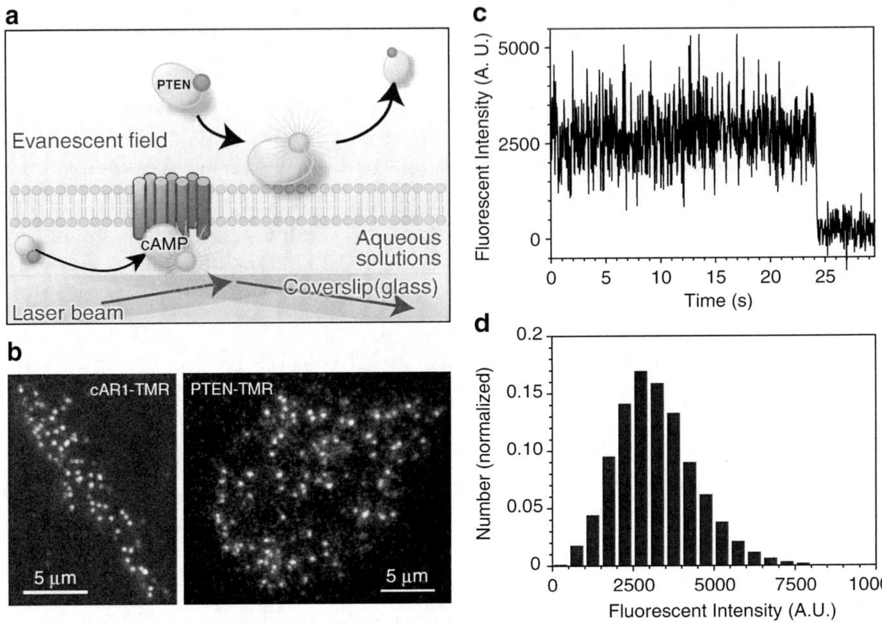

Fig. 2.2 Single molecule imaging in living cells. (**a**) Schematic drawing of single molecule imaging. The basal membrane of a *Dictyostelium* cell is illuminated by evanescent fields generated by the total internal reflection of an incident laser. Diffusible fluorescent molecules in the cytoplasm or extracellular aqueous solution, such as PTEN-TMR and Cy3-cAMP, are not seen due to their rapid diffusions. However, their fluorescence can be seen when these molecules bind to the plasma membrane or membrane-bound receptors. (**b**) Typical images of single-molecules of cAR1-TMR (*left panel*) and PTEN-TMR (*right panel*). (**c**) Typical time course of the fluorescence intensity arising from a single cAR1-TMR spot showing single-step photobleaching. (**d**) Distribution of cAR1-TMR fluorescence intensities.

Target molecules for single-molecule imaging by TIRFM should be labeled with a fluorescent dye, which must be selected with care. In order to conduct kinetic analysis through single-molecule imaging, we need fluorophores that do not interfere with the function of the molecule. Also, fluorescent dyes that have a bright fluorescent intensity and are stable over a long observation time are preferred. Commonly used labels include green fluorescent protein (GFP or its variants) or a tag protein such as HaloTag [17, 27, 28]. We have applied several fluorescent-labeling techniques to *Dictyostelium* cells, having examined the photobleaching time of the label and signal-noise ratio of the acquired images. Currently, we find that HaloTag labeling is best for the single molecule imaging of signaling molecules in *Dictyostelium* cells.

In the HaloTag labeling method, the gene encoding target protein is fused to the HaloTag gene and transferred into the cells. There, the HaloTag-conjugated proteins are labeled specifically by treating with a fluorescent HaloTag ligand like tetramethylrhodamine (TMR). Using our equipment, TMR has both better brightness and a longer photobleaching time than fluorescent proteins such as GFP or YFP. In fact, we can observe individual single molecules in living cells for tens of seconds or more under continuous excitation. Moreover, the labeling efficiency can be controlled to a level suitable for single-molecule imaging by adjusting the amount of HaloTag ligand. Labeling efficiency is sometimes problematic for single-molecule analysis because a high density of labeled molecules makes it difficult to identify single molecules. Therefore, the labeling efficiency should be such that overlapping between fluorescent spots when the molecules exhibit rapid diffusion is avoided. Empirically, the densities of the fluorescent spots should be limited to less than ~0.1 molecule/μm^2 for molecules with diffusion coefficients of ~0.2 $\mu m^2/s$ and less than ~0.7 molecule/μm^2 for molecules with diffusion coefficients of ~0.02 $\mu m^2/s$.

The simplest way to verify the functionality of the labeled molecules is to substitute these with native molecules, which can be easily done in *Dictyostelium* cells [15, 21, 25]. Using this haploid organism, functional fusion proteins for chemotactic signaling molecules including cAMP receptor, subunits of trimeric G protein, PI3K and PTEN have been successfully generated. Halo-tagged cAR1, $G\alpha_2$, $G\gamma$ and PTEN molecules can be confirmed as functional by examining whether the Halo-tagged genes for these molecules can rescue phenotypic defects in cells lacking the corresponding genes [28]. For instance, PTEN knockout cells exhibit severe defects in cell division, morphology and chemotactic movement. When HaloTag-conjugated PTEN (PTEN-Halo) is introduced, the transformed *pten*-null cells behave like wild type cells. PTEN-Halo was visualized by staining with TMR (PTEN-TMR), revealing a uniform localization on the membrane, as expected for functional PTEN molecules. When a fluorescent analogue of an extracellular ligand like chemoattractant is used for single molecule imaging analysis, it is also important to confirm whether the labeled analogue can activate the corresponding signaling pathways. We have generated a fluorescent analogue for cAMP (Cy3-cAMP) to monitor the signal inputs for chemotaxis [18, 54]. Cy3B-cAMP, which is commercially available (Amersham LKB, a component of cAMP Fluorescence Polarization Immunoassay kit, RPN3595), can also be used.

Figure 2.2b shows a typical example of fluorescent spots arising from TMR-labeled cAR1-Halo (cAR1-TMR) on the basal membrane of living *Dictyostelium* cells. To verify that the fluorescent spots represent single cAR1-TMR molecules, some characteristic features of a single molecule should be confirmed. One is single-step photobleaching, which occurred upon continuous excitation (Fig. 2.2c), and is consistent with the photobleaching of single fluorescent molecules. The histogram for cAR1-TMR fluorescence intensity exhibited a Gaussian distribution with a single mean value (Fig. 2.2d), which is also characteristic of single fluorescent molecules. For a mixture containing dimer and monomer molecules, the distribution of the fluorescence intensities should exhibit a sum of two Gaussian distributions.

Using the same experimental setup, PTEN-TMR on the basal membrane of living *Dictyostelium* cells can also be observed (Fig. 2.2b). PTEN molecules undergo dynamic shuttling between the membrane and cytoplasm. PTEN-TMR fluorescent spots behave differently from those of cAR1-TMR. Free PTEN-TMR molecules in the cytosol cannot be imaged clearly as fluorescent spots at video frame rates because of rapid Brownian motion. When PTEN-TMR binds to membrane, however, diffusible movements slow due to its association, which allows for the visualization of clearly resolved TMR fluorescence spots (Fig. 2.2b). Following PTEN-TMR release from the membrane, the fluorescent spot suddenly disappears. Thus, association/dissociation events of a single signaling molecule on the membrane can be visualized in living cells. Additionally, the lateral diffusion of PTEN-TMR molecules can be observed simultaneously with association/dissociation reactions. Signaling molecules shuttling between the membrane and cytoplasm should also have a fluorescence intensity that exhibits a single-step disappearance and a quantized distribution. However, the disappearance of fluorescent spots should not necessarily be regarded as photobleaching because molecules may dissociate from the membrane before photobleaching occurs.

By analyzing the behaviors of individual molecules on the membrane statistically, their kinetic information such as dissociation rates of signaling complex can be obtained. Statistical analysis of the lateral diffusion is described in detail in Chapter 12 [29]. Here we explain briefly single-molecule kinetic analysis of signaling molecules for a simple case [28, 31]. First, the time duration between the appearance and disappearance of individual molecules on cells is measured. Then, the distribution of the time duration is constructed by counting the fraction of the duration. The number at $t = 0$ is the total number of molecules measured, which decays with time because of dissociation. Figure 2.3 shows a normalized distribution, also known as a cumulative probability distribution. Cumulative probability distributions of the time duration can be fitted with the following equation,

$$f(t) = a_1 \exp(-k_1 t) + a_2 \exp(-k_2 t) + \cdots \tag{2.1}$$

where k_i represents the dissociation rates and a_i represents the relative amount of the ith component. Summing a_i over all i equals 1. Inversing k_i gives the lifetimes, τ_i, of the molecules.

When membrane-integral proteins such as receptors are observed and analyzed, the distribution of the time duration reflects the photobleaching events because membrane-integral proteins do not dissociate from the membranes. Figure 2.3a show the cumulative probability distributions of cAR1-TMR and cAR1-YFP. The distribution of cAR1-TMR is well fitted to a single exponential curve with the time constant $\tau = 11$ s ($k_{bleaching} = 0.09\ \mathrm{s}^{-1}$), which represents the photobleaching time constant of TMR in living cells under our experimental system. On the other hand, cAR1-YFP has two time constants, $\tau_1 = 0.52$ s (86%) and $\tau_2 = 2.2$ s (14%). The multiple time constants suggests that YFP frequently exhibits photoblinking [30] (Fig. 2.3b). The shorter observable time and complicated photo-stability of YFP may cause misinterpretations of data for YFP-tagged target proteins.

For molecules dynamically shuttling between the membrane and cytosol, the cumulative probability distribution of the molecules reflects the release from the membrane if the dissociation rates of the molecules are relatively faster than

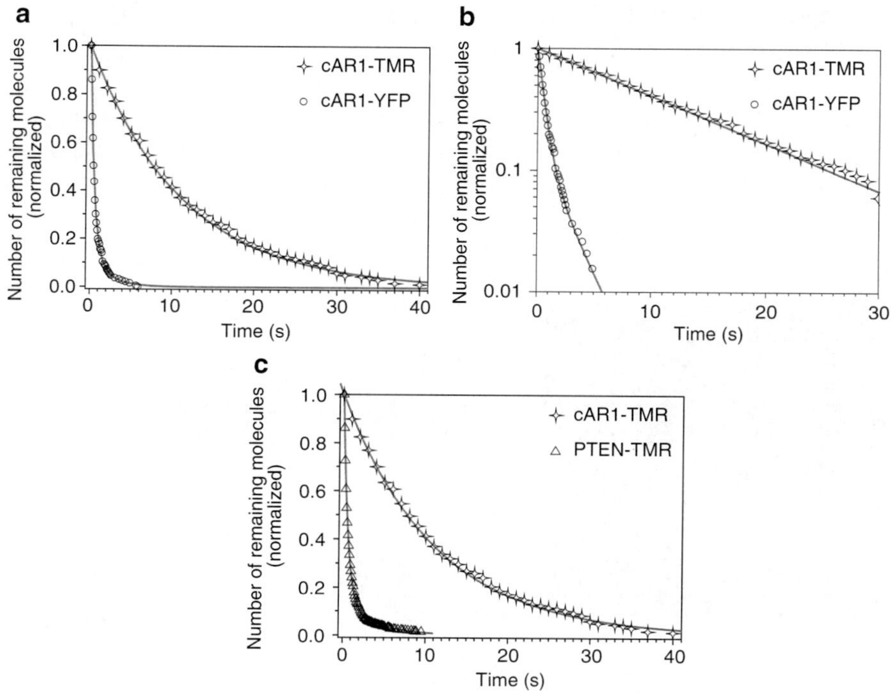

Fig. 2.3 Lifetime analysis of cAR1 and PTEN. (**a**) Cumulative probability distributions of cAR1-TMR (*stars*) and cAR1-YFP (*circles*). *Solid lines* represent the fitting curves of the data to a single exponential function with a time constant of 11 s (cAR1-TMR) and sum of two exponential functions with time constants of 0.52 s (86%) and 2.2 s (14%) (cAR1-YFP). (**b**) Semi-logarithmic plots of cumulative probability distributions facilitate the distinction between multiple and single time constant in the reaction. (**c**) Comparison of the lifetimes of cAR1-TMR (*stars*) and PTEN-TMR (*triangles*). Shorter lifetime of PTEN-TMR indicates the dissociation of PTEN-TMR from the membrane prior to the photobleaching of TMR.

the photobleaching rates of the fluorescence dyes. For example, the lifetimes of PTEN-TMR (0.6 s, 87% and 4.1 s, 13%) without correction are significantly shorter than that of cAR1-TMR (Fig. 2.3c), meaning that PTEN molecules dissociate from the membrane before photobleaching. Because the dissociation rates k_i obtained from Eq. 2.1 for experimental data are apparent values, they should be corrected by incorporating the photobleaching rate $k_{bleaching}$. The actual dissociation rates are obtained by subtracting the photobleaching rate from the apparent dissociation rates: $k = k_1 - k_{bleaching}$. The lifetime analysis of membrane-integral proteins such as receptors is sometimes performed as a control for signaling molecules that exhibit dynamic shuttling between the membrane and cytosol. Lifetime analysis has been successfully applied to kinetic studies of ligand-receptor complexes and down-stream molecules including G-protein, PTEN and PH domain-containing proteins, as described below [27, 28, 54, 57].

2.4 Single-Molecule Kinetic Analysis of Stochastic Signal Inputs for Chemotaxis

To elucidate how *Dictyostelium* cells sense the chemoattractant cAMP, fluorescently-labeled cAMP (Cy3-cAMP) was prepared and its binding to the receptors on the cells was observed by using TIRFM (Fig. 2.4a) [54]. When Cy3-cAMP solution

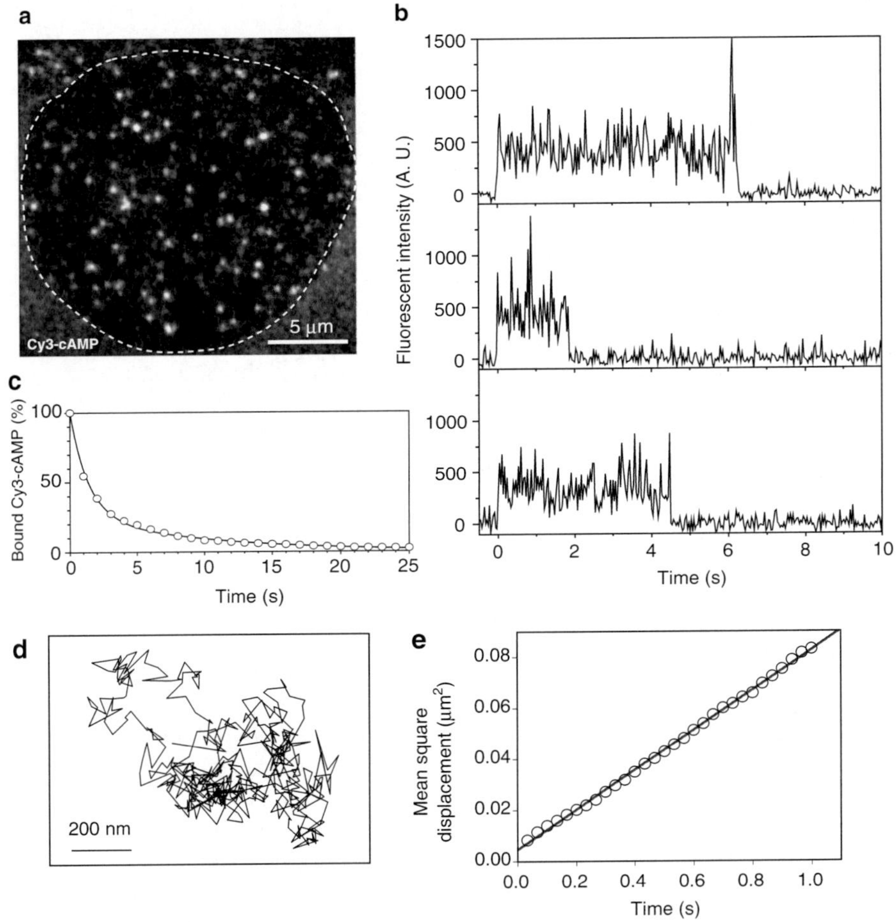

Fig. 2.4 Single molecule imaging analysis of signal inputs for chemotactic responses. (**a**) Typical image of Cy3-cAMP single-molecules. Cell contour is shown by *broken line*. (**b**) Typical time courses of fluorescence intensity from single Cy3-cAMP spots showing association and dissociation events between Cy3-cAMP and cAR1. Cy3-cAMP molecules bound to the receptor at time 0. (**c**) Cumulative probability distribution of Cy3-cAMP. (**d**) Typical single-molecule trajectory of Cy3-cAMP bound to its receptor on a living cell. Time length of the trajectory, 15.4 s. (**e**) Plot of the mean-squared displacement versus time, indicating that the receptor diffuses by simple Brownian motion with a diffusion coefficient of ~ 0.02 $\mu m^2/s$.

was applied to the cells, the fluorescence arising from the Cy3-cAMP molecule became visible as a single spot on the membrane, followed by the spot undergoing lateral diffusion and then suddenly disappearing (Fig. 2.4b). The cumulative probability distribution of Cy3-cAMP lifetimes were well fitted to a sum of exponential curves with constants of 0.93 s^{-1} (61%) and 0.11 s^{-1} (39%), which correspond to lifetimes of 1.1 and 8.2 s, respectively (Fig. 2.4c). These constants represent the dissociation rates of Cy3-cAMP-receptor complexes. In a simple kinetic scheme, ligand-binding reactions can be written as $R + L \leftrightarrow RL$, where R and L represent a receptor and ligand, respectively. The dissociation rates represent the rates for the state transition RL to R. Multiple dissociation rates indicate the receptor takes multiple kinetic states. Exponential distributions of the ligand-binding lifetimes demonstrate that the ligand dissociates from the receptor randomly. The majority of Cy3-cAMP-receptor complexes adopted the faster-dissociation state, meaning that the cells sense the chemoattractant within a 1 s sampling time. Lateral diffusion also can be analyzed simultaneously for individual Cy3-cAMP-bound receptors by calculating the mean square displacement (Fig. 2.4d, e) [29]. The diffusion coefficient of the individual Cy3-cAMP-receptor complex was ~0.02 $\mu m^2/s$. cAR1-Halo expressed in wild-type cells had a diffusion coefficient of ~0.02 $\mu m^2/s$ irrespective of cAMP being present, suggesting that ligand-binding to the cAR1 receptor does not affect lateral mobility on the membrane.

Single-molecule imaging of cAMP binding to the receptor in living cells can be also used to examine the receptor occupancy for chemotaxis. By counting the number of Cy3-cAMP complexes bound to a chemotaxing cell, the cell's receptor occupancy was determined, although the region measured was restricted to the cell's basal surface. Figure 2.5a shows an example of the temporal changes in

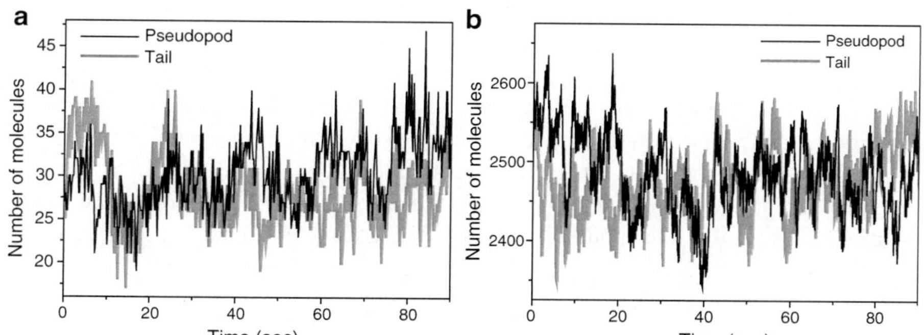

Fig. 2.5 Fluctuations in chemotactic signaling inputs. (**a**) Temporal changes in the number of bound cAMP molecules were measured by counting Cy3-cAMP spots on the anterior pseudopod (*black line*) and the posterior tail (*gray line*). (**b**) Numerical simulations of signal inputs. Temporal changes in the receptor occupancy in the anterior (*black line*) and posterior (*gray line*) halves of a single cell were simulated. It is assumed that the receptor adopt two states: fast- (dissociation rate, 1.0 s^{-1}; association rate, 4 × 10^{-6} $M^{-1}s^{-1}$; 78,000 molecules/cell) and slow-dissociating (dissociation rate, 0.04 s^{-1}; association rate, 4 × 10^{-6} $M^{-1}s^{-1}$; 4,000 molecules/cell).

receptor occupancy of a chemotaxing cell. The time series of the receptor occupancy at the anterior and posterior halves of the cell occasionally reverse against the direction of the cAMP gradient due to stochastic fluctuations in the ligand-receptor binding reaction. In other words, the posterior end may have higher occupancy despite the gradient being higher at the anterior. Thus, signal inputs on the basal surface of a chemotaxing cell are quite noisy. But we should note that the time series of the receptor occupancy does not represent the total inputs made by the chemotactic signals. This means we cannot exclude the possibility that such reversals are rare events when considering the cell's entire surface. Next, we performed a numerical simulation of ligand binding to the receptors (Fig. 2.5b). For simplicity, we assume that the receptors adopt slower- or faster-dissociating states stably without state transitions and that they are uniformly localized on the membrane. Also, we assume a Poisson process for the ligand-binding reaction. When the cAMP gradient across the cell body is 2% at 10 nM, differences in receptor occupancy between anterior and posterior halves of a cell can sometimes be negative, meaning that the ligand-binding patterns on the cell are reversed against the ligand concentration gradients. For direct measurements of total receptor occupancy over the whole surface of chemotaxing cells, further technical developments are required including numerical simulations [36, 37].

2.5 Stochastic Signal Transduction and Processing by Chemoattractant Receptors

As described in Section 2.4, chemoattractant receptors have been revealed to operate stochastically in living cells. Here we discuss the application of the *gain-fluctuation relation* (GFR) to transmembrane signaling initiated by chemoattractant receptors [46, 53]. A detailed explanation of the GFR is described in Chapter 13. The GFR can describe noise generation, amplification and propagation along a signaling cascade based on the kinetic properties of signaling molecules.

Let us consider a simple model for transmembrane signaling in which ligands bind to receptors stochastically to activate the receptors and initiate second messengers that act as stochastic signal outputs (Fig. 2.6a). The scheme is as follows:

$$\mathrm{R} + \mathrm{L} \underset{k_{off}}{\overset{k_{on}}{\leftrightarrow}} \mathrm{R}^*, \quad \mathrm{R}^* + \mathrm{X} \xrightarrow{k_p} \mathrm{R}^* + \mathrm{X}^*, \quad \mathrm{X}^* \xrightarrow{k_d} \mathrm{X}, \tag{2.2}$$

where R, R* and L represent inactive receptors, active receptors and ligands, respectively. X and X* are inactive and active forms of second messengers. k_{on} and k_{off} represent ligand-association and dissociation rates, respectively. k_p and k_d represent production and degradation rates for the second messenger, respectively. The average number of active receptor R^* and second messenger X^* per one cell can be given by,

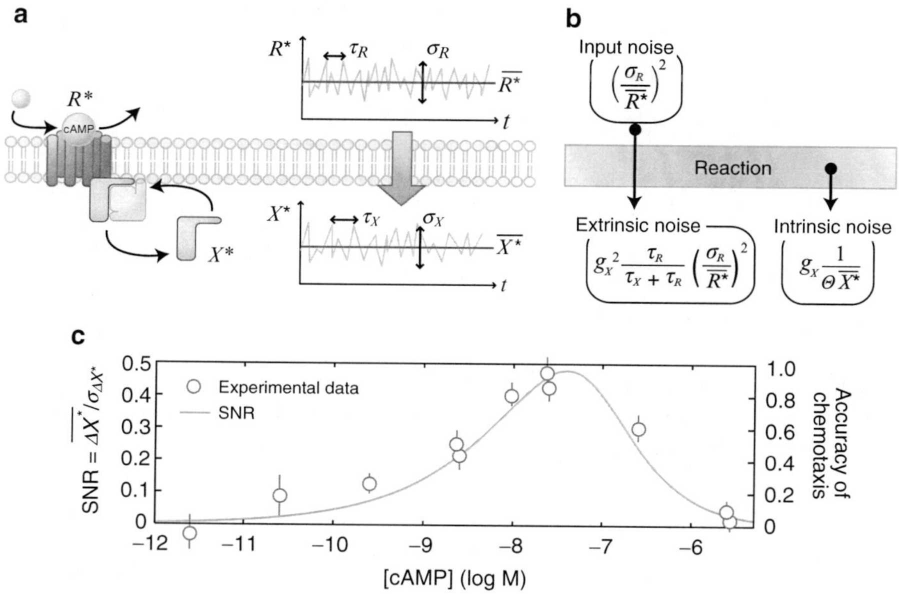

Fig. 2.6 Noise propagation and generation in chemotactic responses. (**a**) Simple model for transmembrane signaling. Ligand bound receptor becomes active (R*) and then activates second messenger (X*). Both input and output signal should have noise (σ_R and σ_X, respectively) around their averages ($\overline{R^*}$ and $\overline{X^*}$, respectively). (**b**) Noise generation and propagation. Input noise is propagated through the reaction as an extrinsic noise. Moreover, the reaction itself generates intrinsic noise. (**c**) Comparison of the theoretically obtained SNR (*line*) with experimental data (*circles*) for chemotactic accuracy of Dictyostelium cells.

$$\overline{R^*} = R_{total} \cdot L \cdot (K_R + L)^{-1}, \quad \overline{X^*} = X_{total} \cdot \overline{R^*} \cdot \left(K_X + \overline{R^*}\right)^{-1}, \tag{2.3}$$

where R_{total} is the total molecular number of receptors per single cell, $K_R = k_{off}/k_{on}$ is the affinity for the ligand-receptor binding reaction, X_{total} is the total molecular number of second messengers per cell, $K_X = k_d/k_p$ is the molecular number of active receptors when the activation of the second messenger reaches its half-maximum. Eq. 2.3 represents the input-output relationship between the average number of active receptor $\overline{R^*}$ and the average number of active second messenger $\overline{X^*}$ at ligand concentration L. According to the GFR [46], the relationship between the noise of the active receptor $({\sigma_X}^2)$ and the noise of the active second messenger $({\sigma_X}^2)$ is given by,

$$\frac{{\sigma_X}^2}{\overline{X^*}^2} = g_X \frac{1}{\Theta \overline{X^*}} + {g_X}^2 \frac{\tau_R}{\tau_X + \tau_R} \frac{{\sigma_R}^2}{\overline{R^*}^2}, \tag{2.4}$$

where Θ is a factor to adjust the dimension. $\Theta = 1$ when X^* is measured as a molecular number [46]. τ_R and τ_X are the chara?cteristic time constants of the fluctuations in the ligand-binding and G protein activation reaction, respectively. According to scheme (2.2), the time constants are given by,

$$\tau_R = (k_{on}L + k_{off})^{-1}, \qquad \tau_X = \left(k_p\overline{R^*} + k_d\right)^{-1} \tag{2.5}$$

The gain, g_X, quantifies the amplification rate of the output responses to small changes in input, and is given by,

$$g_X = \frac{\Delta X/\overline{X}}{\Delta R^*/\overline{R^*}} = \frac{d\log X}{d\log R^*} = K_X \cdot \left(K_X + \overline{R^*}\right)^{-1} \tag{2.6}$$

Equation 2.4 describes the input-output relationship between noise in the active receptor concentration and noise in the active second messenger concentration (Fig. 2.6b). The second term on the right hand side of Eq. 2.4 represents the noise propagated from ligand-receptor binding reactions, which is modulated by the time constants τ_R and τ_X and the gain g_X. These parameters can be obtained from the rate constants k_{on}, k_{off}, k_p, and k_d of the signal transduction reaction. In some cases (e.g., k_{off}), the rate constants can be obtained directly by single-molecule experiments. The first term on the right hand side of Eq. 2.4 represents noise generated intrinsically by the second messenger production reaction itself. So even if noise propagated from the ligand-receptor binding reactions is negligible, intrinsic noise is still present. Thus, Eq. 2.4 describes how the kinetic properties of signaling molecules determine both signal and noise transmitted in the corresponding signaling reaction.

We next discuss chemotactic signaling based on the GFR. It is evident that eukaryotic amoeboid cells detect differences in chemoattractant concentrations across their cell body by the fact that they can extend pseudopods toward the chemical gradient without any cellular translocations [4, 34, 63]. This type of gradient sensing is known as spatial-sensing. Chemotactic signals in using a spatial-sensing mechanism originate from differences in the receptor occupancy between the anterior and posterior halves of a cell under a chemical gradient. Recently, the local excitation global inhibition (LEGI) mechanism was proposed as an extended spatial-sensing mechanism. The LEGI mechanism argues that spatial differences in the ligand concentration are sensed by comparing the excitatory signals derived from the receptor occupancy in a local area with the inhibitory signals derived from the average level of receptor occupancy over the entire surface of the cell [16, 24]. This mechanism does not entail direct comparison of the ligand concentration between different points over the cell surface. Instead, differences in receptor occupancy between a local area and the entire surface of the cell determine the chemotactic signal.

Here we consider the spatial-sensing mechanism for simplicity. We assume no polarity in the transmembrane signaling process along the length of a chemotaxis cell. In other words, the kinetic properties of the signaling molecules in the anterior and posterior halves of the cell are the same. A concentration difference in ligands produces a difference in the receptor occupancy, ΔR^*, which then leads to a difference in active second messengers, ΔX^*, between the anterior and posterior region of the chemotactic cells. ΔR^* and ΔX^* should fluctuate with time and include noise $\sigma_{\Delta R}$ and $\sigma_{\Delta X}$, respectively. From the GFR, we obtain the following relation between $\sigma_{\Delta k}/\overline{\Delta R^*}$ and $\sigma_{\Delta x}/\overline{\Delta X^*}$,

$$\left(\frac{\sigma_{\Delta X}}{\overline{\Delta X}}\right)^2 = \frac{1}{g_X \Theta \overline{X^*}}\left(\frac{\overline{R^*}}{\overline{\Delta R^*}}\right)^2 + \frac{\tau_R}{\tau_X + \tau_R}\left(\frac{\sigma_{\Delta R}}{\overline{\Delta R^*}}\right)^2, \tag{2.7}$$

where the first term on the right hand side of Eq. 2.7 is derived from the noise generated intrinsically in the second messenger production reaction, while the second term is derived from the noise propagated from the ligand-receptor binding reaction [53].

We defined the SNR of the chemotactic signals as $\overline{\Delta X^*}/\sigma_{\Delta X}$, which is obtained from Eq. 2.7,

$$SNR = \frac{\overline{\Delta X^*}}{\sigma_{\Delta X}} = \frac{1}{\sqrt{4\left(\frac{L+K_R}{K_R}\right)\left\{\frac{(L+K_{XR})^2}{V_X \cdot K_{XR}} + \frac{\tau_R}{\tau_X+\tau_R}\left(\frac{L+K_R}{R_{total}}\right)\right\}}}\left(\frac{\Delta L}{\sqrt{L}}\right), \tag{2.8}$$

where $K_{XR} = K_X \cdot K_R \cdot (R_{total} + K_X)^{-1}$ is the ligand concentration when X^* is at half maximum and $V_X = R_{total} \cdot X_{total} \cdot (R_{total} + K_X)^{-1}$. By using parameter values obtained experimentally for *Dictyostelium* cells, we obtained the dependence of the SNR on cAMP concentration with a constant 2% gradient (Fig. 2.6c) ([53] for details of parameter values). We found a good agreement between our theoretical SNR and experimental data on the chemotactic accuracy of *Dictyostelium* cells [7, 53]. This suggests that chemotactic accuracy is determined mostly by the SNR of chemotactic signals at the most upstream reactions of the chemotactic signaling system. The SNR is proportional to $\Delta L/\sqrt{L}$ when ligand concentration is much smaller than $K_R$and K_{XR}. Supposing that cells can sense the gradient if the SNR of chemotactic signals is larger than the threshold SNR, or $SNR_{threshold} \leqslant C(\Delta L/\sqrt{L})$ with constant C, we find $\Delta L \geqslant \sqrt{L}\,(SNR_{threshold}/C)$for chemotaxis. Therefore, the minimum difference in ligand concentration necessary for chemotaxis is proportional to the square root of L, $\Delta L_{threshold} \propto \sqrt{L}$. In fact, Van Haastert and a colleague [56] reported the relation between the average concentration of ligand and the corresponding threshold gradient, estimating α in the relation $\Delta L_{threshold} \propto L^{\alpha}$ to be 0.5. We should note that Eqs. 2.7 and 2.8 are applicable to other cell types. In fact, similar dependency for chemotactic accuracy has been observed in mammalian leukocytes and neurons [39, 64].

Equations 2.7 and 2.8 can reveal the regulatory mechanisms used for signal improvements [53]. The SNR of the chemotactic signals is determined by the kinetic parameters of both the receptor and the downstream second messenger (k_{on}, k_{off}, k_p, k_d, K_X and K_R) along with their respective quantities (R_{total} and X_{total}). Therefore, changing a number of parameters individually potentially reveals a great

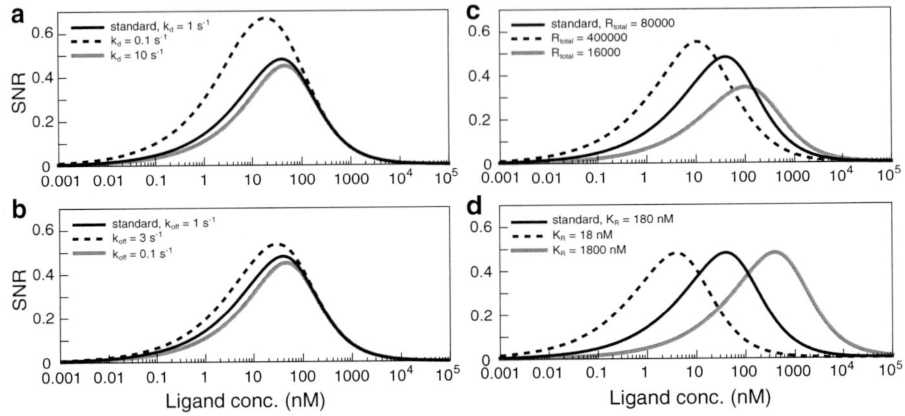

Fig. 2.7 Signal improvements. (**a**) Time-averaging effects. The SNR was improved by increasing the time constants of the second messenger. The degradation rates (k_d) were changed and the corresponding SNR was calculated: (*gray line*) 10 s^{-1}; (*black line*) 1 s^{-1}; (*broken line*) 0.1 s^{-1}. (**b**) Receptor fluctuation-dependent signal improvements. The dissociation rates of the ligand (k_{off}) were: (*gray line*) 0.1 s^{-1}; (*black line*) 1 s^{-1}; (*broken line*) 3 s^{-1}. (**c**) Dependency of the SNR on receptor expression levels. The receptor numbers are 16,000 (*gray*), 80,000 (*black*), 400,000 (*broken*) molecules/cell. (**d**) Effects of affinity modulation on the SNR. Ligand-binding affinity: 18 nM (*broken*), 180 nM (*black*), and 1,800 nM (*gray*).

deal about the relationship between the kinetic properties of signaling molecules and the corresponding cellular response.

The first example is how the SNR can be improved by increasing τ_X (Fig. 2.7a). The second term on the right hand side of Eq. 2.7 describes how the noise of the active receptor, $R*$, propagates into the noise of the second messenger X^*. The term $\tau_R/(\tau_X + \tau_R)$ is known as the time-averaging factor of the noise. When the time constants of the second messenger production are faster than those of the active receptor ($\tau_X \ll \tau_R$), the noise of the active receptor propagates more efficiently into the noise of the second messenger because the term $\tau_R/(\tau_X + \tau_R)$ increases gradually with a decrease in τ_X. In this case, the second messenger reaction can follow rapid temporal changes of the active receptor. On the other hand, in the case of $\tau_X \gg \tau_R$, the second messenger reaction cannot follow the noise of the active receptor. Instead, the noise of the active receptor is averaged temporally. Thus, in order to reduce the noise generated at the ligand-binding reaction, a relatively slower reaction is required during second messenger production. A longer second messenger lifetime, which corresponds to slower degradation, causes noise reduction more effectively by time-averaging effects (Fig. 2.7a). This means that the regulatory mechanism for degradation or inactivation of second messengers has a pivotal role on signal improvements for chemotaxis. In the case of G-protein, lifetimes of its active form on the membrane affect the SNR of the chemotactic signal, which is determined by both GTPase activity and its membrane-binding stability. In our preliminary single-molecule analysis of $G\alpha_2$ and $G\beta\gamma$, receptor stimulation causes an enhancement in the membrane-binding stability of both subunits, suggesting that the SNR of the chemotactic signal is improved by modulating the lifetimes of G-protein on the membrane.

Second, the SNR can be improved by decreasing τ_R (Fig. 2.7b). When ligand concentration is increased, the ligand-binding rate, $k_{on}L$, accelerates, resulting in a decrease in the time constant τ_R (Eq. 2.5). Moreover, signal improvements are possible by increasing the on-rate (k_{on}) and/or off-rate (k_{off}). This means that faster transitions between the ligand-binding and -unbinding states of the receptors can produce chemotactic signals with higher SNR. That is, an increase in ligand concentration results in better efficiency of chemotactic signals not only by increasing the average concentration of the active receptor but also by decreasing the characteristic time of the active receptor fluctuations. As described below in Section 2.6, chemotactic cells that exhibit a polarity in receptor kinetic states in which ligand dissociation rates k_{off} at the pseudopod region are faster than at the tail region suggest that the SNR is higher at the pseudopod than at the tail [54].

The third and final way to improve SNR discussed in this section involves modulating the total amount of receptor expressed in cells (Fig. 2.7c). Increasing or decreasing the receptor number improves the SNR in the lower and higher concentration ranges of the chemoattractant, respectively. Receptor internalization too can contribute to SNR improvements in the higher concentration ranges. Also, receptor affinity for the chemoattractant is an important factor for adjusting the concentration ranges in chemotaxis (Fig. 2.7d). It is well known that cAMP receptors in *Dictyostelium* cells are phosphorylated upon cAMP stimulation, leading to a three- to approximately sixfold decrease in ligand-binding affinity [60]. Such an affinity shift of receptors contributes to an SNR increase in the higher-ligand-concentration range, which extends the response range to higher ligand concentrations. Thus, we can reveal the molecular mechanisms for both noise reduction and amplification in a stochastically operating signal transduction system by analyzing how the SNR is modulated depending on the kinetics properties of the signaling molecules.

2.6 Kinetic Heterogeneity of Signaling Molecules and Cellular Polarity

As described in Section 2.4, single-molecule imaging analysis of ligand-binding in cells has revealed multiple receptor kinetic states. For most G-protein-coupled receptors (GPCRs), the dissociation rates depend on the interaction of receptors with the coupled G-proteins [19, 22, 58], meaning ligand-binding kinetics can be used to elucidate the coupling of the receptors with G proteins. In membrane fractions that include the receptors and G proteins, ligand-binding kinetics is sensitive to the presence of GTP. In the absence of GTP, the receptors adopt slow-dissociating states, where the receptors bind to G proteins but cannot facilitate their activation. The addition of GTP causes a shift in the receptor kinetic states from slow-dissociating to fast-dissociating states. This shift reflects the dissociation of G protein from the receptor via G protein activation. That is, the receptors adopt a slow-dissociating state with binding to G protein, while they adopt a fast-dissociating state when free from G protein. When G protein is efficiently activated, the receptor undergoes state transitions between the two.

For the case of cAMP receptors, ligand-binding kinetics using radiolabelled cAMP has identified at least two receptor kinetic states and that GTP addition to the membrane fractions increases fast-dissociating one [19]. We have performed single-molecule analysis of cAMP-binding kinetics on a membrane fraction to examine GTP sensitivity (Fig. 2.8a) [54]. In the absence of GTP, slow-dissociating states with dissociation rates of 0.4 s^{-1} or 0.08 s^{-1} were dominantly (78%) observed. GTP addition increased the fast-dissociating state with a dissociation rate of 1.3 s^{-1} up to 66%. When the membrane fractions prepared from the mutant cells lacking either functional G protein subunits Gα_2 or Gβ were used for this assay, GTP sensitivity was not observed, indicating that GTP-induced alterations in the cAMP-binding kinetics reflects altered interactions between the receptors and

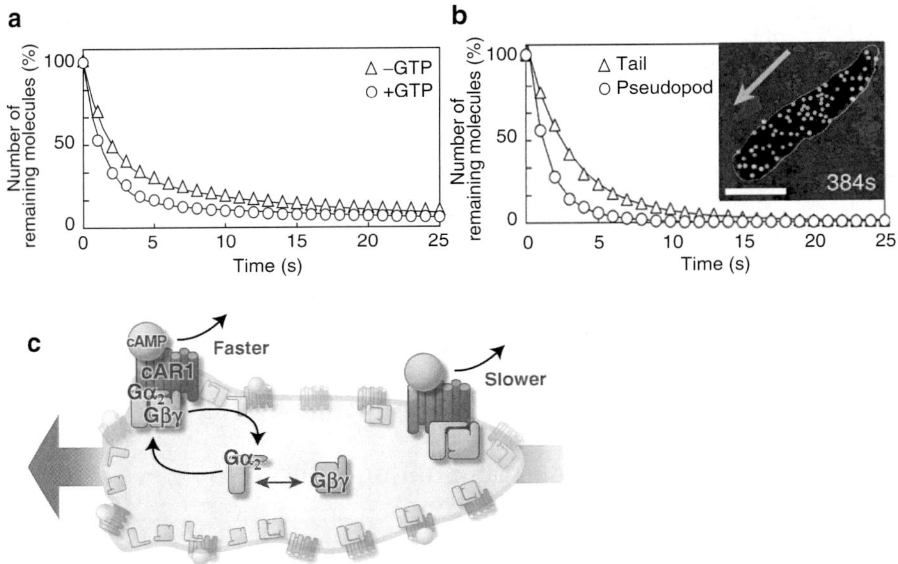

Fig. 2.8 Kinetic heterogeneity of cAMP bindings. (**a**) Cumulative probability distribution of Cy3-cAMP bound to the membrane fraction in the absence (*triangles*) and presence (*circles*) of GTP. (**b**) Cumulative probability distribution of Cy3-cAMP bound to the anterior pseudopod (*circles*) and posterior tail (*triangles*). (*inset*) Receptor occupancy in a chemotaxing cell. The positions of Cy3-cAMP bound to the receptor are shown by the *white dots*. (**c**) Schematic illustration of kinetic heterogeneity in Cy3-cAMP bindings. Faster and slower-dissociation of Cy3-cAMP in the anterior pseudopod and posterior tail, respectively, suggest high and low G protein activation.

their coupled G proteins. These results agree well with those obtained previously by conventional ligand-binding kinetics methods.

Based on these observations, we next examined the kinetic states of the receptor in chemotaxing cells by conducting single-molecule analysis of the cAMP-binding. Under a chemical gradient, *Dictyostelium* cells adopt a polarized shape with a pseudopod at the leading edge and a tail at the rear end (Fig. 2.8b). The cumulative probability distribution of Cy3-cAMP lifetimes was obtained separately for Cy3-cAMP molecules bound to the anterior and posterior halves of a polarized cell. As shown in Fig. 2.8b, the Cy3-cAMP lifetimes exhibited a polarity along the length of the cell. The lifetimes of Cy3-cAMP in the anterior region were about three times shorter than those in the posterior region. In the anterior region, the majority of the receptors adopted a fast-dissociating state with a dissociation rate of 1.1 s^{-1} (71%), while the remaining population of the receptor adopted a slow-dissociating state with a dissociation rate of 0.4 s^{-1} (29%). These dissociation rates resemble those observed in cAMP binding to membrane fractions in the presence of GTP, implying that receptors in the anterior region undergo cycling between G protein-bound and free states, and thus G proteins are efficiently activated at the anterior region. On the other hand, receptors in the posterior region adopted a slow-dissociating state with dissociation rates of 0.4 s^{-1} (76%) and 0.16 s^{-1} (24%), which resemble those observed in cAMP-binding kinetics on membrane fractions in the absence of GTP. This suggests that these receptors are somehow suppressed in their ability to activate G protein. Such polarity in the receptor kinetic states was not found in mutant cells lacking the $G\alpha_2$ or $G\beta$ subunits, suggesting that differences in the receptor states reflect on differences of the coupling with G-protein between the anterior and posterior regions. These results suggest anterior-posterior polarity affect the efficiency of G protein activation along with the length of a chemotaxing cell (Fig. 2.8c). Thus, single-molecule kinetic analysis can reveal the kinetic heterogeneity of signaling molecules within the context of a cell.

Anterior-posterior polarity in the receptor kinetic states may provide a molecular basis for noise-robust signal processing against the stochastic fluctuations seen in receptor occupancy. As described in Section 2.4, receptor occupancy in a cell undergoing chemotaxis towards a chemoattractant source is reversed spontaneously and transiently with respect to the direction of the gradient. If an anterior-posterior polarity affects the efficiency of G protein activation along the length of a cell, the ligand binding at the anterior pseudopod will preferentially affect cell behavior, while the transmembrane signaling at the posterior would be inefficient. Furthermore, polarity in receptor states may explain a polarity in the response to cAMP observed previously in the *Dictyostelium* cells [49]. Specifically, when a cell is stimulated locally with cAMP, it forms a pseudopod. At the anterior region of the cell, a pseudopod forms within a few seconds of cAMP stimulation, while if at the posterior region the formation takes about 40 s or more. This is consistent with the idea that transmembrane signaling by cAMP receptors is somehow inefficient at the posterior.

Such polar kinetic properties has been found in other signaling molecules such as the PtdIns(3, 4, 5)P_3-binding protein Crac and PTEN [27, 57]. Crac molecules stably localized at the pseudopod regions of a polarized cell undergoing chemotaxis

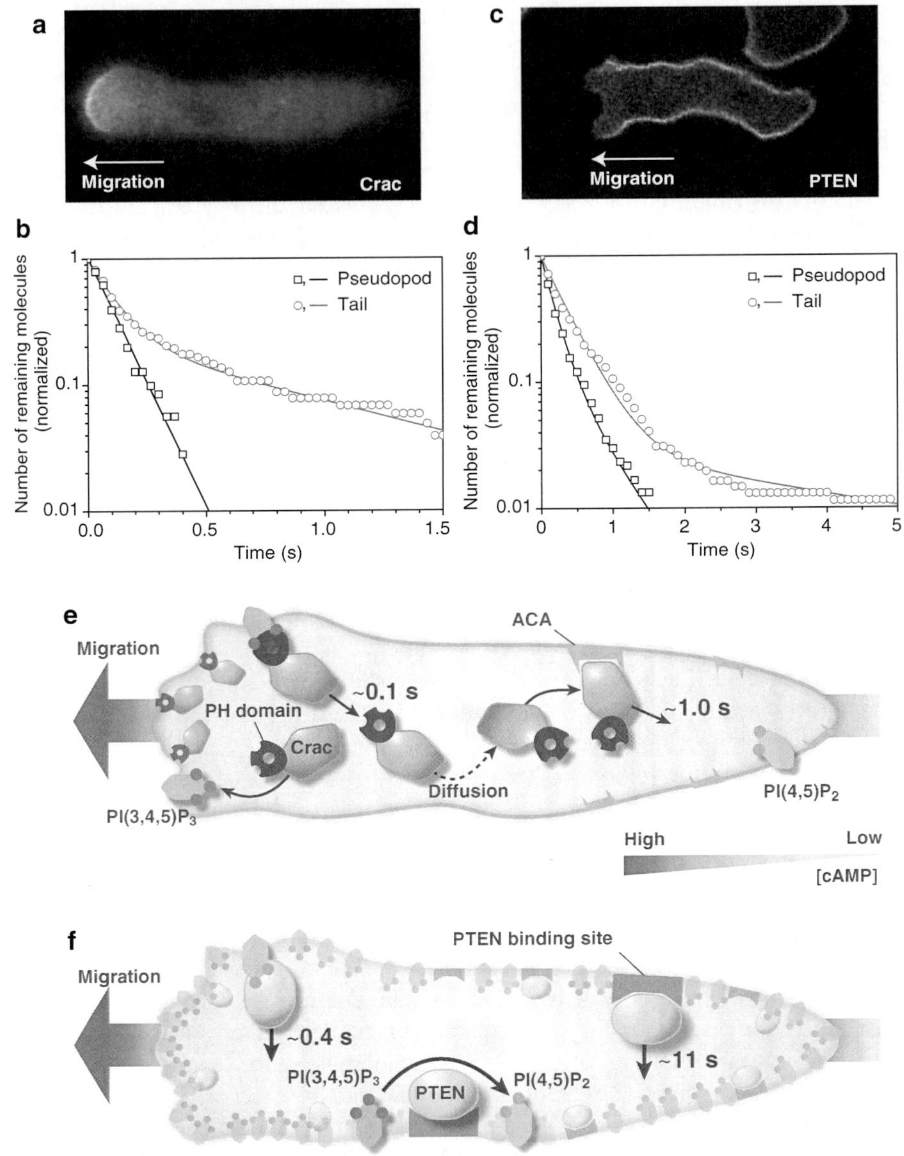

Fig. 2.9 Kinetic heterogeneities of downstream signaling molecules along the length of chemotaxing cells. (**a**) Localization of Crac-GFP at the anterior pseudopods of a polarized cell during chemotaxis. (**b**) Cumulative probability distributions of Crac-GFP in the anterior (*black squares*) and posterior halves (*gray circles*) of chemotaxing cells shown on a semi-logarithmic plot. (**c**) Localization of PTEN-TMR at the lateral and rear regions of a chemotaxing cell. (**d**) Cumulative probability distributions of PTEN-TMR in the anterior (*black squares*) and posterior halves (*gray circles*) of chemotaxing cells shown on a semi-logarithmic plot. (**e**) Schematic illustration of

when observed by conventional epi-fluorescence microscopy (Fig. 2.9a). The cumulative probability distribution of Crac, which was obtained by single-molecule observation, show that almost all Crac molecules dissociated from the anterior membrane within a lifetime of 110 ms (Fig. 2.9b). That is, stable localization of Crac at the pseudopod is maintained by rapid exchange of individual Crac molecules via dynamic shuttling between the membrane and cytosol. Crac molecules were also observed in the posterior regions by single-molecule imaging, which could not be seen when using conventional methods. The number of Crac molecules in the posterior region was ~30 times lower than that observed in the anterior region. The lifetimes of Crac molecules at the posterior regions were 110 ms (76%) and 890 ms (24%). Thus, fast-dissociation and slow-dissociation sites were preferentially localized at the leading edge and at the rear end of the polarized cells, respectively (Fig. 2.9e).

PTEN molecules also exhibit kinetic heterogeneity and dynamic shuttling between the membrane and cytosol. PTEN localized predominantly at the lateral and rear regions of a chemotaxing cell (Fig. 2.9c). Such stable localization is maintained by the shuttling of individual PTEN molecules. Lifetimes of PTEN molecules at the lateral and rear regions were 0.4 s (97%) and 11 s (3%), respectively, while the lifetime at the pseudopod regions was 0.2 ~0.6 s (Fig. 2.9d). Thus, even if the same signaling molecules are present at the anterior and posterior regions of a polarized cell, the signaling molecules adopt different kinetic states between the anterior and posterior regions (Fig. 2.9f).

The kinetic heterogeneity of downstream signaling molecules with relation to cellular polarity has the potential to modulate the quality of the transmitted signals and thus affect the corresponding molecular and cellular responses. As described in Section 2.5, downstream signaling reactions with faster kinetic rates can follow rapid temporal changes of upstream reactions, and thus cause effective noise propagation. On the other hand relatively slower reactions cannot follow such rapid temporal changes in the upstream reactions, leaving the noise from the upstream reactions to be averaged temporally. Thus, polarity in the kinetic heterogeneity of signaling molecules suggests that the anterior region may follow and respond to relatively faster and smaller changes in the environmental signals, while the posterior region may observe relatively slower and larger changes. That is, it is likely that the anterior region of chemotactic cells is more sensitive to environmental changes than the posterior one. The polarity in kinetic properties of signaling molecules downstream of the cascade may also provide a molecular basis for the robustness

Fig. 2.9 (continued) Crac membrane recruitment. Fast and slow-dissociation binding sites for Crac have been identified as PI(3,4,5)P_3 and adenylyl cyclase A (ACA) dependent, respectively. PI (3,4,5)P_3 and ACA-dependent sites are localized at the pseudopod and the tail of chemotaxing cells, respectively. (**f**) Schematic illustration of the kinetic heterogeneity of PTEN. PTEN molecules show fast-dissociation bindings, which depend on PIP_2, and slow-dissociation bindings, which are observed at the lateral and rear regions of chemotaxing cells.

against stochastic fluctuations seen in receptor occupancy and thus explain the polar responses made to chemotactic stimulations observed in *Dictyostelium* cells [49]. As a result, heterogeneity in the signaling molecules with relation to cellular polarity may further establish and maintain cellular polarity.

2.7 Conclusion Remarks

Here we take the chemotactic signaling of *Dictyostelium* cells as a typical example for stochastic signal processing and transduction in living cells. Single-molecule imaging analysis has revealed the kinetic properties of the signaling molecules responsible for chemotaxis. The GFR can be used to evaluate how these kinetic properties affect cellular chemotactic responses. Our analyses revealed that the stochastic properties of chemoattractant receptors most upstream of the chemotactic signaling cascade determine the chemotactic accuracy of the cells. The noise within transmembrane signaling by the receptors limits the precision of the directional sensing, suggesting that receptor-G protein coupling and its modulation have an important role on chemotaxis efficiency in cells. The noise generated at the ligand-receptor complex can be reduced through the slow dynamics of signaling molecules by using time-averaging effects. Overall, polarity within the kinetic heterogeneity of signaling molecules may provide a molecular basis for noise-robust signal processing.

Acknowledgments The authors thank all of the members of Stochastic Biocomputing Group in Osaka University for discussion and also thank Peter Karagiannis for critical reading of the manuscripts.

References

1. Axelrod D (2001) Total internal reflection fluorescence microscopy in cell biology. Traffic 2:764–774
2. Berg HC, Purcell EM (1977) Physics of chemoreception. Biophys J 20:193–219
3. Bialek W, Setayeshgar S (2005) Physical limits to biochemical signaling. Proc Natl Acad Sci USA 102:10040–10045
4. Devreotes PN, Zigmond SH (1988) Chemotaxis in eukaryotic cells: a focus on leukocytes and Dictyostelium. Annu Rev Cell Biol 4:649–686
5. Dormann D, Weijer CJ (2006) Imaging of cell migration. EMBO J 25:3480–3493
6. Eichinger L, Pachebat JA, Glockner G et al (2005) The genome of the social amoeba Dictyostelium discoideum. Nature 435:43–57
7. Fisher PR, Merkl R, Gerisch G (1989) Quantitative analysis of cell motility and chemotaxis in Dictyostelium discoideum by using an image processing system and a novel chemotaxis chamber providing stationary chemical gradients. J Cell Biol 108:973–984
8. Fornel F (2001) Evanescent waves. Springer, Berlin, Heidelberg, New York
9. Funatsu T, Harada Y, Tokunaga M et al (1995) Imaging of single fluorescent molecules and individual ATP turnovers by single myosin molecules in aqueous solution. Nature 374:555–559

10. Gerisch G (2009) Imaging actin cytoskeleton dynamics in Dictyostelium chemotaxis. Method Mol Biol 571:385–400
11. Graf R, Brusis N, Daunderer C et al (2000) Comparative structural, molecular, and functional aspects of the Dictyostelium discoideum centrosome. Curr Top Dev Biol 49:161–185
12. Hereld D, Jin T (2009) Chemotaxis. Methods Mol Biol 571. Humana Press, New York
13. Hibino K, Watanabe TM, Kozuka J et al (2003) Single- and multiple-molecule dynamics of the signaling from H-Ras to cRaf-1 visualized on the plasma membrane of living cells. Chemphyschem 4:748–753
14. Hibino K, Shibata T, Yanagida T et al (2009) A RasGTP-induced conformational change in C-RAF is essential for accurate molecular recognition. Biophys J 97:1277–1287
15. Howard PK, Ahern KG, Firtel RA (1988) Establishment of a transient expression system for Dictyostelium discoideum. Nucleic Acids Res 16:2613–2623
16. Iglesias PA, Devreotes PN (2008) Navigating through models of chemotaxis. Curr Opin Cell Biol 20:35–40
17. Iino R, Koyama I, Kusumi A (2001) Single molecule imaging of green fluorescent proteins in living cells: E-cadherin forms oligomers on the free cell surface. Biophys J 80:2667–2677
18. Janetopoulos C, Ma L, Devreotes PN et al (2004) Chemoattractant-induced phosphatidylinositol 3, 4, 5-trisphosphate accumulation is spatially amplified and adapts, independent of the actin cytoskeleton. Proc Natl Acad Sci USA 101:8951–8956
19. Janssens PM, Van Haastert PJ (1987) Molecular basis of transmembrane signal transduction in Dictyostelium discoideum. Microbiol Rev 51:396–418
20. Kitamura K, Tokunaga M, Iwane AH et al (1999) A single myosin head moves along an actin filament with regular steps of 5.3 nanometres. Nature 397:129–134
21. Kuwayama H, Yanagida T, Ueda M (2008) DNA oligonucleotide-assisted genetic manipulation increases transformation and homologous recombination efficiencies: Evidence from gene targeting of Dictyostelium discoideum. J Biotechnol 133:418–423
22. Levitzki A, Marbach I, Bar-Sinai A (1993) The signal transduction between beta-receptors and adenylyl cyclase. Life Sci 52:2093–2100
23. Lu HP, Xun L, Xie XS (1998) Single-molecule enzymatic dynamics. Science 282:1877–1882
24. Ma L, Janetopoulos C, Yang L et al (2004) Two complementary, local excitation, global inhibition mechanisms acting in parallel can explain the chemoattractant-induced regulation of PI(3, 4, 5)P3 response in dictyostelium cells. Biophys J 87:3764–3774
25. Manstein DJ, Titus MA, De Lozanne A et al (1989) Gene replacement in Dictyostelium: generation of myosin null mutants. EMBO J 8:923–932
26. Mato JM, Losada A, Nanjundiah V et al (1975) Signal input for a chemotactic response in the cellular slime mold Dictyostelium discoideum. Proc Natl Acad Sci USA 72:4991–4993
27. Matsuoka S, Iijima M, Watanabe TM et al (2006) Single-molecule analysis of chemoattractant-stimulated membrane recruitment of a PH-domain-containing protein. J Cell Sci 119:1071–1079
28. Matsuoka S, Miyanaga Y, Yanagida T et al (2008) Single-molecule imaging of stochastic events in living cells. In: Selvin PR, Ha T (eds) Single-molecule techniques. Cold Spring Harbor Laboratory Press, New York
29. Matsuoka S, Shibata T, Ueda M (2009) Statistical analysis of lateral diffusion and multistate kinetics in single-molecule imaging. Biophys J 97:1115–1124
30. McAnaney TB, Zeng W, Doe CF et al (2005) Protonation, photobleaching, and photoactivation of yellow fluorescent protein (YFP 10C): a unifying mechanism. Biochemistry 44:5510–5524
31. Miyanaga Y, Matsuoka S, Ueda M (2009) Single-molecule imaging techniques to visualize chemotactic signaling events on the membrane of living Dictyostelium cells. Method Mol Biol 571:417–435
32. Noegel AA, Schleicher M (2000) The actin cytoskeleton of Dictyostelium: a story told by mutants. J Cell Sci 113:759–766
33. Parent CA (2004) Making all the right moves: chemotaxis in neutrophils and Dictyostelium. Curr Opin Cell Biol 16:4–13
34. Parent CA, Devreotes PN (1999) A cell's sense of direction. Science 284:765–770

35. Postma M, Bosgraaf L, Loovers HM et al (2004) Chemotaxis: signalling modules join hands at front and tail. EMBO Rep 5:35–40
36. Rappel WJ, Levine H (2008a) Receptor noise limitations on chemotactic sensing. Proc Natl Acad Sci USA 105:19270–19275
37. Rappel WJ, Levine H (2008b) Receptor noise and directional sensing in eukaryotic chemotaxis. Phys Rev Lett 100:228101
38. Ridley AJ, Schwartz MA, Burridge K et al (2003) Cell migration: integrating signals from front to back. Science 302:1704–1709
39. Rosoff WJ, Urbach JS, Esrick MA et al (2004) A new chemotaxis assay shows the extreme sensitivity of axons to molecular gradients. Nat Neurosci 7:678–682
40. Sakmann B, Neher E (1995) Single channel recording. Plenum, New York
41. Sako Y, Yanagida T (2003) Single-molecule visualization in cell biology. Nat Rev Mol Cell Biol (Suppl):SS1–5
42. Sako Y, Minoghchi S, Yanagida T (2000) Single-molecule imaging of EGFR signalling on the surface of living cells. Nat Cell Biol 2:168–172
43. Sasaki AT, Firtel RA (2006) Regulation of chemotaxis by the orchestrated activation of Ras, PI3K, and TOR. Eur J Cell Biol 85:873–895
44. Sato MJ, Kuwayama H, van Egmond WN et al (2009) Switching direction in electric-signal-induced cell migration by cyclic guanosine monophosphate and phosphatidylinositol signaling. Proc Natl Acad Sci USA 106:6667–6672
45. Shaffer BM (1975) Secretion of cyclic AMP induced by cyclic AMP in the cellular slime mould Dictyostelium discoideum. Nature 255:549–552
46. Shibata T, Fujimoto K (2005) Noisy signal amplification in ultrasensitive signal transduction. Proc Natl Acad Sci USA 102:331–336
47. Song L, Nadkarni SM, Bodeker HU et al (2006) Dictyostelium discoideum chemotaxis: threshold for directed motion. Eur J Cell Biol 85:981–989
48. Swaney KF, Huang CH, Devreotes PN (2010) Eukaryotic chemotaxis: a network of signaling pathways controls motility, directional sensing, and polarity. Annu Rev Biophys 39:265–289
49. Swanson JA, Taylor DL (1982) Local and spatially coordinated movements in Dictyostelium discoideum amoebae during chemotaxis. Cell 28:225–232
50. Takagi H, Sato MJ, Yanagida T et al (2008) Functional analysis of spontaneous cell movement under different physiological conditions. PLoS One 3:e2648
51. Tokunaga M, Kitamura K, Saito K et al (1997) Single molecule imaging of fluorophores and enzymatic reactions achieved by objective-type total internal reflection fluorescence microscopy. Biochem Biophys Res Commun 235:47–53
52. Tranquillo RT, Lauffenburger DA, Zigmond SH (1988) A stochastic model for leukocyte random motility and chemotaxis based on receptor binding fluctuations. J Cell Biol 106:303–309
53. Ueda M, Shibata T (2007) Stochastic signal processing and transduction in chemotactic response of eukaryotic cells. Biophys J 93:11–20
54. Ueda M, Sako Y, Tanaka T et al (2001) Single-molecule analysis of chemotactic signaling in Dictyostelium cells. Science 294:864–867
55. Van Haastert PJ (1997) Transduction of the chemotactic cAMP signal across the plasma membrane. In: Maeda Y, Inouye K, Takeuchi I (eds) Dictyostelium. Universal Academy Press, Tokyo
56. Van Haastert PJ, Postma M (2007) Biased Random Walk by Stochastic Fluctuations of Chemoattractant-Receptor Interactions at the Lower Limit of Detection. Biophys J 93:1787–1796
57. Vazquez F, Matsuoka S, Sellers WR et al (2006) Tumor suppressor PTEN acts through dynamic interaction with the plasma membrane. Proc Natl Acad Sci USA 103:3633–3638
58. Waelbroeck I (1999) Kinetics versus equilibrium: the importance of GTP in GPCR activation. Trends Pharmacol Sci 20:477–481

59. Wazawa T, Ueda M (2005) Total internal reflection fluorescence microscopy in single molecule nanobioscience. Adv Biochem Eng Biotechnol 95:77–106
60. Xiao Z, Yao Y, Long Y et al (1999) Desensitization of G-protein-coupled receptors. agonist-induced phosphorylation of the chemoattractant receptor cAR1 lowers its intrinsic affinity for cAMP. J Biol Chem 274:1440–1448
61. Yanagida T, Ueda M, Murata T et al (2007) Brownian motion, fluctuation and life. Biosystems 88:228–242
62. Yumura S, Uyeda TQ (2003) Myosins and cell dynamics in cellular slime molds. Int Rev Cytol 224:173–225
63. Zigmond SH (1974) Mechanisms of sensing chemical gradients by polymorphonuclear leukocytes. Nature 249:450–452
64. Zigmond SH (1977) Ability of polymorphonuclear leukocytes to orient in gradients of chemotactic factors. J Cell Biol 75:606–616

Chapter 3
Single-Molecule Analysis of Molecular Recognition Between Signaling Proteins RAS and RAF

Kayo Hibino and Yasushi Sako

Abstract Single-molecule kinetic and dynamic analyses of the biochemical reactions in living cells are one of the most useful methods of investigating the molecular mechanisms of cellular reactions, especially those involved in signal transduction on the plasma membrane. Here, we focus on single-molecule analyses of the intracellular signaling from RAS to RAF, which occurs on the plasma membrane. RAS and RAF are cell signaling proteins that regulate various cellular behaviors, including cell fate determination for proliferation and differentiation. Using single-molecule techniques in living cells, including single-pair Förster resonance energy transfer (FRET) detection and single-molecule tracking, we directly measured the elementary processes in the reactions between RAS and RAF, including the dissociation of RAS and RAF, the conformational changes in RAF, and the diffusion of RAS and RAF along the plasma membrane. Based on the results of these measurements, we discuss how the mutual molecular recognition between RAS and RAF facilitates accurate signal transduction.

Keywords Conformational change · Cysteine-rich domain (CRD) · Diffusion coefficient · Dissociation · Effector protein · Epidermal growth factor (EGF) · Fluorescence emission spectrum · GDP–GTP binary molecular switch · Green fluorescent protein (GFP) · HeLa cell · Induced conformational change · Initial binding state · Intracellular signaling · Lateral diffusion · Membrane ruffle · Mutual molecular recognition · Objective-type total internal reflection fluorescence (TIRF) microscopy · Plasma membrane · RAF · RAS · RAS-binding domain (RBD) · Serine/threonine kinase · Single-molecule kinetic and dynamic analyses · Single-molecule tracking · Single-pair Förster resonance energy transfer (FRET) · Small GTPase · Translocation · Yellow fluorescent protein (YFP)

K. Hibino (✉) and Y. Sako
Cellular Informatics Laboratory, RIKEN ASI, 2-1 Hirosawa, Wako 351-0114, Japan
e-mail: kayo_hibino@riken.jp; sako@riken.jp

Y. Sako and M. Ueda (eds.), *Cell Signaling Reactions: Single-Molecular Kinetic Analysis*,
DOI 10.1007/978-90-481-9864-1_3,

3.1 Intracellular Signaling from RAS to Its Effectors

The cellular signaling protein RAS participates in diverse biological processes, including cell proliferation, differentiation, apoptosis, and survival [1]. RAS plays roles in these processes through its direct interactions with various species of effector proteins [2]. The importance of RAS–effector signaling in cells is obvious from the fact that its disruption has been observed in many types of human cancers [1–3]. Because of the biological and pharmacological significance of RAS, its structure and function have been studied intensively. However, it is unclear how RAS recognizes the appropriate effectors in the context of different extracellular stimuli and turns on the effector functions. Here, we describe a single-molecule kinetic and dynamic analysis of the mutual molecular recognition between RAS and one of its effectors, C-RAF (RAF).

RAS is a small GTPase with a molecular weight of 21 kDa, which binds and hydrolyzes guanine nucleotides [4, 5]. In living cells, RAS cycles between the GDP-bound inactive form (RASGDP) and the GTP-bound active form (RASGTP) [4, 5]. Active RASGTP undergoes high-affinity interactions with effector molecules [2, 4, 5]. Through these interactions, RAS is thought to trigger the activation of the effector molecules. Because of this function, RAS is often called a "GDP–GTP binary molecular switch" [2]. The cyclic turnover of active RASGTP and inactive RASGDP is coupled to the hydrolysis of bound GTP to bound GDP, and the exchange of bound GDP for free GTP, which predominates numerically over GDP in the cytoplasm [4, 5]. This cycle is regulated by activators called "GDP/GTP exchange factors (GEFs)" and inactivators called "GTPase-activating proteins (GAPs)" [4, 5]. The roles of GEFs and GAPs are crucial because the intrinsic reaction rates of GTP hydrolysis and nucleotide exchange on RAS are too slow to have physiological meaning [5].

When cells are stimulated with extracellular signals, such as epidermal growth factor (EGF), RAS is activated by GEF and interacts with effectors, including RAF, a cytoplasmic serine/threonine kinase [6, 7]. Regardless of its nucleotide status, RAS is anchored to the inner leaflet of the plasma membrane through lipid modifications in its C-terminal region [8]. Therefore, the interaction between RAF and RASGTP results in the translocation of RAF from the cytoplasm to the plasma membrane [9, 10]. At the plasma membrane, RAF is activated by an unknown kinase and transduces signals downstream [1]. RAS induces the similar translocation of other effectors, including phosphatidylinositol-specific phospholipase Cε (PI–PLCε) and Ral guanine nucleotide dissociation stimulator (RALGDS), to the plasma membrane after the cells are stimulated by growth factors [11, 12].

All known RAS effectors contain one or two RAS-binding site(s) [2]. Two RAS-binding sites occur in the N-terminal domain of the RAF molecule: the primary RAS-binding site or RAS-binding domain (RBD), and the secondary site or cysteine-rich domain (CRD) [13]. The affinity between RAS and RBD depends on the nucleotide status of RAS [14–16]. We can easily confirm this phenomenon, i.e., that RBD strongly prefers RASGTP to RASGDP as its binding partner, in an in vitro pull-down assay [17]. Therefore, it has been believed that RAS induces activation of downstream effectors when the affinity between RAS and the RAS-binding sites increases after RAS is activated (GDP/GTP exchange). In parallel with this in vitro observation, several studies of cells have reported that RAF is translocated from the cytoplasm to the plasma membrane and accumulates over vast areas of the plasma membrane after stimulation induces RAS activation [9, 10]. We have also observed the translocation of green fluorescent protein (GFP)-tagged whole molecules of RAF (GFP–RAF) in living HeLa cells stimulated with EGF (Fig. 3.1a) [18].

However, in our experiments, the translocation of the RBD and RBDCRD fragments of RAF tagged with GFP did not depend markedly on RAS activation (Fig. 3.1b, c) [19]. After RAS activation, most of the RBD molecules remained in the cytoplasm, although a proportion of GFP–RBD molecules accumulated on the plasma membrane (Fig. 3.1b). On the contrary, RBDCRD–GFP localized entirely at the plasma membrane, even before RAS activation, and remained there after RAS activation (Fig. 3.1c). These results indicate that, contrary to the suggestion of in vitro experiments, a simple increase in the affinity between RAS and one or both of the two RAS-binding sites in RAF when RAS is activated is insufficient for the

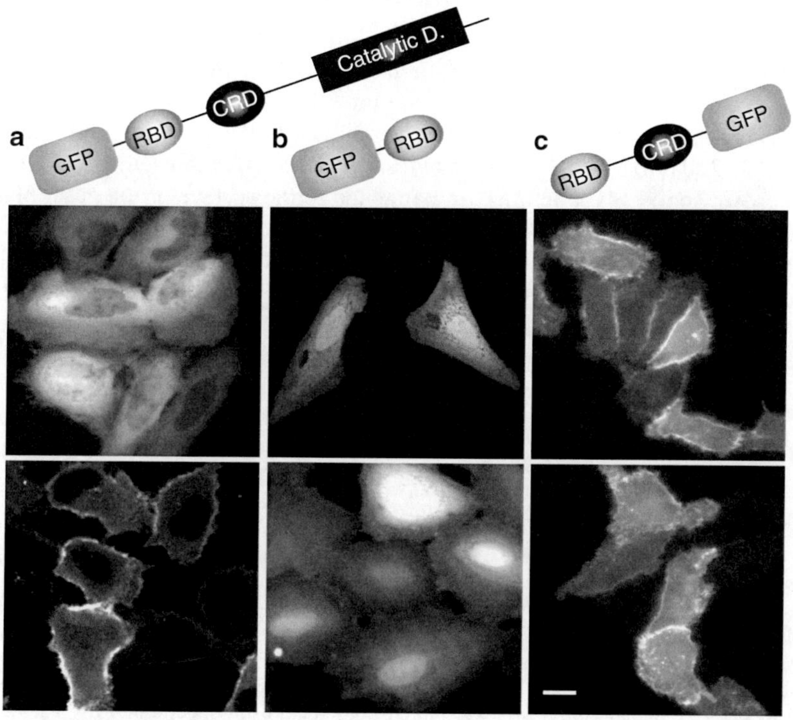

Fig. 3.1 Intracellular distribution of RAF and fragments of RAF containing the RAS-binding domains (RBD and CRD). GFP–RAF (**a**), GFP–RBD (**b**), or RBDCRD–GFP (**c**) was transiently coexpressed with RAS in HeLa cells. The intracellular distributions of the proteins were observed by scanning confocal microscopy before (*upper panels*) and after (*lower panels*) stimulation with EGF to induce RAS activation. GFP–RAF in quiescent cells (a, *upper*) and GFP–RBD (b) were predominantly distributed in the cytoplasm. GFP–RAF in cells after EGF stimulation (a, *lower*) and RBDCRD–GFP (c) were localized on the plasma membrane. Scale bar: 10 μm. Catalytic D: catalytic domain.

accurate and efficient signaling from RAS to RAF in living cells. We thought that the efficient association between RBD and RASGTP observed in pull-down assays was caused by the dense accumulation of the proteins on the microbeads. How then does the molecular switch RAS recognize RAF and turn on the RAF function in living cells? The behaviors of the RBD and RBDCRD fragments of RAF suggest that the C-terminal domain of RAF, in addition to the RBD and CRD in the N-terminal part of the molecule, is required for RAF to distinguish accurately between RASGDP and RASGTP. It has been reported that CRD interacts with the C-terminal catalytic domain of RAF [20]. A dynamic interaction between the RAS-binding sites and the C-terminal domain of RAF is presumably related to the accurate signaling from RAS to RAF. To confirm this hypothesis, we examined the precise kinetics of the interaction between RAS and RAF using single-molecule measurements in living cells.

3.2 Single-Molecule Kinetics of the Interactions Between RAS and RAF in Living Cells

GFP–RAF, in which RAF is tagged with GFP at its N-terminus, was monitored in living HeLa cells using objective-type total internal reflection fluorescence (TIRF) microscopy focused on the basal plasma membrane. Because of the limited excitation depth (~200 nm) of TIRF microscopy, single GFP–RAF molecules are detected as fluorescent spots only when they are attached to the surface of the basal plasma membrane [18]. Free GFP–RAF molecules in the cytoplasm within the excitation depth of TIRF microscopy increase the background fluorescence, but are not detected as spots because the diffusion of the free GFP–RAF in the cytoplasm is too rapid for them to be detected as spots at the video rate (30 frames s^{-1}) used in our experiments. Even in quiescent cells, a small number of spots were observed, representing GFP–RAF molecules bound to inactive RASGDP (Fig. 3.2a, left). After the cells were stimulated with EGF to induce RAS activation, the number of GFP–RAF spots increased on the plasma membrane (Fig. 3.2a, right). This reflects

Fig. 3.2 Single-molecule kinetics of dissociation between RAS and RAF. (**a**) HeLa cell transiently expressing GFP–RAF was observed by TIRF microscopy before (*left*) and after (*right*) stimulation with EGF to induce RAS activation. Single molecules of GFP–RAF bound to the basal cell membrane were detected as fluorescent spots (scale bar: 5 μm). The *inset* (*right panel*) is a magnified view of single GFP–RAF molecules (scale bar: 1 μm). (**b–g**) On-time distributions of GFP–RAF (b and c), GFP–RBD (d and e), and RBDCRD–GFP (f and g) bound to RAS on the plasma membrane before (b, d, and f) and after (c, e, and g) stimulation with EGF. N indicates the number of spots examined. *Solid lines* were fitted to the data using kinetic models. The best-fit values of the rate constants (s^{-1}) obtained from these fittings are shown with the error ranges. *Dotted lines* in b, d, and e are the results of fitting to a two-component exponential function. The best-fit values and fractions of the fittings to a two-component exponential function are 4.4 ± 0.27 s^{-1} (97%) and 0.66 ± 0.24 s^{-1} (3%) for b, 4.8 ± 0.49 s^{-1} (94%) and 0.85 ± 0.23 s^{-1} (6%) for d, and 3.2 ± 0.11 s^{-1} (98%) and 0.50 ± 0.19 s^{-1} (2%) for e. These results indicate that the minor components are insignificant. k is the dissociation rate constant. k_1, k_2, and k_3 are the rate constants for the dissociation from the initial binding state, for the formation of the kinetic intermediate, and for the dissociation from the kinetic intermediate, respectively.

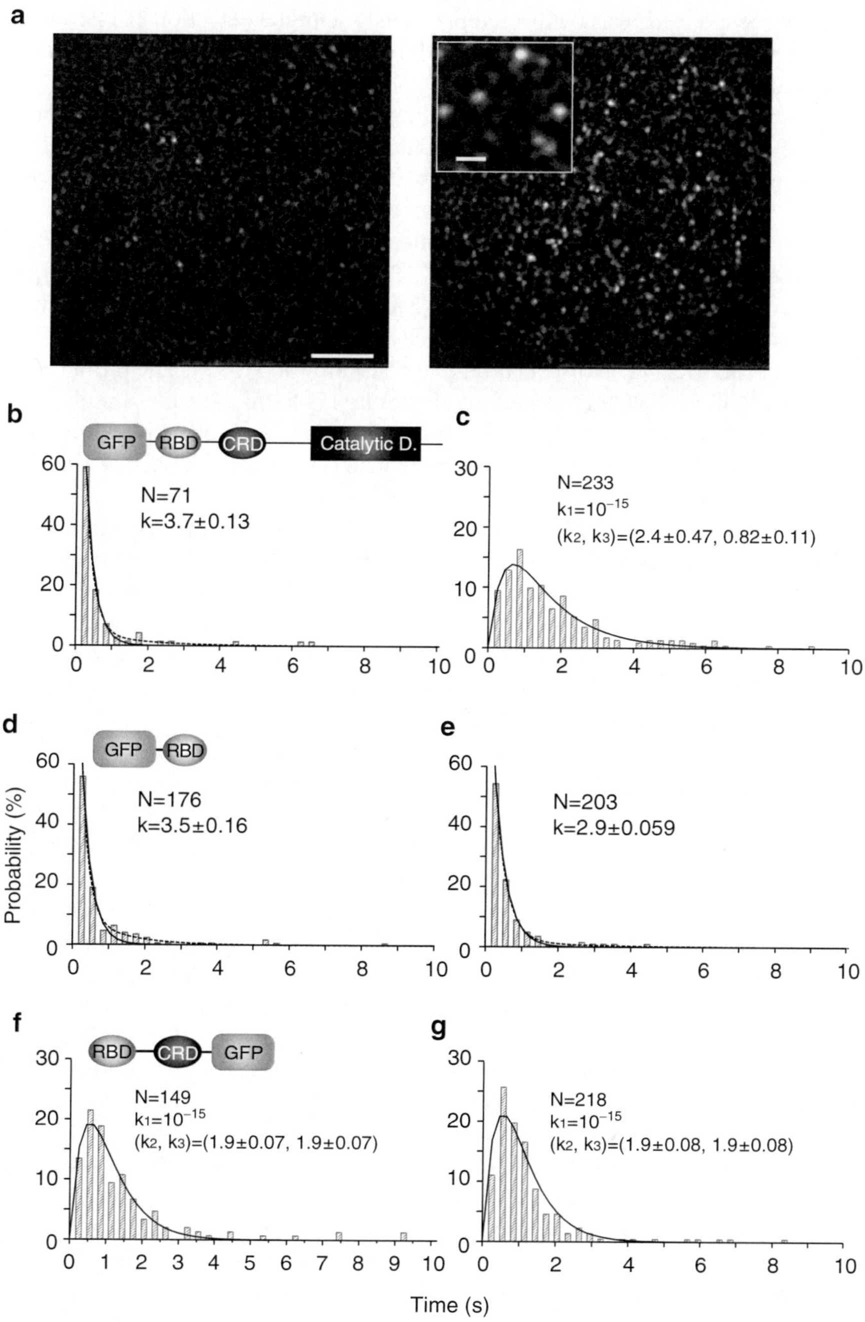
a
b
GFP RBD CRD Catalytic D.
N=71
k=3.7±0.13
c
N=233
k1=10^−15
(k2, k3)=(2.4±0.47, 0.82±0.11)
d
GFP RBD
N=176
k=3.5±0.16
e
N=203
k=2.9±0.059
f
RBD CRD GFP
N=149
k1=10^−15
(k2, k3)=(1.9±0.07, 1.9±0.07)
g
N=218
k1=10^−15
(k2, k3)=(1.9±0.08, 1.9±0.08)
Probability (%)
Time (s)

the translocation of RAF as individual protein molecules. The direct interaction between RAS and RAF was confirmed previously with the detection of intermolecular Förster resonance energy transfer (FRET) between yellow fluorescent protein (YFP)–RAS and GFP–RAF [18].

The fluorescent spots of the RAF molecule diffused laterally along the plasma membrane during its association with RAS, and then dissociated into the cytoplasm. We measured the duration of the RAF association with the plasma membrane as "on-time" (Fig. 3.2b–g). The length of the on-time is related to the dissociation kinetics of RAF from RAS. The distribution of on-times was constructed from single-molecule of multiple RAF molecules. The on-time distribution of RAF interacting with RASGDP could be described by a single exponential function (Fig. 3.2b). This means that the dissociation rate between RASGDP and RAF was constrained by a single stochastic event. The dissociation rate constant was 3.7 s^{-1}. The profile of the on-time distribution changed to a peaked distribution 2–5 min after the stimulation of the cells with EGF to induce RAS activation (Fig. 3.2c), which implies that RAF dissociated from RASGTP through an intermediate state. Direct dissociation of RAF from the initial state of association with RASGTP was negligible (the best-fit value of the rate constant was $<10^{-15}$ s^{-1}) [19]. The rate constant of the transition to the intermediate (0.8 or 2.4 s^{-1}) was similar to or smaller than the rate constant of the direct dissociation of RAF from RASGDP (3.7 s^{-1}). Therefore, absence of the direct dissociation was not caused by competition between dissociation (in a similar way of RAF on RASGDP) and intermediate formation. Instead, the initial binding state between RASGTP and RAF differs from that between RASGDP and RAF, at least in the time resolution of our experiments (~0.1 s).

In similar experiments, the on-time distribution of the RBD fragment of RAF interacting with RAS could be described by a single exponential function, irrespective of the nucleotide status of RAS. The dissociation rate constants of RBD from RASGDP and from RASGTP were 3.5 and 2.9 s^{-1}, respectively (Fig. 3.2d, e). These values are nearly equal to the dissociation rate constant of RAF from RASGDP (3.7 s^{-1}), suggesting that RAF associated with RASGDP via RBD alone. Conversely, the on-time distribution of the RBDCRD fragment interacting with RAS peaked, irrespective of the nucleotide status of RAS (Fig. 3.2f, g), and the kinetic analysis showed that the direct dissociation of RBDCRD from the initial association state was negligible, even though the rate constant of transition to the intermediate (1.9 s^{-1}) was again smaller than that of the direct dissociation of RBD from RAS (2.9 and 3.5 s^{-1}). These results mean that the detected "initial association state" of RBDCRD with RAS differed from the association state of RBD with RAS. The dissociation kinetics of RBDCRD from RAS is similar to those of the whole molecule of RAF from RASGTP in two important points: the presence of an intermediate state and the absence of a direct dissociation from the initial association state. Both RAF and the RBDCRD fragment include the functional RBD and CRD. Therefore, in addition to RBD, CRD seems to be involved in the initial association of RAF with RASGTP and RBDCRD with RAS (both the GDP and GTP forms). Furthermore, the C-terminal domain of RAF, in addition to RBD and CRD, is required for RAF to distinguish RASGTP from RASGDP.

3.3 Conformational Changes in RAF

It is thought that RAF has at least two conformations: a globular "closed form" and a stretched "open form" [20, 21]. In the closed form of RAF, CRD is thought to interact intramolecularly with the C-terminal catalytic domain. We examined the relationship between the conformation and translocation of RAF. To detect the conformation of RAF in living cells, we constructed a FRET-based probe (RAF-FRET-PROBE), in which RAF was tagged with GFP and YFP at its N- and C-termini, respectively (Fig. 3.3a). It was expected that the FRET efficiency would increase when RAF adopts the closed form and decrease when RAF adopts the open form.

We confirmed the basic function of the RAF-FRET-PROBE expressed in HeLa cells. After stimulation of the cells with EGF, the RAF^{WT}-FRET-PROBE (constructed from wild-type RAF) translocated from the cytoplasm to the plasma membrane in the same way as wild-type RAF and GFP–RAF (Fig. 3.3b). We then measured the fluorescence emission spectrum of RAF-FRET-PROBEs constructed from wild-type RAF and two point mutants (C168S and S621A) of RAF in living cells (Fig. 3.4). Both the mutants are reported to have conformations biased to the open form. Compared with the spectrum for RAF^{WT}-FRET-PROBE (Fig. 3.4a), the fluorescence emission spectra of the probes, each containing one point mutation, were depressed around the emission wavelength of YFP (Fig. 3.4b

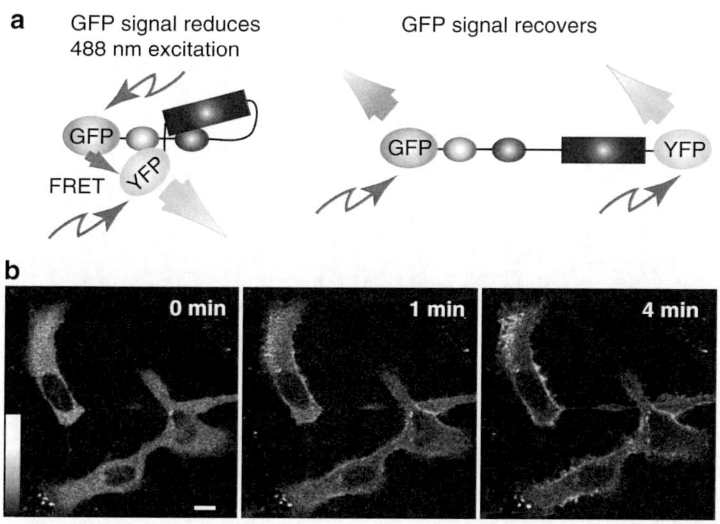

Fig. 3.3 RAF-FRET-PROBE. (**a**) Design of the RAF-FRET-PROBE used to detect conformational changes in RAF. GFP signals were reduced by FRET from GFP to YFP in the closed conformation (*left*) and were recovered in the open conformation (*right*). (**b**) HeLa cells transiently coexpressing RAF^{WT}-FRET-PROBE (the probe constructed from wild-type RAF) and RAS were observed with a confocal microscope. The cells were stimulated with EGF to induce RAS activation, and successive images of the same cells were acquired at the indicated times after stimulation. These images show the summation of GFP and YFP fluorescence. Scale bar: 10 μm.

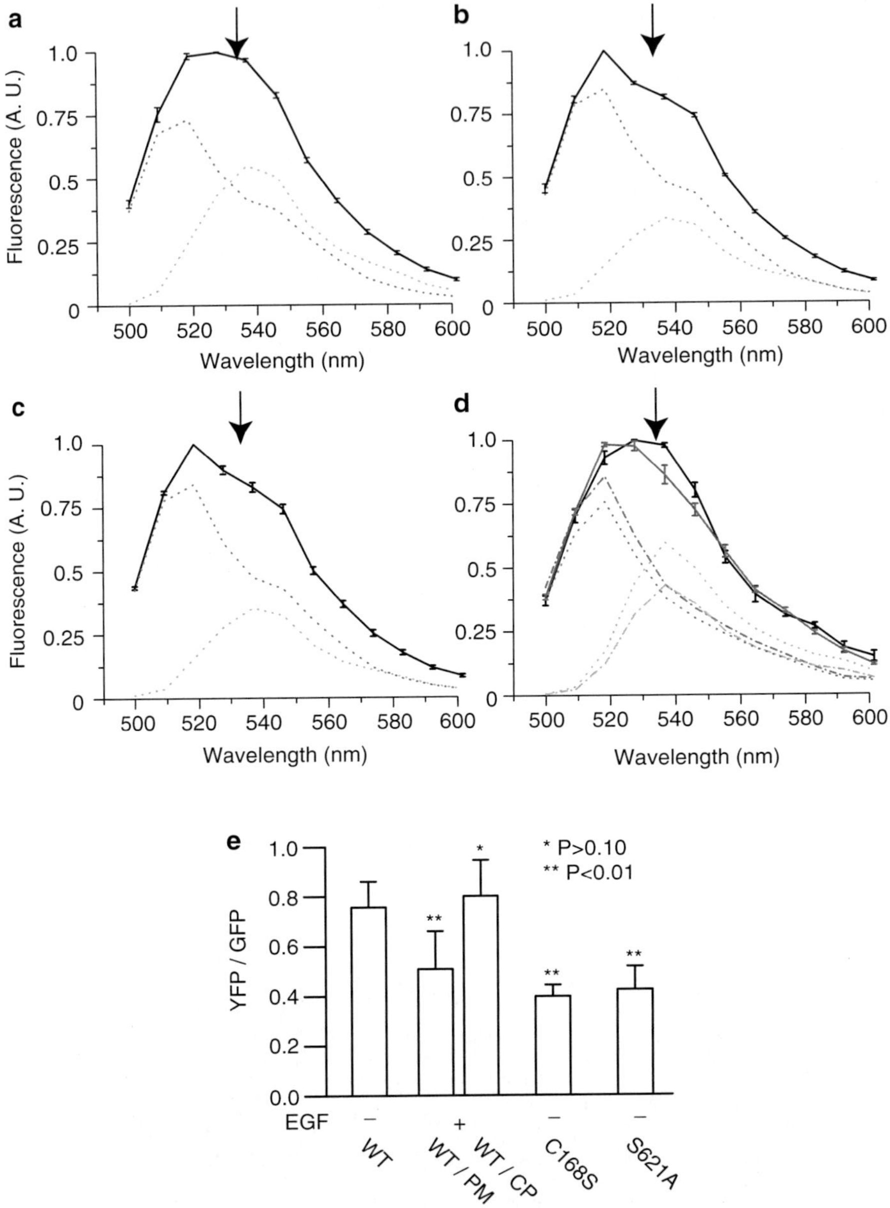

Fig. 3.4 Ensemble detection of FRET in RAF-FRET-PROBEs. Fluorescence emission spectra were measured in single HeLa cells expressing RAF^{WT}-FRET-PROBE (**a** and **d**), RAF^{C168S}-FRET-PROBE (**b**), or RAF^{S621A}-FRET-PROBE (**c**). Each probe was coexpressed with RAS to increase the efficiency of the ensemble FRET detection. Only the cells in (d) were stimulated with EGF. The ensemble average for the spectra of 6–7 cells is shown with the standard error. In (d), the spectra in the cytoplasm (*black*) and at the plasma membrane (*gray*) are shown separately. In the measurements shown in (a) and (b), all the regions of the cells showed similar spectra. The spectra

and c). These results mean that the open-biased mutations induced a reduction in the FRET efficiency, as expected. Thus, the changes in the spectrum of RAF-FRET-PROBE reflect the conformational changes in RAF, as we expected.

The conformational changes in RAF were monitored using the RAF^{WT}-FRET-PROBE (Fig. 3.4d). In cells expressing wild-type RAS and stimulated with EGF, the spectrum of RAF^{WT}-FRET-PROBE that accumulated at the plasma membrane was depressed around the YFP emission, as observed in the spectra of the probes containing the open-biased mutants (C168S and S621A) in quiescent cells. The emission spectrum of the RAF^{WT}-FRET-PROBE that remained in the cytoplasm was not significantly different from that in quiescent cells (Fig. 3.4a, e). These results are consistent with the hypothesis that RAF adopts the closed conformation in the cytoplasm of quiescent cells and, upon stimulation of the cells, the closed RAF changes to the open conformation after its translocation to the plasma membrane, where RAF interacts with RASGTP.

3.4 Single-Molecule Imaging of the RAF Conformation in Living Cells

The conformation of individual RAF molecules associated with RAS was observed using single-pair FRET imaging in living cells. The upper panels in Fig. 3.5a show dual-color images of individual RAF^{WT}-FRET-PROBE molecules on the plasma membrane of a quiescent cell, where the probe molecules were associated with RASGDP. The fluorescent signals in the GFP and YFP channels were observed separately with dual-view optics [19, 22]. The fluorescence emission was suppressed in the GFP channel, indicating a high FRET efficiency from GFP to YFP. In ensemble-molecule FRET detection using confocal microscopy, it was not easy to separate the signals of the molecules on the plasma membrane in quiescent cells from the dominant signals of the molecules in the cytoplasm. Using single-molecule imaging, which can detect a small number of molecules on the plasma membrane, as observed in quiescent cells, we confirmed that RAF molecules binding RASGDP take the closed conformation, like the soluble RAF molecules in the cytoplasm.

In contrast, in cells stimulated with EGF to induce RAS activation, fluorescence emission was detected in both the GFP and YFP channels from the same fluorescent

Fig. 3.4 (continued) in (**c**) were obtained from the plasma membrane region. All the spectra of the FRET probes were normalized to the peak intensity. The *dotted lines* are the unmixed spectra for GFP and YFP, respectively, from the emission spectrum of the FRET probe observed under each condition. In (**d**), the *dotted* and *dashed lines* are unmixed spectra obtained from the cytoplasm and the plasma membrane, respectively. Depression of the spectrum around 530 nm (*arrow*) indicates low FRET efficiency, suggesting the open conformation of RAF. Each emission spectrum in (**a–d**) was separated into GFP and YFP signals, and the averages of the relative YFP signals normalized to the GFP signals are shown, with their standard deviations (**e**). The YFP signals in b (C168S) and c (S621A) and from the plasma membrane in (**d**) (WT/PM) are significantly smaller than that in (**a**).

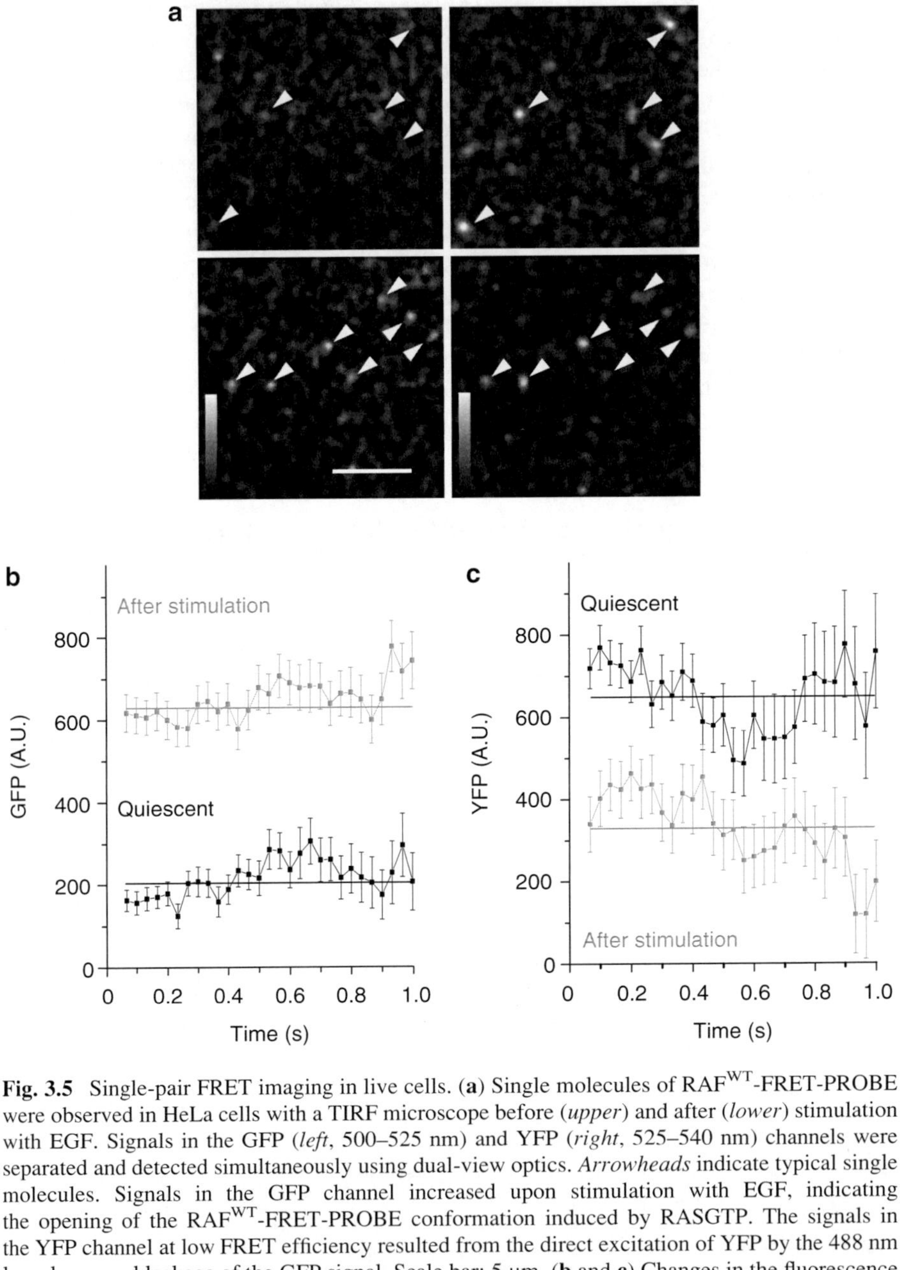

Fig. 3.5 Single-pair FRET imaging in live cells. (**a**) Single molecules of RAF^{WT}-FRET-PROBE were observed in HeLa cells with a TIRF microscope before (*upper*) and after (*lower*) stimulation with EGF. Signals in the GFP (*left*, 500–525 nm) and YFP (*right*, 525–540 nm) channels were separated and detected simultaneously using dual-view optics. *Arrowheads* indicate typical single molecules. Signals in the GFP channel increased upon stimulation with EGF, indicating the opening of the RAF^{WT}-FRET-PROBE conformation induced by RASGTP. The signals in the YFP channel at low FRET efficiency resulted from the direct excitation of YFP by the 488 nm laser beam and leakage of the GFP signal. Scale bar: 5 μm. (**b** and **c**) Changes in the fluorescence intensities of unmixed GFP (b) and YFP (c) are shown for the RAF^{WT}-FRET-PROBE molecules after their association with RAS, when the cells were stimulated (*gray*) or not (*black*) with EGF. The ensemble averages for the GFP signals from 242 (*black*) and 183 (*gray*) molecules from two different cells under each condition are plotted with their standard errors. The *solid lines* are time averages for the fluorescent signals. Because of the limited temporal resolution of the measurements, intensities within 0.1 s could not be determined.

spots of individual probe molecules (Fig. 3.5a, lower panels), indicating that FRET efficiency was reduced after RAS activation, and that RAF molecules bound to RASGTP take the open conformation. (The signal in the YFP channel was detected by the direct excitation of YFP and the leakage of the GFP signal, even in the absence of FRET. The YFP signal leaked negligibly into GFP channel. Therefore, the increased signal in the GFP channel indicated reduced FRET efficiency.) To examine the RAF conformation immediately after its association with RAS, we calculated the unmixed GFP fluorescence of RAF^{WT}-FRET-PROBE as a function of time after it had bound to the plasma membrane in cells before and after RAS activation (Fig. 3.5b). In this experiment, which detected the intramolecular single-pair FRET from GFP to YFP, the intensity of the GFP fluorescence was inversely proportional to the FRET efficiency. The FRET efficiency was reduced immediately after the RAF^{WT}-FRET-PROBE bound to RASGTP (Fig. 3.5b, c), suggesting that RASGTP induced a conformational change in RAF at a very early stage (<0.1 s) in the molecular recognition between individual RASGTP and RAF molecules. This result is consistent with the single-molecule kinetic analysis, suggesting that the initial association state between RAF and RASGDP differs from that between RAF and RASGTP.

3.5 Molecular Recognition Between RAS and RAF Mutants

To understand the relationship between the conformational changes in RAF and the molecular recognition process between RAS and RAF, we examined the distributions of three RAF mutants in cells before and after RAS activation (Fig. 3.6). The first mutant was the open-biased RAF S621A mutant. The S621A mutant adopted the open conformation, as shown above, in the ensemble FRET measurements (Fig. 3.4c), and accumulated at the plasma membrane before RAS activation (Fig. 3.6a). This open-biased mutant cannot distinguish between RASGDP and RASGTP and binds firmly to both forms of RAS. This behavior is similar to that of the functional RBDCRD fragment of RAF (Fig. 3.1c). Considering these data together, it is highly likely that the open-biased RAF S621A mutant associates with RAS (both RASGDP and RASGTP) via RBD and CRD simultaneously, because it always exposes both RBD and CRD to RAS. The S621A mutation in RAF is thought to cause the loss of one of the phosphorylation sites that act as the binding sites for the adaptor/scaffold protein 14-3-3 [23]. The binding of 14-3-3 to phosphorylated S621 seems to function as a stabilizer of the closed conformation of RAF in quiescent cells (see the next section).

The second RAF mutant was the CRD-deficient C168S mutant. As shown in the FRET measurements above, C168S adopts the open form, like S621A, because C168 is crucial not only for CRD function but also for the intramolecular interaction between CRD and the c-terminal catalytic domain. However, C168S did not accumulate at the plasma membrane, even after RAS activation (Fig. 3.6b and c). This behavior of C168S is similar to that of the RBD fragment of RAF (Fig. 3.1b), suggesting that the binding of RAF to RAS via both CRD and RBD is essential for RAF to bind firmly to RAS and accumulate at the plasma membrane.

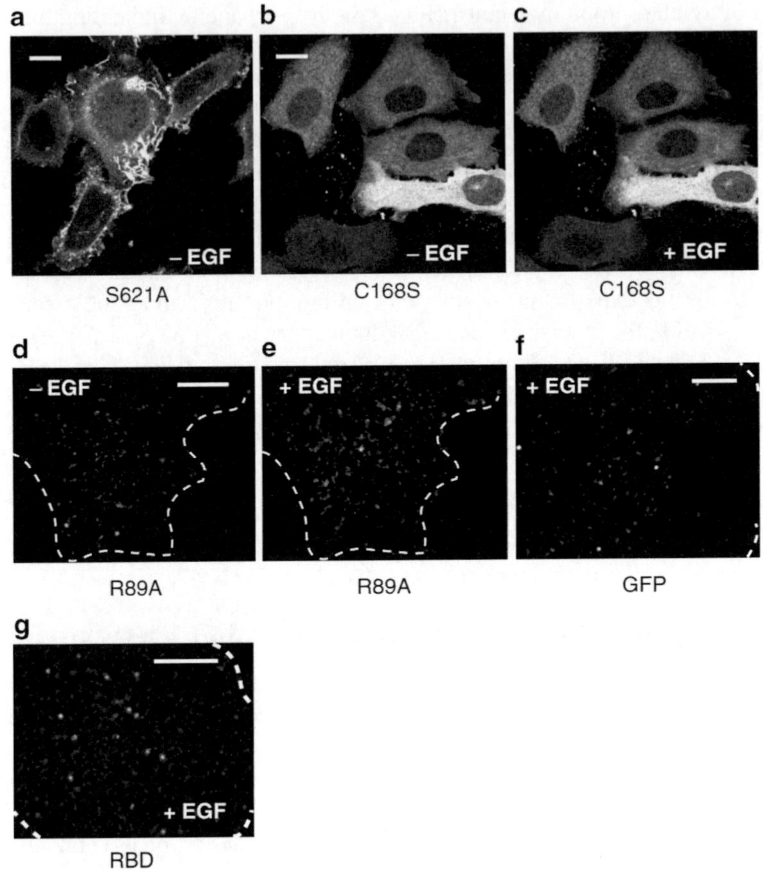

Fig. 3.6 Intracellular distributions of RAF mutants. The S621A (**a**) or C168S (**b**, **c**) mutant of RAF was coexpressed in HeLa cells with RAS. The cells were observed using confocal microscopy before (a, b) and after (c) stimulation with EGF to induce RAS activation. Scale bar: 10 μm R89A tagged with GFP (**d**, **e**), GFP alone (**f**), or GFP-RBD (**g**) was expressed in HeLa cells with RAS and observed using TIRF microscopy before (d) and after (e–g) stimulation with EGF to induce RAS activation. Outlines of the cells are shown as *dashed lines*. Scale bars: 5 μm.

The last RAF mutant was the RBD-deficient R89A mutant, which adopts the closed form. Regardless of the activation status of RAS, the association of the RAF R89A mutant with the plasma membrane was observed much more rarely than that of the RBD fragment and as rarely as that of GFP bound nonspecifically to the plasma membrane (Fig. 3.6d–g), suggesting that the closed form of RAF exposes only RBD to RAS and initially interacts with RAS via RBD.

With these results, we verified that the conformational change in RAF between its closed and open forms is related to the covering and uncovering of the CRD by the C-terminal catalytic domain of RAF, and is essential for accurate molecular recognition between RAS and RAF.

3.6 Mutual Molecular Recognition Between RAS and RAF

The results of our single-molecule measurements provide a new model of the molecular recognition between RAS and RAF. This model suggests that RAS regulates the conformational change in the effector molecule RAF, and accurately turns on the RAF function via this mechanism, as follows. RAF adopts a closed conformation in the cytoplasm and contacts both RASGDP and RASGTP via the RBD, which is the unique RAS-binding site exposed in the closed conformation. In the closed conformation, CRD, the second RAS-binding site of RAF, is covered by the catalytic domain as the result of an intramolecular interaction. When it contacts RASGDP, RAF dissociates directly into the cytoplasm, maintaining its closed conformation, and is not activated by phosphorylation on the plasma membrane. However, when RAF encounters RASGTP, a conformational change from the closed to the open form is immediately induced in the RAF molecule by this newly found activity of RASGTP, and then the RAF molecule is phosphorylated and thus activated. The RASGTP-induced conformational change in RAF is so important that RAF mutants lacking this conformational change cannot distinguish between RASGDP and RASGTP (Fig. 3.6).

To satisfy this scenario, the structure of RAF in quiescent cells must be strongly biased towards the closed conformation, so the spontaneous transition of the RAF structure towards the open conformation is suppressed, and/or the lifetime of the open conformation caused by thermal fluctuations is short. Were this not the case, RAF in the open conformation would bind firmly to both RASGDP and RASGTP, as is true for the RBDCRD fragment and the open-biased S621A mutant, and its activity could not be regulated accurately via signal transduction from RAS. It is thought that 14-3-3 interacts with RAF at the phosphorylated S259 residue following the CRD and phosphorylated S621 in the catalytic domain to crosslink these regions [23]. S621 is constitutively phosphorylated, even in quiescent cells [24], and the mutation S621A induces a shift in the equilibrium of the RAF conformation to the open form (Fig. 3.4c). These observations suggest that 14-3-3 functions in stabilizing the closed conformation of RAF in quiescent cells to avoid mis-signaling from RASGDP to RAF.

3.7 Spatial Heterogeneity of the Reaction Between RAS and RAF in Living Cells

So far, we have described the interactions between RAS and RAF in the early (2–5 min) stage of cell stimulation by EGF. At this stage, RAF accumulates almost homogeneously along the plasma membrane [18]. While more than 20 min after continuous stimulation, RAF forms micrometer-scale accumulation patches scattered over the plasma membrane (Fig. 3.7a, left). Ensemble-molecule imaging of cells at this stage showed that the fluorescence intensity of GFP–RAF per unit area decreased at the bulk membrane, which was outside the patches, whereas GFP–RAF molecules accumulated in local areas on the plasma membrane and formed accumulation patches for a long period of time (>60 min after continuous stimulation by EGF). The fluorescence intensity of GFP–RAF in these patches on the cells increased 35 min after EGF stimulation by a factor of 2.5 compared with that on the bulk membrane. In these local accumulation patches, reorganization of the cytoskeleton and the formation of membrane ruffles were stimulated [18].

Using single-molecule imaging, the on-time distribution of GFP–RAF was measured separately at the bulk membrane and at the local accumulation patches after photo bleaching (Fig. 3.7a, right). The on-time distribution of GFP–RAF on the bulk membrane (Fig. 3.7b) was similar to that of GFP–RAF in quiescent cells (Fig. 3.2b), suggesting that GFP–RAF was interacting with RASGDP on the bulk membrane. The on-time was prolonged in the patches (Fig. 3.7c). The on-time distribution of RAF within the local accumulation patches fitted a two-component exponential function with rate constants of 2.7 s^{-1} (63%) and 0.63 s^{-1} (37%) [18]. However, based on the results of the single-molecule kinetic analysis described in the sections above, it is more likely that both RASGDP and RASGTP were present in the local accumulation patches and that the on-time distribution was a mixture of the distributions of RAF with RASGDP and RAF with RASGTP. Therefore, the distribution was analyzed with a combination of the single exponential function for RAF with RASGDP and the peaked distribution for RAF with RASGTP (Fig. 3.7c). The result of fitting suggests that 22% of RAF molecules appeared within the patches interacted with RASGTP and the remaining (78%) RAF interacted with RASGDP.

Considering the short periods of the on-time of RAF on RASGDP (the median was 0.30 s) compared to those on RASGTP (the median was 1.33 s), 44% of RAF in the local accumulation patches was associating with RASGDP. Therefore, among the 2.5-fold accumulation of RAF in the local patches, 1.1-fold associated with RASGDP, that is, density of RASGDP was almost equal in the patches and the bulk membrane. This means that the local accumulation of RAF was caused by an additional concentration of RASGTP at the patches, not by a local stimulation of RAS activation under the uniform distribution of RAS along the plasma membrane. We anticipate that the heterogeneous concentration of RAS was attributable to dynamic interactions between GEF, GAP, and effector molecules. It is highly likely that the local and prolonged dynamic activation of RAS induced cellular morphological changes.

It is generally accepted that the amount of active RAS in cells, which correlates with the amount of RAF translocated to the plasma membrane, increases to a peak about 2–10 min after cell stimulation and returns to the basal level within 30 min of cell stimulation [25, 26]. However, in our observations, the localized accumulation of RAF to the patches on the plasma membrane was sustained for more than 60 min after the start of cell stimulation. Our observations might seem inconsistent with earlier biochemical reports. However, the fraction of RAF molecules that accumulated in the patches was small ($3 \pm 5\%$ of the total amount) [18], which is similar to the amount of basal RAS activity in quiescent cells [26]. In quiescent cells, this

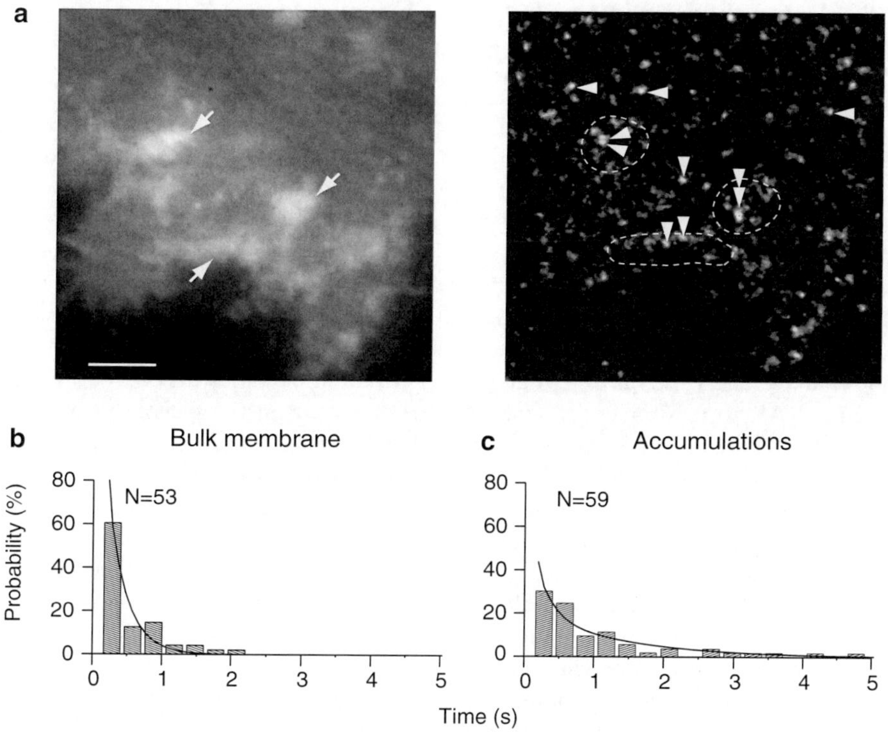

Fig. 3.7 Heterogeneous distribution of RAF on the plasma membrane. (**a**) A cell coexpressing RAS and GFP–RAF was observed with TIRF microscopy 35 min after stimulation with EGF. Dense accumulation patches of fluorescent spots of GFP–RAF were observed (*arrows*). After photobleaching, GFP–RAF was observed as small fluorescent spots of single molecules (*arrowheads*). Regions surrounded by *dashed lines* are the accumulated areas. Scale bar: 5 μm. (**b**) On-time distributions of GFP–RAF outside the accumulation patches. The distributions were fitted to a single exponential function. The best-fit value for the dissociation constant was 2.4 ± 0.06 s^{-1}. (**c**) On-time distributions of GFP–RAF inside the accumulation patches. The distribution was fitted to a combination of the single exponential function with RASGDP (Fig. 3.2b) and the peaked function with RASGTP (Fig. 3.2c). The fractions of the single exponential and peaked functions were 78% and 22%, respectively. N indicates the number of molecules measured.

small amount of activated RAS is dispersed over vast areas of the plasma membrane. However, during the 20–60 min of cell stimulation, RAF molecules recruited by RASGTP were concentrated in specific regions and significant physiological activities including membrane ruffling were observed in these regions, even though the total amount in the cells was small. Such small and local signaling from RAS to RAF is difficult to detect using biochemical techniques. In the membrane ruffles formed from the patches of RAF, activated RAS is likely to be involved in signaling to reorganize the cytoskeleton, which is mediated by the Rho family of small GTPases and phosphatidylinositol 3-kinase (PI3K) [27, 28].

3.8 Lateral Diffusion of RAS–RAF Complexes on the Plasma Membrane

Single-molecule imaging can be used to measure the lateral movements of RAS and RAF along the plasma membrane. In HeLa cells stimulated with EGF, the lateral diffusion coefficients for individual GFP–RAS and GFP–RAF molecules were calculated from the mean square displacements during 33–150 ms time intervals in individual single-molecule trajectories. The distributions of the diffusion coefficients of GFP–RAF within the accumulation patches and on the bulk membrane were not significantly different (Fig. 3.8a, b). The average diffusion coefficient was 0.04 and 0.08 $\mu m^2\ s^{-1}$ within the patches and on the bulk membrane, respectively. Both the average (0.11 $\mu m^2\ s^{-1}$) and the distribution of the diffusion coefficients of GFP–RAS were similar to those of RAF observed on the plasma membrane (Fig. 3.8c), confirming that the RAF molecules interacted with RAS. The diffusion coefficients of the RAS and RAF molecules were distributed in a wide range from 0.01 to several $\mu m^2\ s^{-1}$. The diffusion coefficient of 0.01 $\mu m^2\ s^{-1}$ is indistinguishable from the apparent diffusion coefficient of the GFP molecule fixed onto the plasma membrane, and several $\mu m^2\ s^{-1}$ is in the range of the diffusion coefficients of the molecules freely diffusing along the biological membrane.

The on-time distributions indicated that the half-lives of the RAF molecules within the accumulation patches and on the bulk membrane were 0.51 and 0.18 s, respectively. During these periods of association with RAS, each RAF molecule moved around 0.08 and 0.06 μm^2 on average, respectively. Considering the area of the accumulation patches (~3 μm in diameter), the probability is minimal that the RAF molecules moved between the accumulation patches and the bulk membrane by lateral diffusion along the membrane. Therefore, the association of cytoplasmic RAF molecules with the concentrated binding sites, rather than the lateral transport of RAF molecules from the bulk membrane, is the major mechanism maintaining the accumulation patches of RAF.

3.9 Discussion and Perspectives

Single-molecule kinetic and dynamic analyses have suggested that the signaling from RAS to RAF occurs not by unilateral recognition of the conformational change in RAS by the RAF molecule, but by the active recognition of RAF by RASGTP inducing the opening of RAF conformation. The commonly accepted mechanism for the regulation of signaling from RAS to RAF, in which the affinity between RAS and the RAS-binding sites of RAF increases when RAS is activated, is insufficient for the signal transduction between RAS and RAF in living cells. However, the RASGTP-induced conformational change in RAF is essential for the accurate molecular recognition between RAS and RAF. These intermolecular interactions are inherently stochastic, so a subset of RAF molecules associated

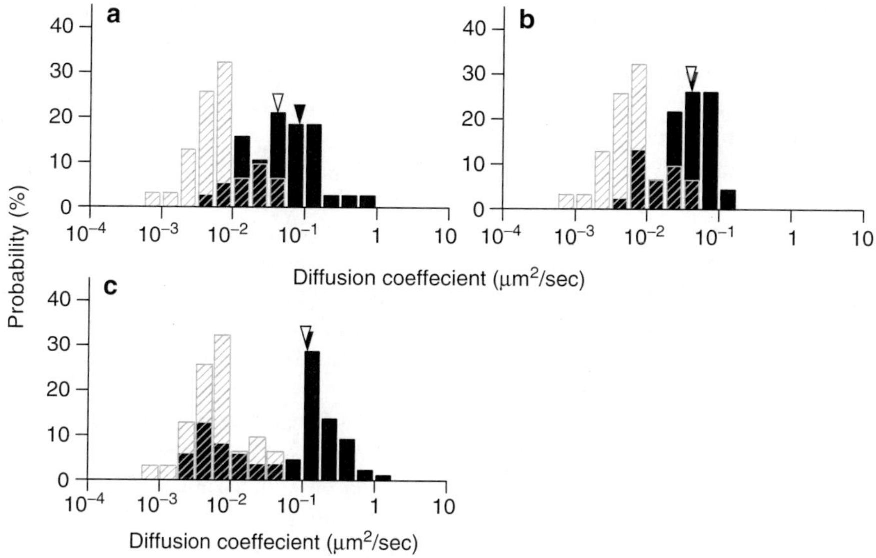

Fig. 3.8 Distributions of the lateral diffusion coefficient of RAS and RAF. Distributions of the lateral diffusion coefficient determined for individual molecule of GFP–RAF outside (**a**) and inside (**b**) the accumulation patches and for GFP–RAS (**c**) in HeLa cells after stimulation with EGF. *Closed* and *open arrowheads* indicate the averages and medians of the distributions, respectively. The distribution of the diffusion coefficient of GFP, measured on the fixed cell membrane, is shown in *gray*.

with RASGDP would encounter the kinase responsible for RAF activation even though the affinity between RAF and RASGDP is low. This would mistakenly activate RAF if RAS–RAF recognition depended only on the affinity change. Actually, RASGTP induces a conformational change in RAF that causes a qualitative change in the interaction between RAS and RAF. This mechanism can function as a safety device to prevent mis-signaling from RAS to RAF.

RAS has various species of effectors involved in different downstream reactions. One unsolved and important question is how such diverse molecular interactions are realized and organized. Single-molecule analysis of RAS–RAF signaling suggests that both the RBD and CRD of RAF associate simultaneously with RASGTP, so RAS presents a broad interface for its molecular recognition of RAF. It is possible that the multilateral molecular recognition between RAS and its effector molecules is important in allowing diverse molecular interactions. Not only RAF, but several other RAS effectors contain multiple RAS-binding domains [2]. Furthermore, our analysis of RAS–RAF recognition suggests that the interface between RAS and RAF changes dynamically depending on the RAS activation status and during the process of the RAS–RAF interaction. The closed form of RAF recognizes RAS using only RBD, and only when RAS is in the active form, CRD is uncovered and binds with RAS. Such complex dynamics of molecular recognition could be common in the complex interactions between RAS and its effectors that have multiple RAS-binding sites and could be used in the spatial and temporal regulation by RAS of the appropriate effector functions in response to different intracellular states and/or extracellular signals. To assess this possibility, we require a technology that precisely measures the structural dynamics of single molecules in living cells with high temporal resolution.

In a later stage in the stimulation of cells by EGF, we observed spatially and temporally heterogeneous signaling from RAS to RAF, even though the cells were stimulated uniformly with EGF, and both the association between EGF and its receptors and the initial translocation of RAF occurred almost uniformly along the cell surface. This heterogeneity appeared on two levels. One was the patch-like distribution of RAF when recruited onto the plasma membrane, observed in ensemble molecules, and the other was the different on-time distributions of RAF as single molecules inside and outside the patches. Although the mechanisms that induce these heterogeneities are not known, some kind of feedback loop in RAS–RAF signaling could be involved in this phenomenon. The presence of feedback loops has been reported in RAF-induced RAF activation via the dimerization of RAF molecules [29] and the negative feedback of RAS activation from extracellular signal-regulated kinase (ERK) activated downstream from RAF [30]. Regardless of the mechanisms, the spatially heterogeneous signaling from RAS to downstream effectors, including RAF, is related to the polarization and directional movements of cells observed after EGF stimulation. It is also possible that RAS controls the signaling to specific effectors using heterogeneous activation.

Despite the importance of the roles of RAS in various cellular reactions, the precise kinetics of the process of effector activation by RAS remains unknown, to the best of our knowledge. There have been many attempts to reconstruct the dynamics of cell signaling networks, including RAS activity, with *in silico* computation. However, in these studies, the step of RAS signaling has been described as an ad hoc process [31–35]. Single-molecule imaging of RAS–RAF signaling in living cells offers a way to undertake the kinetic study of RAS functions. The major questions to be answered in the near future are how RAS selects the appropriate effector molecule, how the spatially heterogeneous signaling of RAS is initiated and maintained, and how it affects cell behaviors. Single molecule kinetic and dynamic analyses are powerful methods with which to approach these questions.

Acknowledgments We would like to thank our collaborators, Tatsuo Shibata (Hiroshima University) and Toshio Yanagida (Osaka University), and the members of our laboratory.

References

1. Wellbrock C, Karasarides M, Marais R (2004) The Raf proteins take center stage. Nat Rev Mol Cell Biol 5:875–885
2. Repasky GA, Chenette EJ, Der CJ (2004) Renewing the conspiracy theory debate: does Raf function alone to mediate Ras oncogenesis? Trends Cell Biol 14:639–647

3. Downward J (2003) Targeting Ras signalling pathways in cancer therapy. Nat Rev Cancer 3:11–22
4. Lowy DR, Willumsen BM (1993) Function and regulation of Ras. Annu Rev Biochem 62:851–891
5. Kaziro Y, Itoh H, Kozasa T, Nakafuku M, Satoh T (1991) Structure and function of signal-transducing GTP-binding proteins. Annu Rev Biochem 60:349–400
6. Avruch J, Zhang X, Kryiakis JM (1994) Raf meets Ras: completing the framework of a signal transduction pathway. Trends Biochem Sci 19:279–283
7. Daum G, Eisenmann-Tappe I, Fries H-W, Troppmair J, Rapp UR (1994) The ins and outs of Raf kinases. Trends Biochem Sci 19:474–480
8. Hancock JF (2003) Ras proteins: different signals from different localizations. Nat Rev Mol Cell Biol 4:373–384
9. Leevers SJ, Paterson HF, Marshall CJ (1994) Requirement for Ras in Raf activation is overcome by targeting Raf to the plasma membrane. Nature 369:411–414
10. Stokoe D, Macdonald SG, Cadwallader K, Symons M, Hancock JF (1994) Activation of Raf as a result of recruitment to the plasma membrane. Science 264:1463–1467
11. Matsubara K, Kishida S, Matsuura Y, Kitayama H, Noda M et al (1999) Plasma membrane recruitment of RalGDS is critical for Ras-dependent Ral activation. Oncogene 18:1303–1312
12. Song C, Hu C-D, Masago M, Ki K, Yamawaki-kataoka Y et al (2001) Regulation of a novel human phospholipase C, PLCε, through membrane targeting by Ras. J Biol Chem 276:2752–2757
13. Avruch J, Khokhlatchev A, Kyriakis JM, Luo Z, Tzivion G et al (2001) Ras activation of the Raf kinase: Tyrosine kinase recruitment of the MAP kinase cascade. Recent Prog Horm Res 56:127–155
14. Vojtek AB, Hollenberg SM, Cooper JA (1993) Mammalian Ras interacts directly with the serine/threonine kinase Raf. Cell 74:205–214
15. Warne PH, Viciana PR, Downward J (1993) Direct interaction of Ras and the amino-terminal region of Raf-1 *in vitro*. Nature 364:352–355
16. Zhang X, Settleman J, Kyriakis JM, Takeuchi-Suzuki E, Elledge SJ et al (1993) Normal and oncogenic p21ras proteins bind to the amino-terminal regulatory domain of c-Raf-1. Nature 364:308–313
17. Jd R, Bos JL (1997) Minimal Ras-binding domain of Raf1 can be used as an activation-specific probe for Ras. Oncogene 14:623–625
18. Hibino K, Watanabe TM, Kozuka J, Iwane AH, Okada T et al (2003) Single- and multiple-molecule dynamics of the signaling from H-Ras to cRaf-1 visualized on the plasma membrane of living cells. Chemphyschem 4:748–753
19. Hibino K, Shibata T, Yanagida T, Sako Y (2009) A RasGTP-induced conformational change in C-RAF is essential for accurate molecular recognition. Biophys J 97:1277–1287
20. Cutler RE, Stephens RM, Saracino MR, Morrison DK (1998) Autoregulation of the Raf-1 serine/threonine kinase. Proc Natl Acad Sci USA 95:9214–9219
21. Terai K, Matsuda M (2005) Ras binding opens c-Raf to expose the docking site for mitogen-activated protein kinase kinase. EMBO Rep 6:251–255
22. Hibino K, Hiroshima M, Takahashi M, Sako Y (2009) Single-molecule imaging of fluorescent proteins expressed in living cells. Method Mol Biol 544:451–460
23. Tzivion G, Luo Z, Avruch J (1998) A dimeric 14-3-3 protein is an essential cofactor for Raf kinase activity. Nature 394:88–92
24. Morrison DK, Heidecker G, Rapp UR, Copeland TD (1993) Identification of the major phosphorylation sites of the Raf-1 kinase. J Biol Chem 268:17309–17316
25. Satoh T, Endo M, Nakafuku M, Akiyama T, Yamamoto T et al (1990) Accumulation of $p21^{ras}$GTP in response to stimulation with epidermal growth factor and oncogene products with tyrosine kinase activity. Proc Natl Acad Sci USA 87:7926–7929
26. Gibbs JB, Marshall MS, Scolnick EM, Dixon RAF, Vogel US (1990) Modulation of guanine nucleotides bound to Ras in NIH3T3 cells by oncogenes, growth factors, and the GTPase activating protein (GAP). J Biol Chem 265:20437–20442

27. Rodriguez-Viciana P, Warne PH, Khwaja A, Marte BM, Pappin D et al (1997) Role of phospoinositide 3-OH kinase in cell transformation and control of the actin cytoskeleton by Ras. Cell 89:457–467
28. Nimnual AS, Yatsula BA, Bar-Sagi D (1998) Coupling of Ras and Rac guanosine triphosphatases through the Ras exchanger Sos. Science 279:560–563
29. Heidorn S, Milagre C, Whittaker S, Nourry A, Niculescu-Duvas I et al (2010) Kinase-dead BRAF and oncogenic RAS cooperate to drive tumor progression through CRAF. Cell 140:209–221
30. Dong C, Waters S, Holt K, Pessin J (1996) SOS phosphorylation and disassociation of the Grb2-SOS complex by the ERK and JNK signaling pathways. J Biol Chem 271:6328–6332
31. Bhalla U, Iyengar R (1999) Emergent properties of networks of biological signaling pathways. Science 283:381–387
32. Brightman F, Fell D (2000) Differential feedback regulation of the MAPK cascade underlies the quantitative differences in EGF and NGF signalling in PC12 cells. FEBS Lett 482:169–174
33. Kholodenko B (2000) Negative feedback and ultrasensitivity can bring about oscillations in the mitogen-activated protein kinase cascades. Eur J Biochem 267:1583–1588
34. Schoeberl B, Eichler-Jonsson C, Gilles E, Müller G (2002) Computational modeling of the dynamics of the MAP kinase cascade activated by surface and internalized EGF receptors. Nat Biotechnol 20:370–375
35. Sasagawa S, Ozaki Y, Fujita K, Kuroda S (2005) Prediction and validation of the distinct dynamics of transient and sustained ERK activation. Nat Cell Biol 7:365–373

Chapter 4
Single-Channel Structure-Function Dynamics: The Gating of Potassium Channels

Shigetoshi Oiki

Abstract The ion channel is a molecular device for electrical signaling in the cell. The high throughput rate of ion permeation through the channel allowed measurements of ionic currents through single channel molecule (the single-channel current) and a demonstration of the kinetic behavior of the opening and closing (gating) of the permeation pathway. Recently, the crystal structure of potassium channels has been elucidated and structural parts relevant to its function were identified. To detect the structural dynamics upon gating, we have established a diffracted X-ray tracking method and traced the conformational changes of the KcsA potassium channel at the single molecule level. During the opening and closing of the pH-dependent gate, the channel rearranged its architecture reversibly, and generated a global twisting motion around the axis of symmetry. Recording the gating dynamics at the single molecule level using structural and functional methods provides complementary information important for constructing an integrative view of the channel gating. Recent advances in the study and understanding of channel dynamics are reviewed and discussed.

Keywords Action potential · Allosteric · Born energy · Bragg condition · Bulge helix · Bundle crossing · Circular symmetry · Concerted model · CPD-truncated channel · Cytoplasmic domain · Diffracted X-ray tracking method; DXT · Diffraction · Diffraction spots · Diffusion constant · Discrete Markovian model · Dwell times · E71A mutant · Electrical signals · Eyring-type kinetics · Gating · Gating particles · Gold nanocrystal · Helical bundle (HB) domain · helix gate · Hodgkin– Huxley Equation · Inactivation · KcsA channel · KNF (Koshland-Nemethy-Filmer) model · Kramers-type kinetics · Log-binned histogram · M1, M2 helix · Markov model · Mean square displacement (MSD) · Membrane current · Memoryless

---

S. Oiki (✉)
Department of Molecular Physiology and Biophysics, University of Fukui Faculty of Medical Sciences, Fukui 910-1193, Japan
e-mail: oiki-fki@umin.ac.jp

Y. Sako and M. Ueda (eds.), *Cell Signaling Reactions: Single-Molecular Kinetic Analysis*, DOI 10.1007/978-90-481-9864-1_4,

· MWC (Monod-Wymann-Changeux) model · Patch-clamp · pH-sensor (PS) domain · Planar lipid bilayer · Pore · Pore helix · Selectivity filter · Sequential model · Single-channel current recordings · Synchrotron radiation facility · Transmembrane domain · Twisting motion · Voltage clamp method · Voltage-gated channel · Voltage sensor · Water-filled pore · White X-rays

List of Abbreviations

DXT	Diffracted X-ray Tracking
KNF model	Koshland-Nemethy-Filmer model
MWC model	Monod-Wymann-Changeux model
HB domain	helical bundle domain
PS domain	pH-sensor domain
MSD curve	mean square displacement curve
CPD-truncated channel	cytoplasmic domain-truncated channel

4.1 Introduction

The ion channel is a membrane protein which elicits electrical signals in the cell. This function is carried out by the essential properties of channel proteins, ion permeation and gating [12]. As a result of conducting the ionic current with a high throughput rate, the channel generates changes in the membrane potential through charging and discharging the electric capacitor of the cell membrane. The opening and closing of the permeation pathway (gating) under the regulation of sensory signals elicit specific patterns of membrane potential (Fig. 4.1), propagating through the membrane. Compared to diffusion and chemical reactions, electrical signaling of

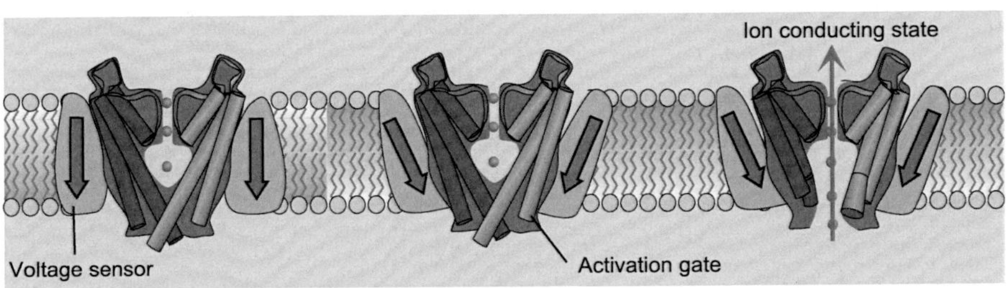

Fig. 4.1 Gating of the channel. The channel is composed of different domains, such as the sensor and the pore domains. For the voltage-gated channel, for example, the voltage sensor detects the changes in the membrane potential and transfers the message to the pore domain, leading to the opening of the gate for ion permeation.

ion channels enables fast and far-reaching conduction. In order to understand the physiological function of the channel, an elucidation of the underlying mechanism of channel gating is crucial.

In ion channels, dynamic activities, including ion permeation and gating, are related to electrical events, and electrophysiological methods have been evolved to inquire the various aspects of the physical properties of the channel molecules. Among these, techniques for single-channel current measurements [32] have revealed the fine mechanisms of the ion permeation and gating of channel proteins. Recent determination of the crystal structures of potassium channels [7] has dramatically advanced the understanding of channel function. To further investigate relationships between the structure and function of channel proteins, we have developed a method for detecting conformational changes in the channel proteins at the single molecule level (Diffracted X-ray Tracking method; DXT) [38]. In this chapter, the gating of ion channels are reviewed from the structural and functional point of view and our recent experimental results on the structural dynamics of the KcsA potassium channel are presented [27].

4.2 The Gating Phenomenon

Among other membrane proteins having ion transporting activity, ion channels have unique features named gating. In response to certain stimuli, such as ligand binding and membrane stretch, ion permeation is turned on and off (gating). One category of these channels, the voltage-gated channel, bears voltage sensors in its molecule and responds to changes in the membrane potential. By acquiring sensitivity to the membrane potential, the channel, having the ability to generate currents, enables free manipulation of the membrane potential as an electrical signal. The gating process, thus, underlies essential functional roles of channel proteins.

To study channel function, measuring ionic current through the channel under controlled membrane potential is fundamental, and electrophysiological techniques, such as the patch-clamp [32], and planar lipid bilayer methods [23], have been developed extensively. On the cell membrane several thousands of channel molecules exist, and ensemble current is a linear sum of single-channel currents. In fact, the membrane current through single species of channel molecules can be expressed as $I(V, t) = n\, p(V, \mathrm{t})\, i(V)$, in which $I(V, t)$ is the macroscopic current at the membrane potential V and time t, n the number of channels, $p(V, \mathrm{t})$ the open probability of the gate at V and t, and $i(V)$ the single-channel current through an open channel. A simple linear system of channel performance makes functional studies both straightforward and quantitative. In parallel to the macroscopic current measurements, the single-channel currents are readily measured, in which opening and closing transitions of the gate and the ionic currents through single molecules are available.

In this chapter, we focus on the potassium channel to examine the gating dynamics, and the KcsA potassium channel is the target of our study. Since first crystallization of the KcsA channel, a huge data have been accumulated on the structural and functional information on gating of the KcsA channel. This chapter is organized as follows. First, the voltage-gated potassium channel is introduced to illustrate the kinetic features of gating. Second, functional studies on the gating are related to the recently determined crystal structure of the potassium channels. The KcsA potassium channel is introduced as a typical channel protein. Then, to focus on the conformational dynamics of the gating transitions, the results of the DXT method for the KcsA potassium channel at the single molecule level are presented. Finally, these single channel dynamics are discussed.

4.3 Gating Kinetics

4.3.1 Kinetics of the Macroscopic Current

One of the most exciting achievements of biological sciences in the twentieth century is the elucidation of the mechanism of the action potential. Hodgkin and Huxley first characterized the kinetic nature of the channel gating of sodium and potassium channels in the neuron with a set of differential equations [13]. This phenomenological description was based on the following electrophysiological experiments. Using the voltage clamp method,[1] the channel current elicited by depolarization steps were recorded. The time course of the macroscopic currents was decomposed into components of Na^+ and K^+ currents (Fig. 4.2a). During activation of the gate, both the Na^+ and K^+ currents exhibited activation after a delay, displaying a sigmoidal shape. As the membrane potential became more depolarized, the time course of activation was accelerated with a shorter delay. This voltage dependent gating is the most fundamental feature of these channels (voltage-gated channel). Hodgkin and Huxley focused on the time course of the current, such as the delay and the shape of the activation. They conjectured that several independent "gating particles" flip to the "on" state upon depolarization and the channel becomes conductive only when all the gates are in the "on" state. Assuming two (on and off) states taken by each gating particle (Fig. 4.2b), a differential equation for each gate was integrated by raising the number of the gating particles involved to an exponent in the kinetic equation (Fig. 4.2c and the Hodgkin–Huxley equation). The time course of the Na^+ and K^+ currents were well fitted by the power number of three for the Na channel and four for the K channel.

$$C_m \frac{dV}{dt} = -g_L(V - E_L) - g_{Na}m^3h(V - E_{Na}) - g_K n^4(V - E_K) + I$$

[1] The gating status varies as a function of the membrane potential (V_m) and the time. If V_m is not fixed under the control of the external circuit, elicited currents lead to changes in V_m, and interpretation of the time evolved currents become complicated.

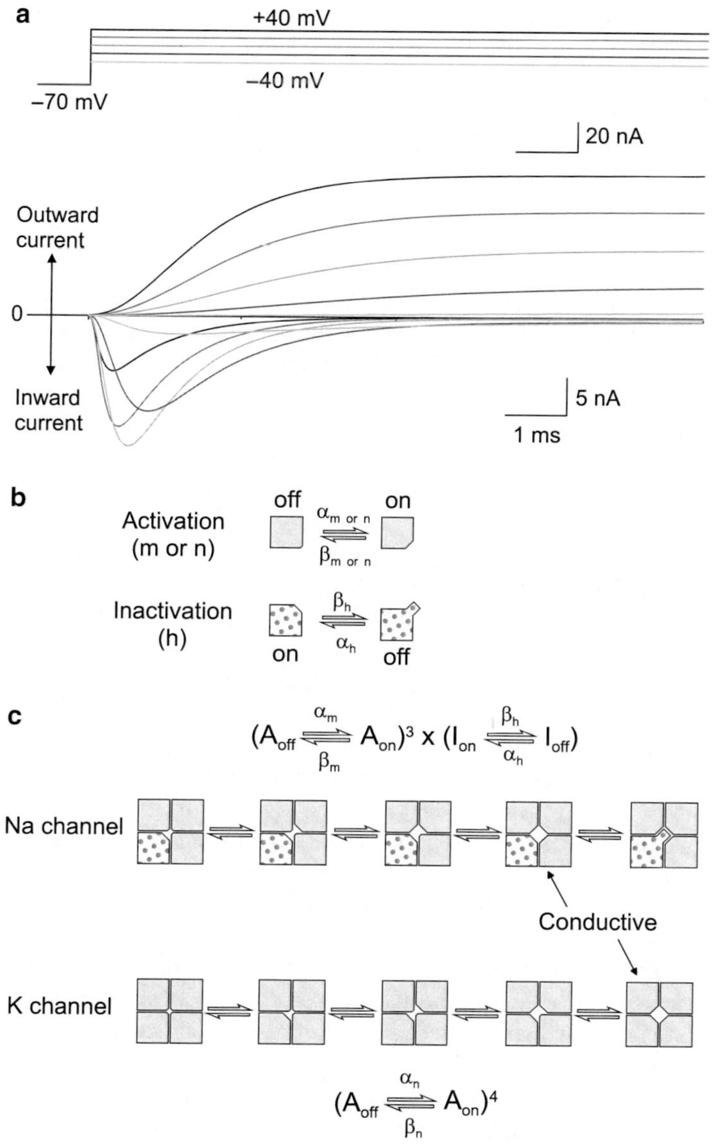

Fig. 4.2 Gating of the Na and K channels. (**a**) Current recordings upon depolarizing voltage steps. The inward currents (*downward*) represent the Na current and the outward currents (*upward*) represent the K current. The membrane potentials were changed from −70 to depolarizing potentials ranging from −40 to +40 mV. These current traces were generated from the Hodgkin–Huxley equation. (**b**) The gating "particles" for the activation and the inactivation in the Hodgkin–Huxley model. (**c**) The Hodgkin–Huxley model for the gating of the Na and K channels. For the Na channel three activation gates and one inactivation gate exist. All the gates are voltage dependent and respond independently. The K channel has four activation gates. The Na channel becomes conductive only when the three activation gates and the inactivation gate open.

$$\frac{dm}{dt} = \alpha_m(1 - m) - \beta_m m$$

$$\frac{dh}{dt} = \alpha_h(1 - h) - \beta_h h$$

$$\frac{dn}{dt} = \alpha_n(1 - n) - \beta_n n$$

C_m: the membrane capacitance, V: the membrane potential, t: time, g: the conductance for L (leak), Na (Na channel) and K (K channel), m: the activation gate for the Na channel, h: the inactivation gate for the Na channel, n: the activation gate for the K channel, E: the equilibrium potential for the leak (L), the Na^+ (Na) and the K^+ (K), α: the on rate for m, h and n gates, β: the off rate for m, h and n gates, I: the current.

For the Na channel, the current exhibited a negative peak after the sigmoidal activation, followed by significant decay, called the inactivation (Fig. 4.2a). This two phase behavior was expressed by rendering an additional gate that opens before the depolarization pulse and closes gradually at depolarized potentials. The kinetics of the inactivation gate is also expressed by a differential equation taking two states. Integrating the differential equation numerically, a typical time course of the membrane potential characterizing the action potential was reproduced.

The historical description of the channel gating touches on certain fundamental issues. One can read the kinetics of the gating from patterns of time-evolved current recordings under voltage-clamped conditions. The generated macroscopic currents receive contributions from individual channel, conducting through the open state channels, and many of the closed states are silently anonymous. However, the existence of multiple closed states before openings (the states with not all the gating particles are in the on state) can be detected indirectly from the kinetics of the macroscopic current. For example, the delay of activation would never occur if the channel had only two states (i.e. either a conductive or non-conductive state), in which mono-exponential time course of the activating is followed. On the other hand, it is a series of non-conductive states preceding the opening that generates the delay of current activation.

The mechanistic gating model of Hodgkin and Huxley is an example of the kinetic schemes. Generally, the gating kinetics can be described by kinetic reaction schemes that give the number of open and closed states, and the transition pathways among the states. The rate constants for the transitions (or the probability of a state transition per unit time) are modulated by the membrane potential and ligand concentration. The rate constant depends only on the current (or present) state, not on past events. In particular, the time elapsed in a state does not affect to the rate constants. This memoryless feature is the basic assumption of the Markov model. Such discrete state Markov models have proven highly useful for describing the underlying gating mechanisms [9, 22].

4.3.2 From Hodgkin–Huxley Equation to Allostery

After success of cloning of channel proteins in 1980s by Numa's group [26], electrophysiological methods, combined with the mutational studies of channel proteins, have elucidated the molecular mechanisms of potassium channels [41].

The amino acid sequences of the channel proteins have provided basic information on the channel structure [15]. The potassium channels are mostly homo-tetramer that are reminiscent of four gating particles for the activation gating in the Hodgkin–Huxley equation. From the amino acid sequences, a region of clustered charged-residues was identified [25], and it was proposed that this segment is located in the membrane electric field and responds to changes in the membrane potential (i.e., a voltage sensor). The identification of the sensing parts suggests that the sensing and gating are carried by independent domains in the channel protein: Changes in the membrane potential are sensed by the voltage sensor domain, which are transferred to the ion permeation (pore) domain for opening and closing of the gate. This scheme is similar to other allosteric proteins with an oligomeric structure, having a ligand binding site and a catalytic site [24].

In the Hodgkin–Huxley model of the potassium channel, four gating particles are moved independently into the "on" position and the channel becomes conductive when four of them are in the on position. This kinetics can be assigned as the KNF (Koshland-Nemethy-Filmer) or the sequential model [18] (Fig. 4.3a). Among the equivalent four gating particles, one particle turned on in random fashion, and the relevant subunit changes to a potentially open conformation. Subsequent transitions of the rest of the gating particles to the on position lead to the last instance that the fourth gating particle turned on, and the channel become conductive.

The independent entities for the sensor and the gate, however, suggest an alternative model, in which motions of the gating particle (or the voltage sensor) and the opening and closing of the gate are separate events. After conformational changes of the voltage sensor, the concerted action of the tetrameric subunits results in the opening of the gate. This type of gating is expressed as the MWC (Monod-Wymann-Changeux) or the concerted model [24] (Fig. 4.3b). These two types of the models are projected on the generalized allosteric framework. The Hodgkin–Huxley model or the KNF model is interpreted as the diagonal path and the MWC model is the parallel paths [8] (Fig. 4.3).

The kinetic model for the voltage-gated potassium channel has been refined significantly [35, 42]. The independent transitions for the voltage sensor and the concerted action of the activation gating are still valid, although both the sensor and gate are now described by transitions among more than two states [35].

4.3.3 Single-Channel Current Recordings

Availability of functional measurements at the single molecule level distinguishes channel proteins from other functional proteins. Historically, single-channel current

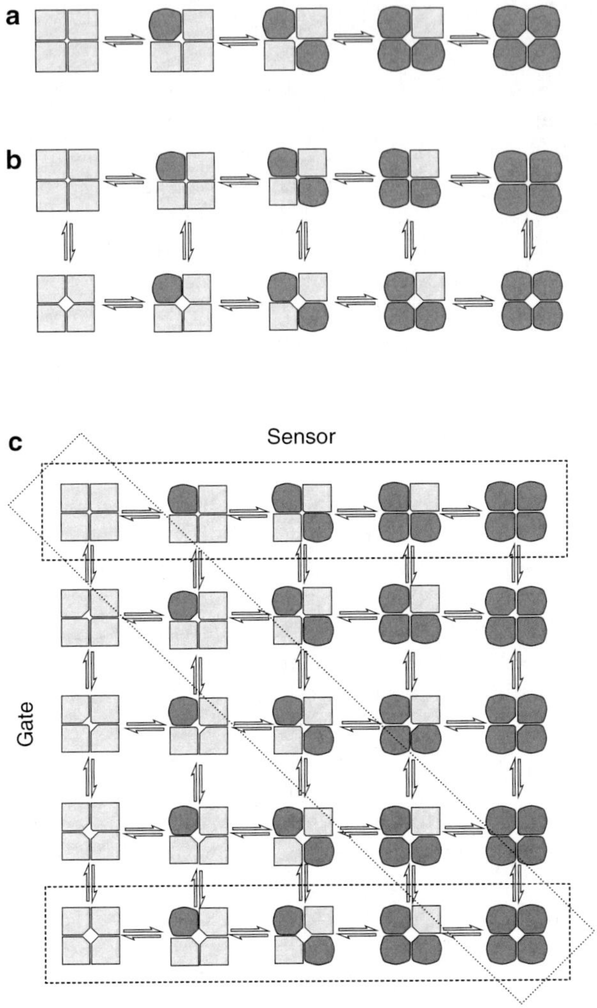

Fig. 4.3 The gating models for the K channels. (**a**) The Hodgkin–Huxley model. In the context of the allostery, this model is expressed as the sequential or the KNF model. (**b**) The concerted or MWC gating model. By introducing independency for the sensor and the gate, the openings of the gate occurs independently from the sensory activity. (**c**) The general allosteric model including all the combinations of the transitions for the sensor and the gate.

measurements were performed using the planar lipid bilayer method and subsequently the patch-clamp method. In the planar lipid bilayer experiments (Fig. 4.4), the purified channels are incorporated into the membrane [14]. Unlike the patch-clamp method, experiments can be performed in well-controlled conditions, such as at varied lipid compositions of the membrane. For some of the channels, lipid compositions are a

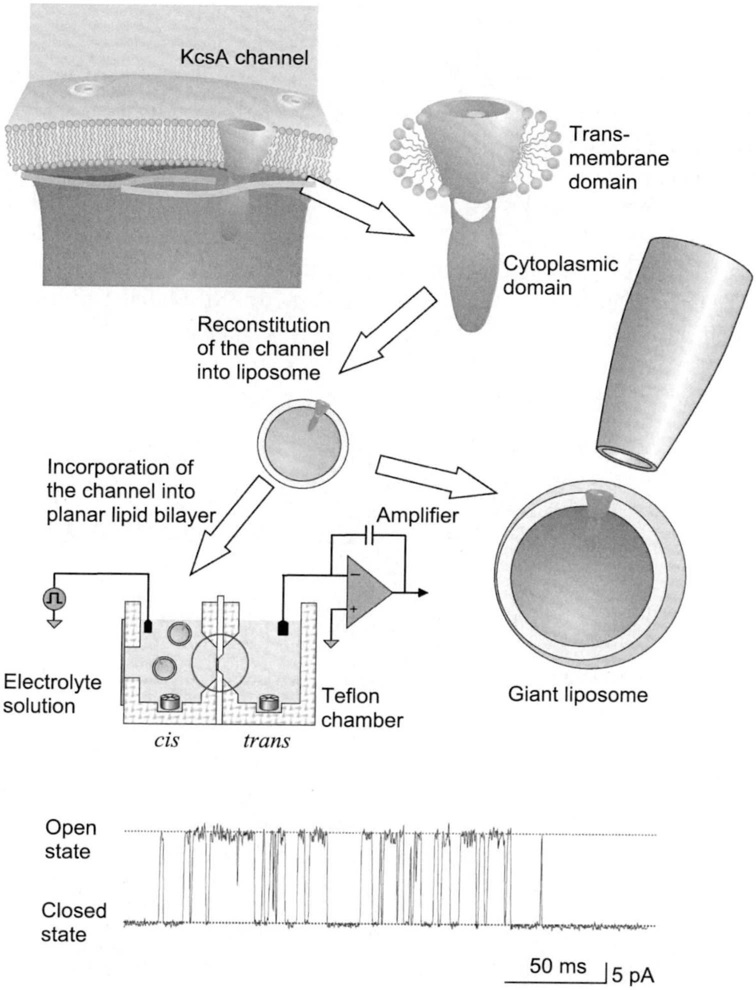

Fig. 4.4 Single-channel current obtained by using the planar lipid bilayer method. The KcsA potassium channel was expressed in *E. coli* and was extracted and purified in the presence of a detergent. The channel was reconstituted into the liposome. The liposome was subjected to fusion to the planar bilayer or was grown as a giant unilamellar vesicle for patch-clamp recordings. This figure is modified from Fig. 1 of the paper published in Journal of the Japanese Society for Synchrotron Radiation Research 22, 183–191.

critical factor to keep the channel active and the planar lipid bilayer method is prerequisite to perform such experiments [21].

Single-channel current recordings provide direct insight into the kinetics. Discrete transitions between zero and a defined level of current amplitude upon the opening and closing of the gate are a common feature in the single-channel current recordings (Fig. 4.4, the lower panel).

4.3.3.1 Single-Molecule Kinetics of the KcsA Channel

Since the first crystallization of the KcsA channel, studies of structure-function relationship of ion channels have been focused on the KcsA channel and huge data have been accumulated. The KcsA channel is a potassium channel originated from the bacteria *Streptomyces lividans* [36]. The ease of mass production of the channel protein enabled both crystallization [7] and spectroscopic investigation [31]. The KcsA channel is one of the simplest and most stable channel proteins. The number of the amino acid residues is 160 and the functional channel is formed as homotetramer. The basic features of potassium channels, such as the high potassium selectivity and high throughput rate, are retained in the KcsA channel. Unlike the voltage-

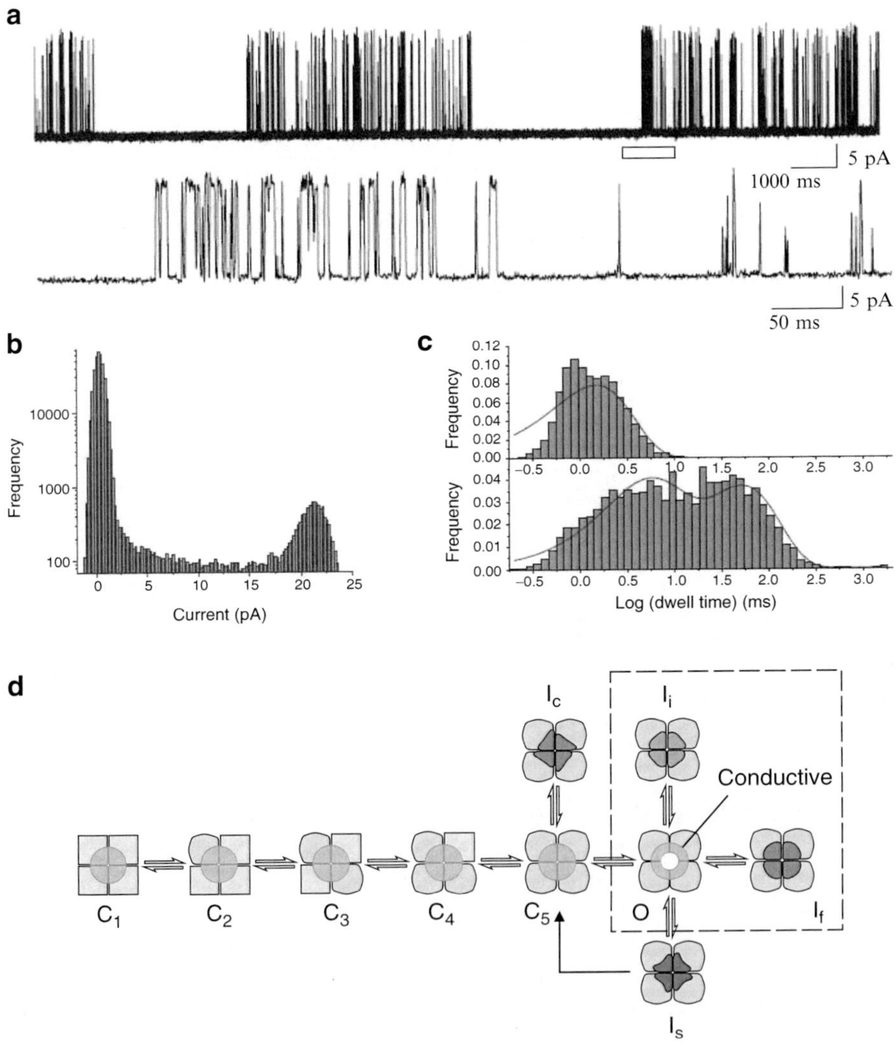

gated potassium channel, the KcsA channel does not have the voltage-sensor domain, and the channel is activated by lowering the pH on the cytoplasmic side.

Figure 4.5a shows representative traces of single channel current from the KcsA channel at pH 4. The opening and closing of the channel are fast and clustered into bursts for a certain period of time. After a long intermission, the channel starts to burst. A histogram of current amplitude is shown (Fig. 4.5b). There are two main peaks corresponding to the closed state (at the 0 current level) and the less frequent open state at around 20 pA. From this histogram by measuring the area under the gaussian curves, the steady-state open probability was obtained.

The kinetics of the channel gating can be characterized by observing the dwell times for the open (upper) and closed (lower) states (Fig. 4.6c). In these histograms, the horizontal scale is in the logarithmic (the log-binned histogram) [39]. The histogram for the open state (upper) was fitted with the single exponential function (Fig. 4.5c, solid line). In the log-binned scale, the exponential function becomes asymmetric bell shape and the peak of the function on the time scale indicates at the time constant.

For the closed dwell time (lower), two components having the time constants of 8 and 30 ms were detected. To describe the gating kinetics, the discrete Markovian model has been used [3, 43], and the numbers of the kinetic components deduced from the dwell-time histogram refer to the number of the discrete states. Thus, these kinetically distinct two non-conducting states correspond to the brief closure within the burst and the long intermission or closure between bursts. However, estimating the rate constants of the kinetic model from the time constants of the histogram needs further considerations [37].

Recently, the gating kinetics of the KcsA channel was examined thoroughly by Perozo's group ([1]). Both structurally and functionally, the closed state and the inactivated state show distinct features, and discriminating these two non-conductive states in the single-channel current recordings are important to characterize the gating. However, the non-conducting states cannot be identified directly as either the closed state of the activation gate or the inactivation gate. To distinguish underlying

Fig. 4.5 Analysis of the gating kinetics for the single-channel current. (**a**) Single-channel current recordings from the KcsA potassium channel at +200 mV in the symmetrical potassium concentration of 200 mM. The pH of the cytoplasmic side was 4.0. The lower trace is the expanded trace for the part indicated by the *white box* below the upper trace. (**b**) The current amplitude histogram. The vertical axis is in the logarithmic scale. The open probability was 25.5%. (**c**) The dwell time histogram for the open (*upper*) and closed (*lower*) states. The time scale is on the logarithmic scale. The *fitted lines* are superimposed. The time constants for the open and the closed states are 1, 8 and 30 ms, respectively. (**d**) A gating model for the KcsA channel proposed by Charkrapani et al. [2]. Each state was shown as a *cartoon*, in which the *outer shape* represents pH-dependent changes (*squares* for non-protonated and *rounded* for protonated) and the *inner circle* represents the gate. The sequence of the closed states (C_1–C_5) was proposed a priori as a sequential activation of the tetrameric channel. From the last closed state (C_5) to the open state (O), it was assumed that the gating parts in each subunit change their conformation cooperatively, which lead to the conductive state. From the pH jump experiment a transition from a closed state to an inactivated state, without passing through the open state, was suggested ($C_5 \rightleftharpoons I_c$). The recovery from the inactivation suggested that there is a pathway shunting the open state ($I_s \rightarrow C_5$). The closed time constants in panel c recorded in the steady-state condition may represent two types of the inactivated state, rather than the closed state (*the states in the box*).

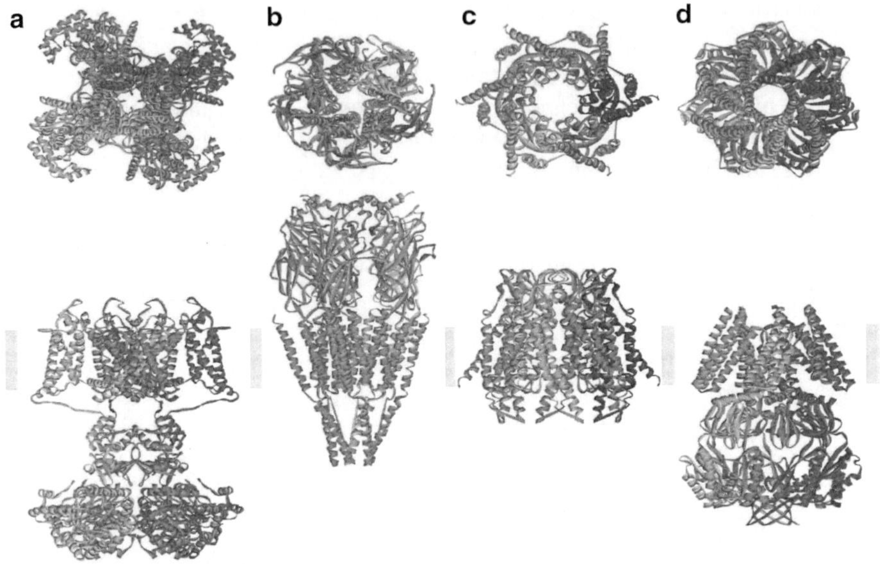

Fig. 4.6 Crystal structures of channel proteins. The *upper panel* represents the top view and the lower the side view, with the *gray bars* representing the membrane thickness. (**a**) The voltage-gated potassium channel (tetramer; pdb code: 2R9R). (**b**) Nicotinic acetylcholine receptor (pentamer; 2BG9). (**c**) The gap junction channel (hexamer; 2ZW3). (**d**) Mechanosensitive channel of small conductance (MscS; heptamer; 2OAU). The K channel is K^+ selective, the acetylcholine receptor is cation selective and the other two channels are permeable to various ion species.

processes, mutations disrupting the inactivation were introduced. Among the mutational sites, Glu71, located on the pore helix, is the most effective site to abolish the inactivation [4, 5]. The single-channel current of the E71A (Glu71 $\rightarrow$ Ala) mutant became fully open at acidic pH. As the cytoplasmic pH was increased, the open probability was decreased and the gating kinetics was slow. This gating kinetics, in the absence of the inactivation, was assigned to those of the activation gate [2]. Differentiating the gating kinetics of the mutant and the wild type elucidated that both non-conductive states observed in the steady-state like those of Fig. 4.5 represent short and long inactivated states (Fig. 4.5d). The closed time (the transition from the C_5 to O) was expected to be slower than those recorded time constants, and the relevant time constant could not be obtained because the number of the slow events was few.

4.4 Structural Dynamics

Functional studies using mutated channels have provided crucial information on the mechanism of gating. On the other hand, the crystal structures of channel proteins have provided unprecedented information relating the functional features of the gating to the relevant structural parts of the channel protein. During gating, dynamic rearrangements of those structural parts are undergoing. Thus, capturing and tracing the dynamic changes in the conformation during gating is prerequisite to understand the gating mechanism. In this section, the crystal structures of channel proteins are reviewed as a background to understand our approach which we have named: "from still pictures to a movie" [27].

4.4.1 *The Crystal Structures of Channel Proteins*

Since 1998 when MacKinnon's group first elucidated the crystal structure of the KcsA potassium channel [7], several crystal structures of channel proteins have been reported [16, 17].

4.4.1.1 The Circular Symmetry of the Channel Structure

Many kinds of channels are composed of identical or similar subunits, and the crystal structures of these channels have revealed that the subunits are assembled around a symmetrical axis. In the circular symmetry of the channel structure, the pore is located at the center of an oligomeric structure along the axis of symmetry (Fig. 4.6). For the small ions abundant in the cellular environment, a high energy barrier of the lipid bilayer is a challenge for the passage of ions. The physical entity of the barrier is the Born energy[2] in the low dielectric medium of the lipid bilayer flanked by the high dielectric medium of the surrounding water [30]. Channel proteins create a water-filled pore in the center of the molecule. This is a rational strategy of design in the channel architecture for the high throughput permeation.

As the top view of the channels illustrates (Fig. 4.6), channels become more selective as the radius of the pore is decreased. Among the various channel proteins, the potassium channels exhibit the highest selectivity (K^+ is more than 1000 times permeable than Na^+), and the crystal structure revealed them to have the narrowest pore diameter of 3Å.

4.4.1.2 Structure of the KcsA Channel

Since the first crystallization of the KcsA channel, crystals have been obtained under various experimental conditions, such as at different ionic conditions [45] and co-crystallized with a blocker [44]. In 2009, the full-length structure was elucidated (Fig. 4.7) [40].

---

[2] The Born energy is the energy required to transfer a charged particle from the high dielectric aqueous phase to the low dielectric membrane interior. This energy, in the range of 100–300 kJ/mol for inorganic monovalent cations, is an immense energy barrier.

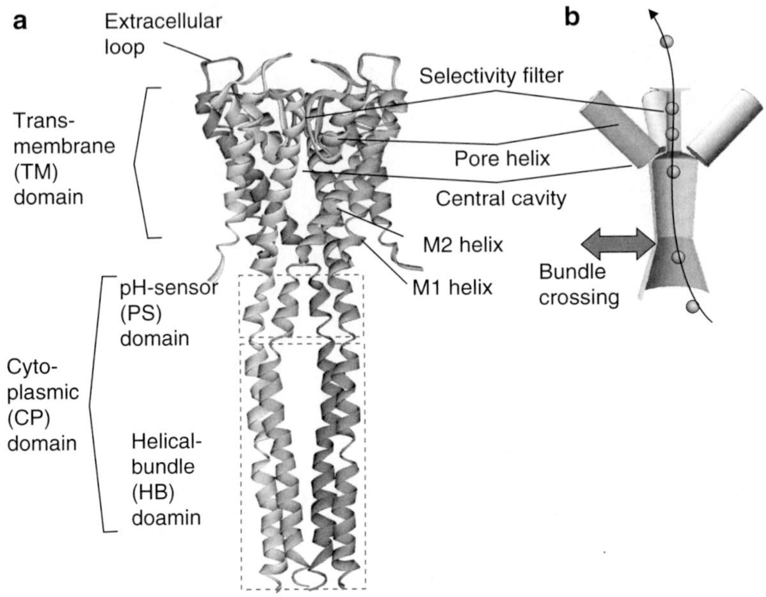

Fig. 4.7 The structure of the KcsA channel. (**a**) The full-length structure and the domains (pdb code: 3EFF). Each subunit of the tetrameric KcsA was colored differently. (**b**) The cut-open pore structure and the pore helix (only three of them are shown). Simplified inner surface of the pore is shown.

The transmembrane (or pore) domain is the most essential part of ion channels since it provides a favorable environment for ion transfer across the membrane. The common architecture of the pore domain for the potassium channels is composed of two transmembrane helices (M1 and M2) and one short helix (the pore helix). The narrowest part of the pore, named the selectivity filter, is lined by four parallel strands.

The diameter of the pore is not uniform (Fig. 4.7b). In the extracellular half of the pore, the radius of the pore is as narrow as 3 Å in diameter, where permeating ions must be dehydrated. At the middle of the pore there is a large space called the central cavity, where ions can stay in fully-hydrated form. The bundle of the M2 helices are crossed at the cytoplasmic end (bundle crossing) and occlude the ion permeation pathway. This part is called the helix gate. Functionally, this gate opens at an acidic pH and is called the activation gate.

In the crystal of the full-length channel, the cytoplasmic domain was divided into two separate domains (Fig. 4.7). The C-terminal end forms a bundle of helices and is called the helical bundle (HB) domain. Between the transmembrane domain and the HB domain there is a short stretch of the α-helix called the bulge helix, in which several charged residues are distributed. Cytoplasmic pH is sensed by these residues and His25 in the M1 helix. Here, the bundle of the bulge helix is called the pH-sensor (PS) domain.

4.4.1.3 Open and Closed Conformations of Potassium Channels

Recently, the crystal structures of several types of potassium channels have been reported [16, 17, 20]. They are activated by specific stimuli, such as changes in the membrane potential and the binding of ligands. They have a sensing domain receiving the stimuli and the message is sent to the pore domain. Thus, the pore domain is the terminal for intra-molecular information transfer, where the message is converted to the opening and closing of the gate.

Some of the channels were crystallized in the closed conformation and others in the open conformation (Fig. 4.8). One may readily find the differences in the architecture for the closed and open states. In the closed conformation, the M2 helix is straight and the bundle is crossed. In contrast, the M2 helix in the open conformation bend at the center and the cytoplasmic half of the helix extends out away from the central axis, allowing a passage of ion along the pore. Both the open and closed conformations retain symmetrical structure.

These exquisite images of the channel architecture in open and closed configurations gave rise to an important question. Does the channel trace conformational trajectory, retaining symmetry during the entire course of the transition? This issue is related to the allosteric mechanism by which the gating follows the KNF model or the MWC model. Such issues of dynamic activity cannot be resolved

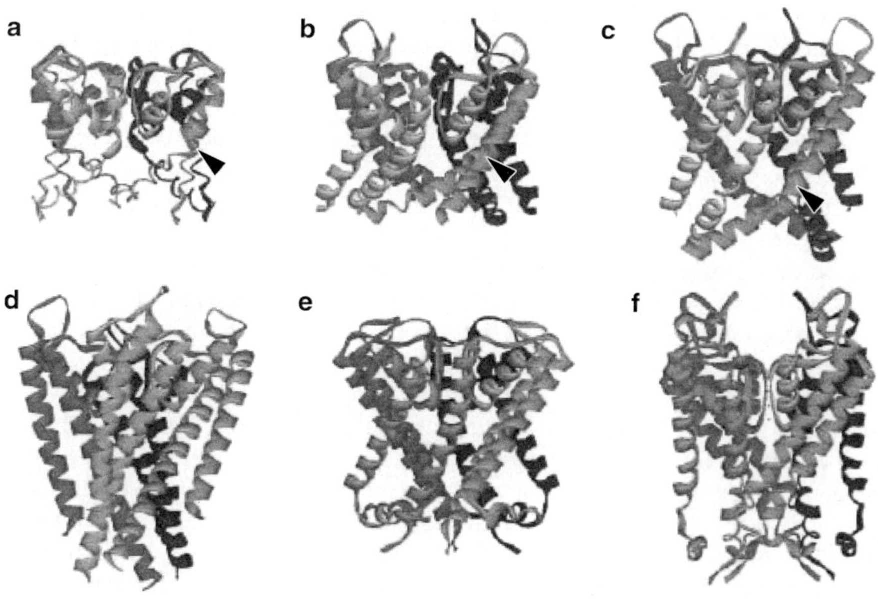

Fig. 4.8 Open (*upper*) and closed (*lower*) conformations of different types of potassium channels. (**a**) MthK (pdb code: 1LNQ), (**b**) KvAP (1ORQ), (**c**) Kv1.2-2.1 chimera (2R9R), (**d**) KcsA (1K4C), (**e**) KirBac3.1 (1XL4), and (**f**) Kir2.2 (3JYC). For the open channel, the hinge point in the M2 helix is indicated by the *arrow heads*.

from the average structure of a huge number of channel proteins. Methods for tracing such conformational changes at the single molecule level are urgently needed.

4.5 Tracing Conformational Dynamics at the Single Molecule Level: The Diffracted X-ray Tracking (DXT) Method

The crystal structure of channel proteins invoked our imagination regarding channel dynamics under gating. Structural changes of proteins have been studied using various methods. Among single molecule measurement techniques, the diffracted X-ray tracking (DXT) method is based on the diffraction principle [29, 33, 34], providing a time series of the geometrical information on the channel structure. Here, the methods and results of the DXT method are briefly described. A complete description of the background, the methods and results are shown elsewhere [27, 38].

4.5.1 The DXT Method

The DXT method is a dynamic Laue method, in which diffraction spots are elicited from a crystal gold bound to a channel protein rather than from a crystalline protein. The gold nanocrystal is of a comparable size (25 nm; [28]) as the channel protein (10 nm in length) (Fig. 4.7). The channel proteins are attached to a glass surface in such a way that the conformational changes are allowed to take place. Focused white X-rays (the energy range: 6–30 keV) from a synchrotron radiation facility (SPring-8: Super Photon ring 8 GeV, Harima, Japan) were irradiated onto the sample and the diffraction spots were recorded by a CCD camera through an image intensifier at the video rate (Fig. 4.9).

Upon gating, for example, the gold nanocrystal changes its orientation along with the conformational changes of the channel, and changes in the incident angle to the gold nanocrystal cast diffraction spot at different locations on the image plane. Thus, the conformational changes of the channel proteins can be monitored by following the trajectory of the diffraction spots. White X-rays cover broad spectrum of the wave lengths, among which there is a beam with relevant wavelength satisfying the Bragg condition[3] for a gold nanocrystal oriented in a certain angle against the beam direction. Even if the gold nanocrystal tilts its orientation and changes the incident angle, a diffraction spot is remained elicited as far as the beam bears a wavelength optimal for the diffraction. This makes the DXT method to keep tracking the diffraction spots. A simple rule of thumb of the DXT principle is that the nanocrystal

[3] $n\lambda = 2d \sin\theta$ where λ: the wave length, n: an integer number, d: the spacing between the planes in the atomic lattice, θ: the angle between the incident ray and the scattering planes.

can be regarded as a tiny mirror. Diffraction, rather than reflection, gives the geometrical information on the conformational change of the channel protein.

Most of the DXT experiments were performed by fixing the channel in the upright orientation (Fig. 4.9a), which allows for free conformational changes during gating. The four reactive sites on the extracellular loop of the KcsA channel (Fig. 4.7a) secured the channel molecule oriented in the upright position, while the flexible extracellular loop allowed the other part of channel to fluctuate. After fixing the channel on the glass plate, a gold nanocrystal was attached at the other end of the KcsA molecule through four cysteine residues from the four subunits. The gold nanocrystal was grown epitaxially on the surface of a NaCl crystal and annealed. The shape of the nanocrystal was plate-like and the size was 25 nm, with a thickness of 10 nm.

In the upright configuration, in which the longitudinal axis of the channel molecule parallels the beam (Fig. 4.9b), the diffraction spots move in a radial

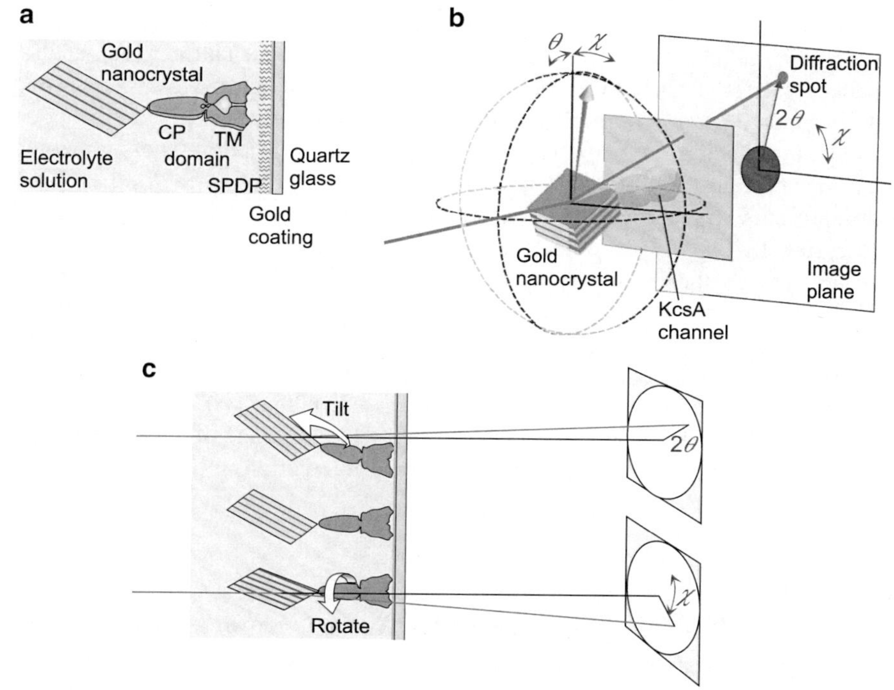

Fig. 4.9 Principles of the DXT method. (**a**) KcsA channels are attached to the quartz glass through a surface modifying reagent. At the C-terminal end of the channel a gold nanocrystal was bound through four cysteine residues of the tetrameric channel. (**b**) Geometrical relationships for the sample and the X-ray beam. *White* X-rays were irradiated perpendicularly to the glass surface. The diffraction spots were recorded by an image-intensified CCD camera at the video rate. Experiments were performed in the beamline BL44B2 of SPring-8. (**c**) The motion of the diffraction spots on the detector plane in relation to the conformational changes of the ion channels on the glass surface. This figure is modified from Fig. 3 of the paper published in Journal of the Japanese Society for Synchrotron Radiation Research 22, 183–191.

direction on the image plane when the channels are tilted from the plane normal (Fig. 4.9c). When channels rotate around the axis of symmetry, the diffraction spots move circumferentially.

4.5.2 Random Brownian Motion of the KcsA Channel in the Closed State

Conformational changes of the KcsA channel were tracked under two different experimental conditions, neutral pH, where the channel stays closed, and the acidic pH of 4, where the channel exhibits active opening and closing transitions. A summary of the results are shown in Fig. 4.10.

First, DXT measurements of the KcsA channel were performed with the channel in the closed state. Figure 4.10a shows some frames of the video images of DXT measurements. The beam was irradiated at the center of the image plane and the direct beam was masked by the beam stopper which cast a shadow as the central disk and a stalk towards it. In the right, image in the boxed region was expanded and only four image frames were picked up. The white arrows indicate the diffraction spot moving back and forth. Generally, multiple diffraction spots emerged on an image plane, and moved randomly and independently of each other. Each spot represents the movements of a single KcsA protein labeled by a single nanocrystal. In Fig. 4.10b, trajectories from approximately 300 molecules were superimposed on the image plane, in which each color represents trajectory of individual molecule. The movements of almost all of the spots were in the radial direction. These movements on the image plane were translated into real space, expressed as the orthogonal coordinates of the rotational and tilt angles (χ and θ). The time courses of the movement was expressed in the 3 D plot in the t-χ-θ coordinate (Fig. 4.10c). The motion was limited within a range of a few degrees in both the χ and θ angles. This motion is a thermal fluctuation of the channel in the closed state.

4.5.3 Global Twisting Motion of the KcsA Channel upon Gating

At acidic pH where the channel undergoes the opening and closing of the gate, we found that diffraction spots moved circumferentially in the image plane. In Fig. 4.10d, representative trajectories were projected on an image plane. Several tens of degrees in the circumferential direction accompanied with augmented movements along the radial direction were observed. The principle of the diffraction indicates that circumferential movements on the image plane correspond to rotational movements around the longitudinal axis of the channel (Fig. 4.11).

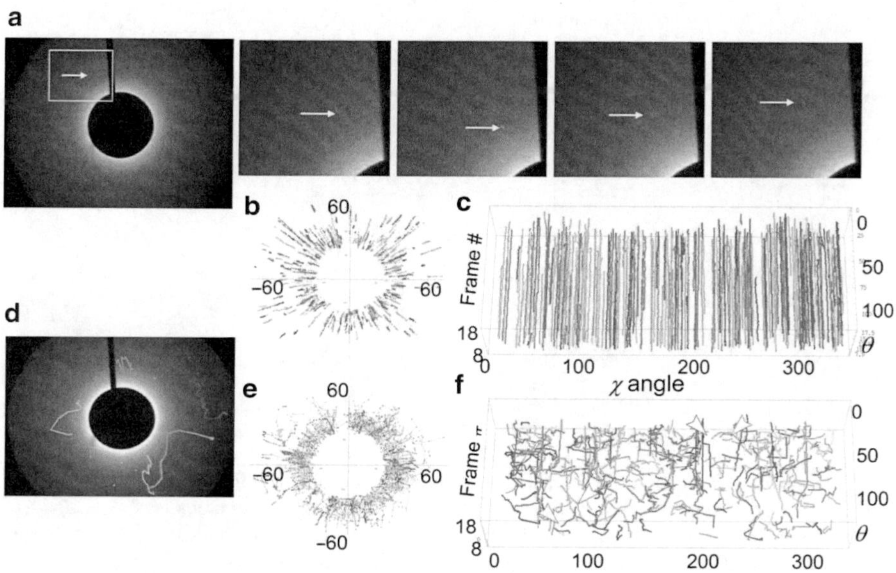

Fig. 4.10 Conformational dynamics of the KcsA channel during the closed (at pH 7.5; panel a to c) and active gating (at pH 4.0; panel d to f) states. (**a**) The video image of the diffraction spots in the closed state. The *boxed area* was expanded and a series of the video frames are shown. (**b**) Trajectories were superimposed on an image plane. Each short stretch of the trajectories along the radial direction represents movement of single channel molecule. Trajectories from 300 molecules are shown. (**c**) The motions of channels in real space on the orthogonal t-χ-θ coordinate. Movement along the χ axis represents rotation and that along the θ axis represents tilt. The scale of the θ axis was expanded. (**d**) Superimposed video image during the gating. The trajectories travel along the circumferential direction. (**e**) Superimposed trajectories. Trajectories from 300 molecules are shown. (**f**) Projection of the trajectories on the χ-θ coordinate. The spots remained on the image plane for short period of time, since the recording range along the θ axis is limited. Thus, trajectories exhibited short lifetime. This figure is slightly modified from Fig. 2 of the paper published in Cell 132: 67–78, 2008.

The range of degrees along the circumferential motion in the image plane represents the range of rotational motion in real space. Thus, the channel twists several tens of degrees, which is as unexpectedly large angle. This was the first observation of channel gating as a twisting motion. Both clockwise and counterclockwise twisting were observed, reflecting random opening and closing events in the steady state. These twisting motions reflect large collective movements among the tetrameric subunits, suggesting concerted conformational changes in the pore domain.

The circumferential movements were vigorous. Many spots skimmed across the image plane and left recorded spots only on a few consecutive frames. Other spots remained in the recording area for longer time but mostly disappeared. Trajectories from approximately 300 molecules are shown in Fig. 4.10e. These spots in the image plane were translated into channel movements in real space as a 3D plot (Fig. 4.10f). Movement along the χ axis predominated over movement along the θ axis. Duration of the observed trajectories are short, and thus, most of

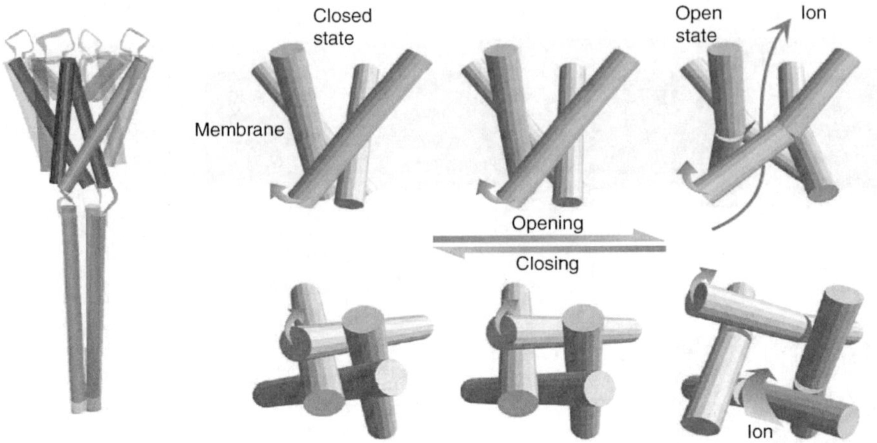

Fig. 4.11 Global twisting motion of the KcsA channel and the underlying conformational changes of the M2 or inner helix. The M2 helices of the TM domain are schematically shown. *Upper*: the side view; *Lower*: the bottom (from the cytoplasmic side) view. The helical bundle crosses at the intracellular end (*bundle crossing*) and prevents ion permeation (closed gate; *left*). Upon gating, the helical bundle unwinds. Then, the helices bend at the center, and the intracellular half of the helices splay out (open gate; *right*). These processes were detected at the single molecular level as global twisting motions around the symmetrical axis. This figure is modified from Fig. 5 of the paper published in Journal of the Japanese Society for Synchrotron Radiation Research 22, 183–191.

the trajectories are fractions of the course of transition between the open and closed conformations.

The random motions were characterized with MSD (mean square displacement) curves (Fig. 4.12). The square of the distance in the rotational angle was plotted as a function of time from the first appearance of the spots on the image plane. The MSD curves for the closed and gating channel are shown (Fig. 4.12a). When the channel was in the closed state, the movements were localized and the vertical scale for the MSD curve was expanded (Fig. 4.12b). The MSD curve was nearly linear and the initial slope gave the diffusion constant of this fluctuating motion.

On the other hand, the MSD curve from the actively gating channel exhibited a large enhancement of the motion and reached saturation. The apparent saturation, however, does not represent the range of the rotational motion. Rather, ensemble of incomplete tracing of the trajectories having different time length of spots on the recording range, and broad distribution of the speed gave the ceiling. The diffusion constant of this movements deduced from the initial slope was approximately 200 times higher than that of the closed state.

Closer inspection of the trajectories revealed that spots often continued to move in the same rotational direction for several steps. This may be a sign of a deviation from random motion along the rotational axis, suggesting that the rotation is driven in one direction for a certain period of time. In contrast to the local fluctuation in the closed state, a path was created in the free energy landscape along the rotational direction during the gating.

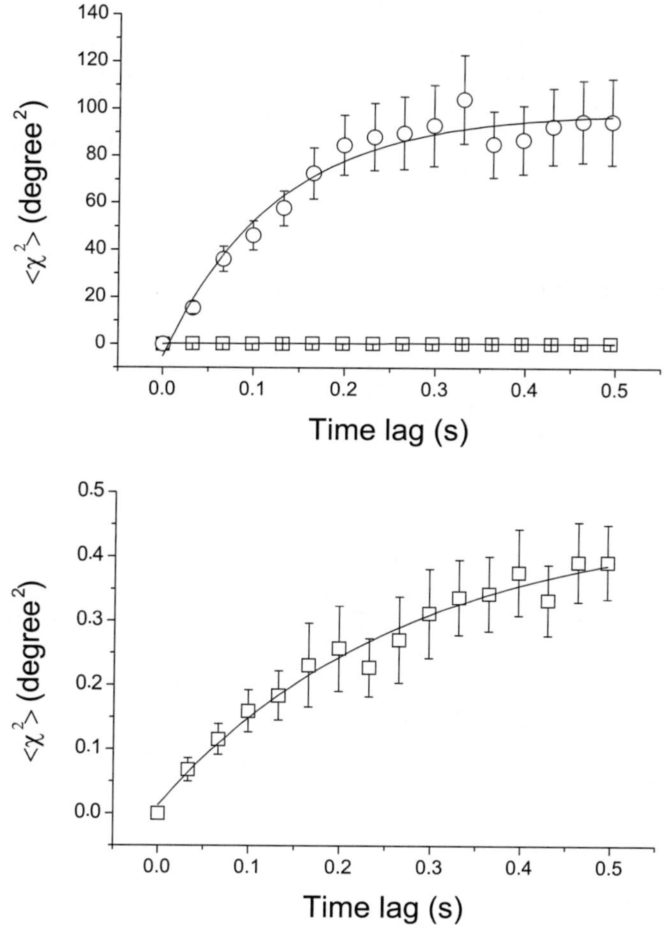

Fig. 4.12 MSD curves for the χ angles. *Upper*: The MSD curves in the closed state (*square symbols*) and during the gating (*circles*). *Lower*: The MSD curve for the closed state with the expanded vertical scale. From the initial slope the diffusion constant was attained: 1.8 degree^2/s for the closed state and 370.5 degree^2/s for the gating state. This figure was slightly modified from Fig. 3 of the paper published in Cell 132: 67–78, 2008.

4.5.4 Cytoplasmic Domain-Truncated Channel

To examine the contribution of the cytoplasmic domain to the twisting motion, the CP domain was enzymatically truncated using chymotrypsin. The cleavage site is located between the pH-sensor domain and the helical bundle domain (Fig. 4.7a), CPD-truncated channel). Thus, the truncated channel retains the pH-sensor domain, and exhibits pH-dependent activity similar to that of the full-length channel [6]. On DXT measurement, the gold nanocrystal was attached at the C-terminal end of the truncated channel. At neutral pH, Brownian motion was observed [38]. At acidic pH, the CPD-truncated channel demonstrated the similar twisting motion. The range of the motion in the χ angle was similar to that of the full-length channel. These results indicate that the twisting motion originated from the TM domain and transferred towards the end of the CP domain. The CP domain does not contribute significantly to the generation of the twisting motion.

4.5.5 Conformational Wave Propagation

The results of the DXT measurements gave fundamental information on the processes of the structural changes upon gating. Here we made an attempt to construct a sequence of events occurring during the gating to provide an integrated picture of the channel gating (Fig. 4.13). Acidic pH is sensed by the PS domain, and the sensitivity is retained even after the truncation of the CP domain. The conformational changes of the PS domain are transferred to the transmembrane helices, allowing the M2 helix to bend at the center. The collective motion of the helices resulted in the twisting motion around the symmetrical axis, which was detected at the C-terminal end of the TM domain. The twisting motion was also detected at the C-terminal end of the full-length channel, meaning that the cytoplasmic domain mediates the twisting motion passively. Since the channel twists and untwists repeatedly at acidic pH, the pH-sensor domain unlocks the reversible rotational motions by lowering the energy barrier between the open and closed conformations.

4.6 Single Molecule Dynamics

The DXT method provided unprecedented information on the structural dynamics of the KcsA channel upon gating. The animated image of the conformational change evokes mechanistic insights of the channel architecture rearranging during the gating. In parallel, single-channel current recordings of the inactivation-free mutant (E71A) revealed the slow kinetics of the helix gate. Here the results of our study on these structural dynamics are discussed in relation to the functional studies.

Before discussing the dynamics of channel proteins, we compared the DXT and single-channel current methods to determine the scope and limitation of these methods. In the DXT method, the gold nanocrystal projects a diffraction spot on a distant image plane upon irradiation with white X-rays, by which small changes in the orientation of the nanocrystal driven by the channel are augmented and can be readily measured at a spatial resolution below 0.1°. On the other hand, in the single-channel current measurements, the pore, allowing fast ion flow through a single channel molecule, enables measurements of gating kinetics in electrophysiological method. The current changes are discrete between the on and off levels, and even tiny changes in the pore structure leads to changes in current amplitude,

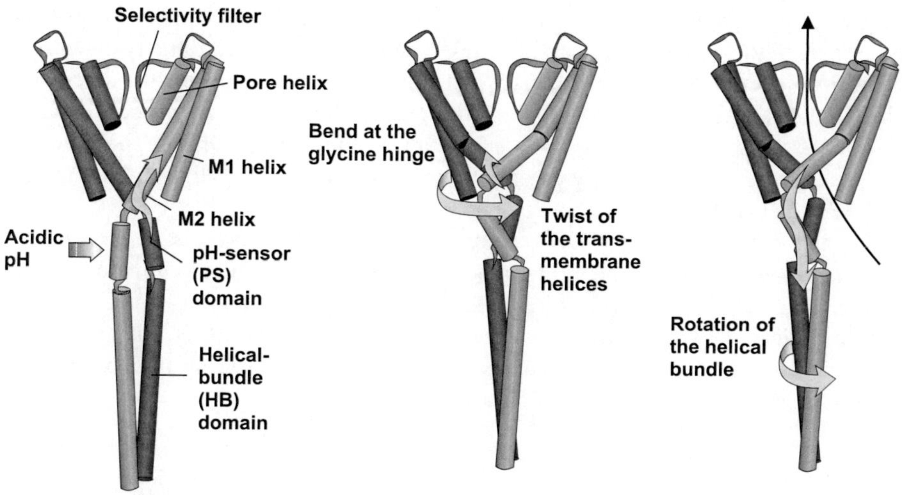

Fig. 4.13 A hypothetical intra-molecular route for conformational changes during the pH-dependent gating. Only two diagonal subunits are shown for clarity. The acidic pH on the cytoplasmic side is sensed by the pH-sensor domain. The relaxed structure of the PS domain transfers to the transmembrane helix. The helix twists and bends at the center and the channel opens. This helical twist propagates towards the C-terminal end of the molecule through the PS and the HB domains. This figure was slightly modified from Fig. 35 of the paper published in Adv. Chem. Phys. 2010 in press.

detected as substates. Thus, similar to the DXT method, fine structural changes of the pore are augmented as changes in the current amplitude.

Upon the pH-dependent gating, the rearrangement of the helical bundle relaxes the bundle crossing. This leads to the pore being permeable, and twists the whole structure, propagating towards the end of the molecule. Single-channel current recordings and the DXT method provide different aspect of the dynamics of the helix gate and complemented each other to integrate the mechanistic view of the gating.

4.6.1 *Discrete vs. Brownian Motions*

At neutral pH, the free energy barrier between open and closed states is very high and the channel stays in the sojourn of the closed conformation (Fig. 4.14). The opening event never occurred in the single-channel current recordings, and the channel exhibited random fluctuation in its structure. When the cytoplasmic pH becomes acidic, the free energy level for the open state approached to that of the closed conformation and the free energy barrier for the gating was significantly lowered. Frequent gating occurred. The gating is a conformational change driven by thermal energy and transition occurred spontaneously and reversibly.

The experimental results of the channel gating from the functional measurements and structural dynamics at the single molecule level shared the common view of the gating dynamics. However, those results gave different impressions of the time scale of the gating kinetics. The single-channel current recording, even with its high time resolution, indicated instantaneous transitions between the open and closed states. Discrete jumps between the current levels do not provide detailed information on the path taken upon surmounting the energy barrier. This kinetics can be expressed as an energy profile having a simple barrier along the transition path (Fig. 4.14, lower left).

On the other hand, the DXT method revealed both clockwise and counter-clockwise twisting, and the fine trajectories of the twisting motion during the gating. This type of motions is a reminiscent of the diffusive motion of the potential profile indicated by the Kramers-type kinetics [11, 19], by which diffusive processes in the condensed system is described, rather than the Eyring-type of kinetics

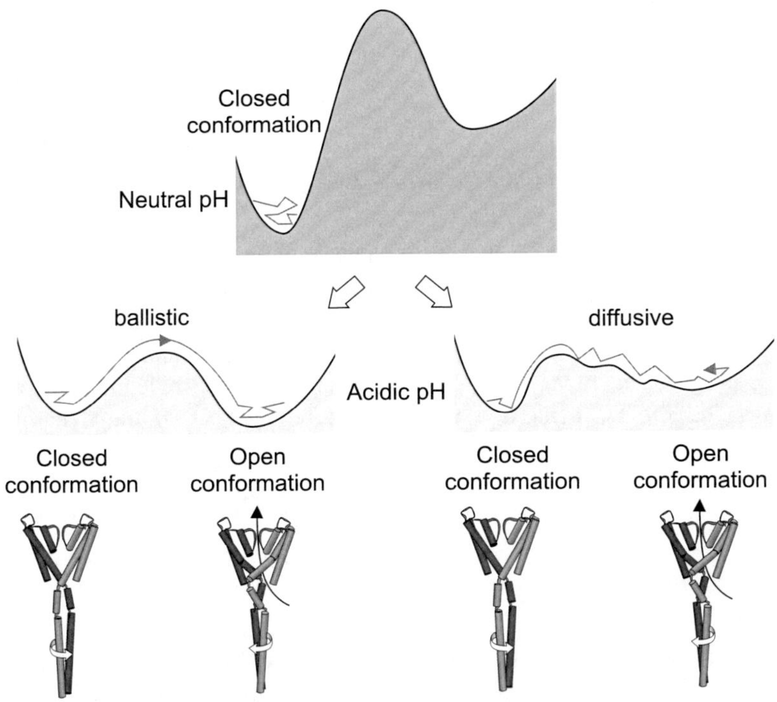

Fig. 4.14 The hypothetical free energy profile for the gating of KcsA channel. The *upper* represents a profile at neutral pH and the lowers for acidic pH. At neutral pH, the free energy barrier towards the open state is so high that the channel stays in the sojourn of the closed state. When the pH becomes acidic, the barrier height is attenuated and the reversible transitions between the open and closed conformations occur frequently. From the single-channel current measurements, the gating transitions were instantaneous and can be characterized by an Eyring-type smooth barrier. On the other hand, the DXT method revealed the directed motions during the gating and can be expressed as a ragged profile. This figure was slightly modified from Fig. 36 of the paper published in Adv. Chem. Phys. 2010 in press.

[10]. The tortuous trajectories suggest that the free energy landscape along the open-close transition is ragged (Fig. 4.14, lower right).

How do these kinetic discrepancies reconciled? During a passage of the trajectory towards the opening, the pore becomes conductive when a critical point is passed. Thus, even when the rotational motion is slow, the time elapsed for crossing the critical point should be instantaneous. To identify the conductive state in the DXT trajectory, simultaneous recordings of the single-channel current and the conformational trajectory at the single molecule level are necessary, which is our technical goal in future work.

In the DXT method, it must be noted that the temporal resolution was limited and the faster motion of the diffraction spots was missed from the recordings. The trajectories captured with our time resolution appeared to be the lowest limit of the speed distribution and thus the recorded trajectories represent a biased population. Also, the trajectories during the gating were incompletely traced because of the limited observation range of the 2θ angle. From these limited observations through a narrow window, what conclusion can be drawn?

This slow motion is, however, not just a result of the limited window of the time resolution. The channel is labeled with a gold nanocrystal that retards the rotational motion by loading it with additional rotational inertia. The perturbation towards the free conformational change, however, gives insights into the mechanistic features of the motion that would otherwise be overlooked without the load. Runs of rotational steps in the same direction along the rotational axis suggest a stochastic process with drift. This mode of motion suggests that the conformational change tracks down the shallow hill on the potential energy surface.

4.7 Conclusion

The structure and function of proteins are central issues in the biological sciences. Among the proteins, the functional phenotype of the channel is unique, allowing for single-channel functional measurements. The behavior of channels, such as the opening and closing of the permeation pathway, however, shares common dynamics that other types of proteins also exhibit. Exploiting the availability of electrophysiological methods providing a quantity of high quality data at the single molecule level, kinetic analysis have been carried out. These analyses are related to the energetics of the gating conformational changes. Along with these functional characterizations, giving the energy landscape for the gating conformational changes, studies of structural dynamics have emerged recently. Our DXT method provides geometrical information, and helps illuminate the mechanistic features of the conformational changes. As the conformational changes spread through the molecule, the channel generated the torque and drove the heavily loaded gold nanocrystal to rotate. These mechanical features of gating dynamics at the single molecule level help elucidate previously unappreciated characteristics of protein dynamics.

References

1. Chakrapani S, Cordero-Morales JF, Perozo E (2007a) A quantitative description of KcsA gating I: macroscopic currents. J Gen Physiol 130:465–478
2. Chakrapani S, Cordero-Morales JF, Perozo E (2007b) A quantitative description of KcsA gating II: single-channel currents. J Gen Physiol 130:479–496
3. Colquhoun D, Hawkes AG (1982) On the stochastic properties of bursts of single ion channel openings and of clusters of bursts. Philos Trans R Soc Lond B Biol Sci 300:1–59
4. Cordero-Morales JF, Cuello LG, Zhao Y, Jogini V, Cortes DM, Roux B, Perozo E (2006) Molecular determinants of gating at the potassium-channel selectivity filter. Nat Struct Mol Biol 13:311–318
5. Cordero-Morales JF, Jogini V, Lewis A, Vasquez V, Cortes DM, Roux B, Perozo E (2007) Molecular driving forces determining potassium channel slow inactivation. Nat Struct Mol Biol 14:1062–1069
6. Cortes DM, Cuello LG, Perozo E (2001) Molecular architecture of full-length KcsA: role of cytoplasmic domains in ion permeation and activation gating. J Gen Physiol 117:165–180
7. Doyle DA, Morais Cabral J, Pfuetzner RA, Kuo A, Gulbis JM, Cohen SL, Chait BT, MacKinnon R (1998) The structure of the potassium channel: molecular basis of K + conduction and selectivity. Science 280:69–77
8. Eigen M (1967) Kinetics of reaction control and information transfer in enzymes and nucleic acids. Almquist & Wiksell, Stockholm
9. Gibb AJ, Colquhoun D (1992) Activation of N-methyl-D-aspartate receptors by L-glutamate in cells dissociated from adult rat hippocampus. J Physiol 456:143–179
10. Glasstone S, Laidler KJ, Eyring H (1941) The theory of rate processes. McGraw-Hill, New York
11. Hanggi P, Talkner P, Borkovec M (1990) Reaction-rate theory: fifty years after Kramers. Rev Mod Phys 62:251–341
12. Hille B (2001) Ion channels of excitable membranes, 3rd edition. Sinauer Associates, Sunderland, MA
13. Hodgkin AL, Huxley AF (1952) A quantitative description of membrane current and its application to conduction and excitation in nerve. J Physiol 117:500–544
14. Iwamoto M, Shimizu H, Inoue F, Konno T, Sasaki YC, Oiki S (2006) Surface structure and its dynamic rearrangements of the KcsA potassium channel upon gating and tetrabutylammonium blocking. J Biol Chem 281:28379–28386
15. Jan LY, Jan YN (1997) Cloned potassium channels from eukaryotes and prokaryotes. Annu Rev Neurosci 20:91–123
16. Jiang Y, Lee A, Chen J, Cadene M, Chait BT, MacKinnon R (2002) The open pore conformation of potassium channels. Nature 417:523–526
17. Jiang Y, Lee A, Chen J, Ruta V, Cadene M, Chait BT, MacKinnon R (2003) X-ray structure of a voltage-dependent K^+channel. Nature 423:33–41
18. Koshland DE Jr, Nemethy G, Filmer D (1966) Comparison of experimental binding data and theoretical models in proteins containing subunits. Biochemistry 5:365–385
19. Kramers HA (1940) Brownian motion in a field of force and the diffusion model of chemical reactions. Physica 7:284–304
20. Kuo A, Gulbis JM, Antcliff JF, Rahman T, Lowe ED, Zimmer J, Cuthbertson J, Ashcroft FM, Ezaki T, Doyle DA (2003) Crystal structure of the potassium channel KirBac1.1 in the closed state. Science 300:1922–1926
21. Marius P, Zagnoni M, Sandison ME, East JM, Morgan H, Lee AG (2008) Binding of anionic lipids to at least three nonannular sites on the potassium channel KcsA is required for channel opening. Biophys J 94:1689–1698
22. McManus OB, Magleby KL (1989) Kinetic time constants independent of previous single-channel activity suggest Markov gating for a large conductance Ca-activated K channel. J Gen Physiol 94:1037–1070

23. Miller C (1986) Ion channel reconstitution. Plenum, New York
24. Monod J, Wyman J, Changeux JP (1965) On the nature of allosteric transitions: a plausible model. J Mol Biol 12:88–118
25. Noda M, Shimizu S, Tanabe T, Takai T, Kayano T, Ikeda T, Takahashi H, Nakayama H, Kanaoka Y, Minamino N et al (1984) Primary structure of electrophorus electricus sodium channel deduced from cDNA sequence. Nature 312:121–127
26. Numa S (1987) A molecular view of neurotransmitter receptors and ionic channels. Harvey Lect 83:121–165
27. Oiki S, Shimizu H, Iwamoto M, Konno T (2010) Single molecular gating dynamics for the KcsA potassium channel. Adv Chem Phys (in press)
28. Okumura Y, Miyazaki T, Taniguchi Y, Sasaki YC (2005) Fabrications of dispersive gold one-dimensional nanocrystals. Thin Solid Films 471:91–95
29. Okumura Y, Oka T, Kataoka M, Taniguchi Y, Sasaki YC (2004) Picometer-scale dynamical observations of individual membrane proteins: the case of bacteriorhodopsin. Phys Rev E Stat Nonlin Soft Matter Phys 70:021917
30. Parsegian A (1969) Energy of an ion crossing a low dielectric membrane: solutions to four relevant electrostatic problems. Nature 221:844–846
31. Perozo E, Cortes DM, Cuello LG (1998) Three-dimensional architecture and gating mechanism of a K + channel studied by EPR spectroscopy. Nat Struct Biol 5:459–469
32. Sakmann B, Neher E (1995) Single-channel recording. Prenum, New York
33. Sasaki YC (2004) Single protein molecular dynamics determined with ultra-high precision. Biochem Soc Trans 32:761–763
34. Sasaki YC, Suzuki Y, Yagi N, Adachi S, Ishibashi M, Suda H, Toyota K, Yanagihara M (2000) Tracking of individual nanocrystals using diffracted x rays. Phys Rev E Stat Phys Plasmas Fluids Relat Interdiscip Topics 62:3843–3847
35. Schoppa NE, Sigworth FJ (1998) Activation of Shaker potassium channels. III. An activation gating model for wild-type and V2 mutant channels. J Gen Physiol 111:313–342
36. Schrempf H, Schmidt O, Kummerlen R, Hinnah S, Muller D, Betzler M, Steinkamp T, Wagner R (1995) A prokaryotic potassium ion channel with two predicted transmembrane segments from Streptomyces lividans. EMBO J 14:5170–5178
37. Shelley C, Magleby KL (2008) Linking exponential components to kinetic states in Markov models for single-channel gating. J Gen Physiol 132:295–312
38. Shimizu H, Iwamoto M, Konno T, Nihei A, Sasaki YC, Oiki S (2008) Global twisting motion of single molecular KcsA potassium channel upon gating. Cell 132:67–78
39. Sigworth FJ, Sine SM (1987) Data transformations for improved display and fitting of single-channel dwell time histograms. Biophys J 52:1047–1054
40. Uysal S, Vasquez V, Tereshko V, Esaki K, Fellouse FA, Sidhu SS, Koide S, Perozo E, Kossiakoff A (2009) Crystal structure of full-length KcsA in its closed conformation. Proc Natl Acad Sci USA 106:6644–6649
41. Vandenberg CA, Bezanilla F (1991) Single-channel, macroscopic, and gating currents from sodium channels in the squid giant axon. Biophys J 60:1499–1510
42. Zagotta WN, Hoshi T, Aldrich RW (1994) Shaker potassium channel gating III: evaluation of kinetic models for activation. J Gen Physiol 103:321–362
43. Zheng J, Vankataramanan L, Sigworth FJ (2001) Hidden Markov model analysis of intermediate gating steps associated with the pore gate of shaker potassium channels. J Gen Physiol 118:547–564
44. Zhou M, Morais-Cabral JH, Mann S, MacKinnon R (2001) Potassium channel receptor site for the inactivation gate and quaternary amine inhibitors. Nature 411:657–661
45. Zhou Y, Morais-Cabral JH, Kaufman A, MacKinnon R (2001) Chemistry of ion coordination and hydration revealed by a K + channel-Fab complex at 2.0 A resolution. Nature 414:43–48

Chapter 5
Immobilizing Channel Molecules in Artificial Lipid Bilayers for Simultaneous Electrical and Optical Single Channel Recordings

Toru Ide, Minako Hirano and Takehiko Ichikawa

Abstract There has been much interest in imaging single drug bindings to ion channel proteins while simultaneously recording single channel current. We developed an experimental apparatus for simultaneous optical and electrical measurement of single channel proteins by combining the single molecule imaging technique and the artificial bilayer technique. However, one major problem is that single molecule imaging of drug bindings is limited by the innate thermal diffusion of channel proteins in the artificial bilayer. Therefore, immobilizing channel proteins in the bilayers is imperative for stable measurements of channel-drug interactions. For future studies on channel-drug interactions, we describe here three different methods for simultaneous optical and electrical observation of single channels in which channel proteins are immobilized. (i) Membrane binding protein annexin V reduces the lateral diffusion of single channel proteins in a concentration-dependent manner. (ii) Channel proteins are immobilized by anchorage through a polyethylene glycol (PEG) molecule to the glass substrate. (iii) Channels immobilized on a gel bead can be directly incorporated into artificial bilayers.

Keywords Annexin · Bilayer · BK-channel · Channel · Co^{2+} affinity gel · Conductance · Detergent · Diffusion · Histidine tag · Immobilization · KcsA · MthK · Phospholipid · Polyethyleneglycol (PEG) · Ryanodine receptor (RyR) · Single channel · Supported bilayer · Total internal reflection fluorescence (TIRF) microscope · Vesicle fusion

T. Ide (✉) and M. Hirano
Graduate School of Frontier Biosciences, Osaka University, 1-3 Yamadaoka Suita, Osaka 565-0871, Japan
e-mail: ide@phys1.med.osaka-u.ac.jp; mina37@phys1.med.osaka-u.ac.jp

T. Ichikawa
Laboratory for Spatiotemporal Regulations, National Institute for Basic Biology, Nishigonaka 38, Myodaiji, Okazaki, Aichi 444-8585, Japan
e-mail: ichi@nibb.ac.jp

Y. Sako and M. Ueda (eds.), *Cell Signaling Reactions: Single-Molecular Kinetic Analysis*,
DOI 10.1007/978-90-481-9864-1_5,

5.1 Introduction

The application of single molecule imaging techniques has spread through many biological fields including the study of cytoplasmic protein and nucleotide dynamics [1, 2]. These powerful techniques have enabled the finding of several biological phenomena which could not have been identified by using multimolecular methods. These techniques can also be applied to study the dynamic properties of ion channel proteins. Ion channels are membrane spanning proteins that regulate the ion permeability of cell membranes. They can be described as transducers since they convert a stimulus such as a ligand binding or membrane voltage change into changes in ion-permeability. Single ion channel current recordings have been performed using the patch-clamp technique or lipid bilayer systems from 40 years ago [3, 4]. However, imaging of single ion-channel proteins in lipid bilayers is a far less established technique. This is because channel proteins express their function only when incorporated into a membrane, which has a very fragile nature. The purpose of our present study is to establish simultaneous electrical and optical observations of single ion channels and elucidate the correlation between ligand binding and current. Although single channel current recordings can directly measure current through a single channel pore, the binding events of a single ligand have been inferred indirectly from multi-molecule experiments. The ability to simultaneously measure single ligand binding events and single channel current fluctuations is of great interest as it would enable us to open a new field of "single molecule pharmacology".

In 1999, we developed an experimental apparatus for simultaneous optical and electrical observation of single ion-channels by combining planar lipid bilayer technique and single molecule imaging technique [5]. There had been numerous studies on single channel recordings by the planar bilayer technique [6] and on single molecule imaging of membrane lipids in self-standing and solid supported bilayer membranes [7, 8]. However, we were the first to report the combination of these two technologies. Using this apparatus, we imaged several types of channels at the single molecule level while simultaneously recording single channel current [9, 10]. For those studies, we developed a method for forming lipid bilayers on an agarose layer through which we could observe single fluorophores in the membrane with an objective type total internal reflection fluorescence (TIRF) microscope.

A major technological obstacle in carrying out simultaneous imaging ligand binding of single-channel and recording electrical change is the lateral diffusion of the channel protein within the lipid bilayer. With our membrane, we could constantly image the whole bilayer. However, because of thermal diffusion by channel proteins in the membrane, it was very difficult to detect interactions between a single ligand molecule and a single ion-channel. Therefore, immobilizing channel proteins in the bilayers is imperative for stable measurements of channel-ligand interactions. For studies on ligand-channel interactions at the single molecule level, we describe here several methods for simultaneous optical and electrical observation of single ion-channels in which channel proteins are immobilized.

5.2 Annexin V Immobilizes Membrane Proteins [11]

Annexins are a family of calcium-dependent proteins that preferentially bind to negatively charged lipids [12–14]. It is known that the binding of annexin V to phospholipids influences the fluidity of the lipid bilayer. It has been reported that annexin V decreased the diffusion of the lipid bilayer using the multi-molecular methods NMR and FRAP [15, 16]. Annexin V is characteristic in its membrane binding manner. Several microscopic studies have revealed that annexin V self-assembles and forms two-dimensional (2D) crystals [17–19]. Peng et al. reported that annexin XII reduced the diffusion of single ryanodine receptor channel type 2 (RyR2) by tracking the center of Ca^{2+} flux [20]. RyR2 is the major calcium release channel of the sarcoplasmic reticulum (SR) in cardiomyocytes. A high calcium concentration in the cell triggers the muscle contraction. We tracked single RyR2 channel proteins in the artificial planner lipid bilayers containing phosphatidylserine (PS). Unlike other annexins the purified form of annexin V is commercially available since it is widely used in testing for apoptosis. We demonstrate that annexin V reduces the lateral diffusion of single RyR2 channel proteins in a concentration-dependent manner.

Figure 5.1a shows a schematic overview of the lipid bilayer formation apparatus. This apparatus consists of two chambers. Lipid bilayers were formed in a small hole (about 100 μm diameter) at the bottom of the upper chamber. The lipid bilayers touched the agarose-coated cover glass at the bottom of the lower chamber. Bilayers were observed with an objective lens type TIRF microscope according to the method of Tokunaga et al. [21, 22] Fig. 5.1b is a schematic representation of the procedures for forming bilayers on the glass coverslip. (1) First, appropriate recording solution was added to the chambers. (2) Lipids dissolved in organic solvent were added to the underside of the bottom of the upper chamber. A thick lipid membrane covered the aperture on the plastic film. (3) The upper chamber was plunged into the lower solution and moved downward until the thick membrane came into contact with the agarose modified glass. (4) By slightly increasing the pressure in the upper chamber by adding a small amount of solution, the membrane began to expand resulting in the center of the membrane becoming a bilayer by contacting the two monolayers. Alternatively, it is possible to form a bilayer on a glass by pressing a bilayer membrane that was pre-formed in an aqueous environment (3'). Plunging a thick membrane into the lower solution, precisely regulating the hydrostatic pressure of the upper side of the membrane, and holding the membrane at a fixed height, the membrane thinned spontaneously to form a bilayer. Since there must be a water layer on the agarose coated glass surface, the two procedures are exactly the same from a microscopic viewpoint. There were no significant differences in channel properties such as single channel conductance and voltage dependency.

Fluorescence spots representing single RyR2 and 1,2-dihexadecanoyl-sn-glycero-3-phosphoethanolamine (DHPE) molecules are shown in Fig. 5.2a. The decay in fluorescence intensity for each spot of DHPE occurred in a single step manner indicating that the observed fluorescence signal was caused by a single fluorophore

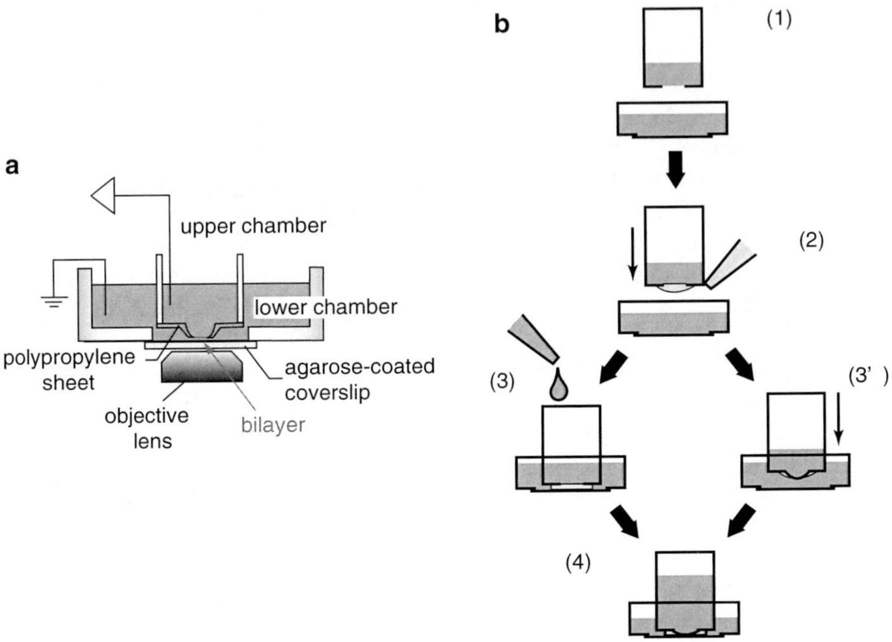

Fig. 5.1 Overview of the lipid bilayer apparatus. (**a**) The apparatus consists of two chambers. A lipid bilayer is formed across a small hole (100–200 μm diameter) at the bottom of the upper chamber. The bottom surface of the lipid bilayer touches the agarose-coated cover glass at the bottom of the lower chamber. (**b**) Schematic representation of the procedures for forming bilayers on the agarose-modified glass. The modified coverslip was fixed with adhesive over the hole on the bottom of the chamber. Bilayers were formed as described in the text.

(Fig. 5.2b). In Fig. 5.2c, the cumulative fluorescence intensity of each spot was plotted into a histogram. The distribution of the fluorescence intensities for each spot showed a single peak which was well fitted with a single Gaussian distribution. Taking into account of these observations we may conclude that each fluorescent spot represented a single DHPE molecule and that single DHPE molecules were conjugated to BODIPY with a dye to lipid ratio of 1:1. The number of channel in a lipid bilayer can be counted by measuring electrical current. Figure 5.2d shows single channel current of the RyR2 channel. There was only one fluorescent spot seen in the bilayer, showing fluorescent spot of RyR2 was observed. Figure 5.2e is the histogram of the current amplitude distribution of Fig. 5.2d. This distribution showed two peaks indicating open and closed state. These data means the spot does not consist of cluster but only one channel exist in lipid bilayer.

Lateral diffusion of single DHPEs and RyR2s were measured in the absence and presence of 0.125, 0.25, 0.5 and 1 μM annexin V. Figure 5.3a is typical trajectories recorded at 30 Hz (1 s). Upper and lower traces are trajectories of DHPE and RyR2, respectively. Cumulative data of the diffusion coefficient and displacement distribution of DHPE and RyR2 at each annexin V concentration is shown in Fig. 5.3b, c,

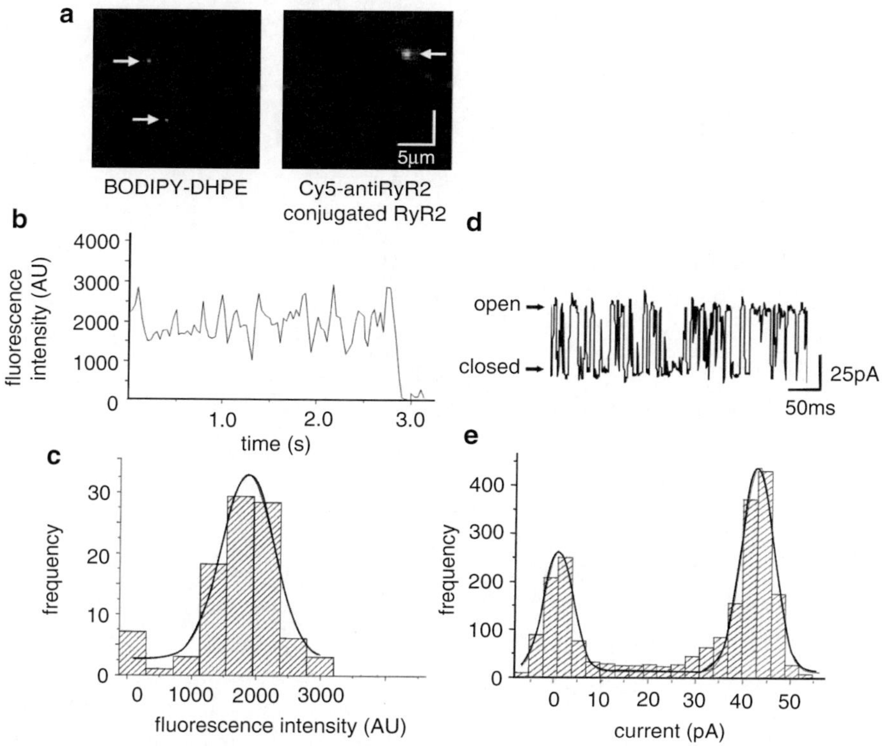

Fig. 5.2 Single molecule imaging of PE and RyR2. (**a**) Imaging single BODIPY-DHPE and Cy5-antiRyR2 conjugated RyR2 (Cy5-RyR2) molecules. *Arrows* indicate the position of each molecule. (**b**) A fluorescence intensity change of single BODIPY-DHPE in a lipid bilayer. The spot was photobleached in a single step, suggesting a single molecule. (**c**) Distributions of cumulative photobleaching step sizes. (**d**) Single channel recording while one spot was observed at +25 mV and (**e**) the distributions of cumulative current amplitude.

respectively. The shape of the distribution became sharp with an increasing concentration of annexin V. This means annexin V reduces the number of large displacements. Annexin V decreased the diffusion coefficients of DHPE and RyR2 in a concentration-dependent manner. In the absence of annexin V the diffusion coefficients of DHPE and RyR2 were $4.81 \pm 2.05 \times 10^{-8}$ cm^2/s (n = 25) and $2.13 \pm 0.809 \times 10^{-8}$ cm^2/s (n = 24), respectively. In the presence of 1 μM annexin V the diffusion coefficients of DHPE and Cy5-RyR2 were $2.23 \pm 2.41 \times 10^{-10}$ cm^2/s (n = 18) and $0.95 \pm 1.49 \times 10^{-10}$ cm^2/s (n = 25), respectively. It was revealed that 1 μM annexin V decreases the diffusion of DHPE and RyR2 molecules to 1/200th of their values in the absence of annexin V. Additionally, in the presence of 1 μM annexin V, only in one chamber did we simultaneously observe two types of diffusion of BODIPY-DHPE: stopped and rapidly moving spots. On the other hand, in the presence and absence of 1 μM annexin V in both chambers, we observed only one type of diffusion, either stopped or rapidly moving spots, respectively. This result indicates that annexin V can access only monolayers, consistent with Isas et al. [23].

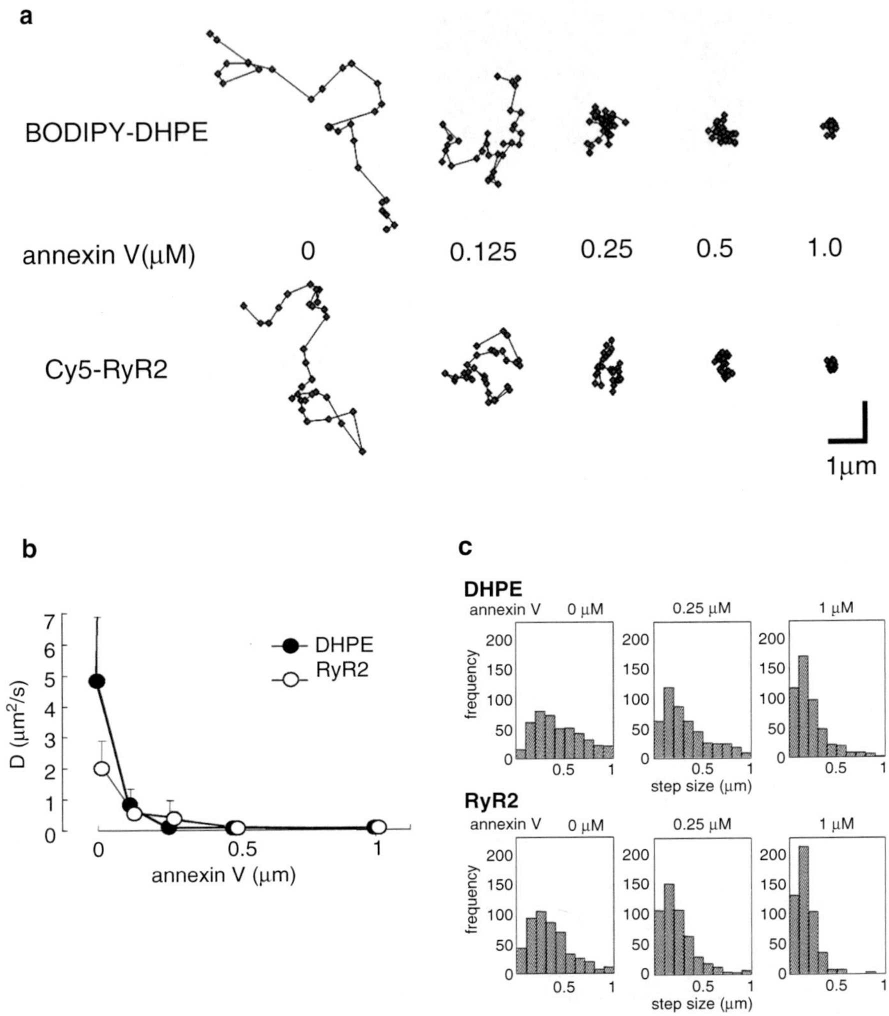

Fig. 5.3 Cumulative data showing effects of annexin V on the lateral diffusion of single PE and RyR2 molecules in bilayers. (**a**) Typical trajectories of single BODIPY-DHPE and Cy5-RyR2 molecules in the lipid bilayers. Thirty data points were recorded at 33 ms intervals. *Upper* and *bottom traces* show BODIPY-DHPE and Cy5-RyR5 trajectories respectively. Diffusion of BODIPY-DHPE and Cy5-RyR2 decreases with the increase of annexin V. (**b**) The diffusion coefficient of PE and RyR2 at each annexin V concentration. *Open* and *closed circles* show the diffusion coefficients of Cy5-RyR2 and BODIPY-PE, respectively. Annexin V decreases the diffusion coefficient of both PE and RyR2 in a concentration-dependent manner. (**c**) The displacement distribution for 33 ms of PE and RyR2 at each annexin V concentration. This graph shows that annexin V reduces the number of large displacements with an increasing concentration of annexin V.

Electrical current traces of single RyR2s in the absence and presence of 1 μM annexin V were shown in Fig. 5.4. In our settings we could not find any significant influence of annexin V on the function of the RyR2. Thus, annexin V can be used in our experimental system which aims at revealing the relationship of ligand binding and the change of current.

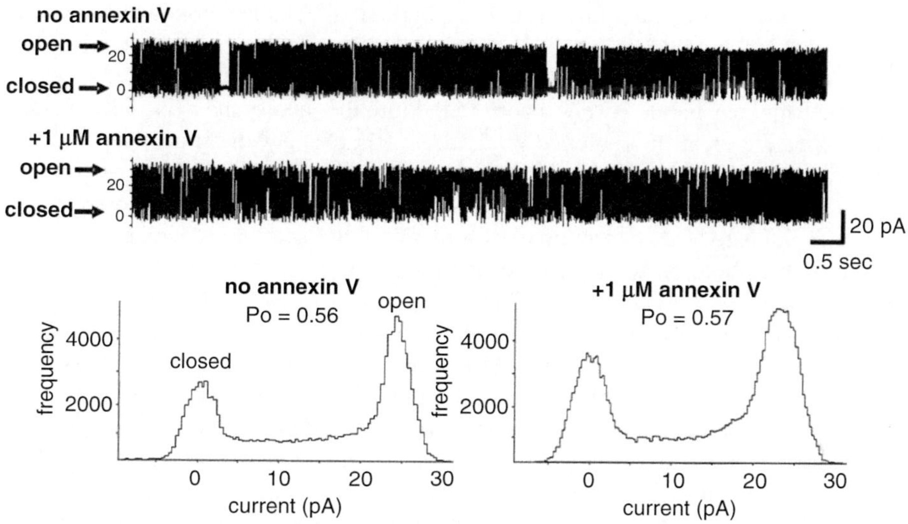

Fig. 5.4 Typical traces of single RyR2 channel recordings with and without annexin V and the histograms of each trace. Experiments were performed at a holding potential of +25 mV. The histograms of each trace are shown below. Open probabilities (Po) of RyR2 with and without 1 μM annexin V were 0.56 and 0.57, respectively. These figures show that annexin V does not alter activation of RyR2 under the conditions used in the present studies.

We propose the following model as a cause for which annexin V decreases the lateral diffusion of lipids and channel proteins in lipid bilayers in a concentration-dependent manner. Lipids and channel proteins diffuse in the area between the complexes of annexin V and the lipid. When the concentration of annexin V increased, the area in which lipids and channel proteins can diffuse is narrowed because the area of the complexes of annexin V and lipid increases. Consequently, the diffusion coefficients of the lipid and channel protein decreased in annexin V concentration-dependent manner. From a recent study with AFM, it was seen that annexin V expands gradually encompassing a grater area [19], which is consistent with our model.

5.3 PEG Supported Bilayers [24]

We describe here a novel method for simultaneous optical and electrical observation of single ion-channels in which channel proteins are immobilized by polyethylene glycol (PEG) molecules. In this method, we used the same apparatus for the annexin-method shown in Fig. 5.1, except that the coverslip at the bottom of the chamber was covered with PEG instead of agarose. The membrane was formed as did in Fig. 5.1b.

BK-channels, which are known to have high single channel conductance, were fluorescently labeled using anti-NH_2-terminus antibodies that had been conjugated with cy5-dye molecules. Bilayer membranes were formed on PEG-coated glass as described above. Channels were incorporated into the membrane by vesicle fusion. Figure 5.5a shows a fluorescence image of a BK-channel in an artificial bilayer membrane that had been conjugated with antibodies at a conjugation ratio of 1.0. Figure 5.5b shows the time course of fluorescence intensity of the spot shown in 5A. The total intensity was approximately four times the stepwise decrease that was caused by the photo bleaching of dyes conjugated to the antibody, which indicates that the channel was labeled with four antibodies because the dye/protein of antibody was 1.0 and four homologous subunits assembled to form a channel. Figure 5.5c shows BK-channel immobilization upon anchoring to the glass surface through a PEG molecule. This figure represents thermal diffusion of a BK-channel. Line segments in the figure show the trajectory of a channel in a bilayer membrane formed on a PEG-coated glass coverslip. Bilayer membranes were formed on PEG spacers with (+) and without (−) anti-BK-channel antibodies. The diffusion coefficient, calculated from single channel trajectories, was $D = 0.64 \pm 0.57\ \mu m^2/s$ without anchoring to the glass via PEG. By contrast, the diffusion coefficient in Fig. 5.5c(+) was much smaller ($D < 0.01\ \mu m^2/s$), showing that the channel was immobilized by anchorage through a PEG (Fig. 5.5e).

Single channel current records shown in Fig. 5.5d indicate that the natural properties of BK-channels can also be measured using immobilized channels. Voltage dependence, which is the signature feature of BK-channels, was maintained when the channel was incorporated into an artificial bilayer and immobilized by a PEG anchor. The PEG-immobilized channel was activated more at lower membrane potential, i.e. it was activated at depolarizing membrane potential, in agreement with native channel behavior. From the slope of the I–V curve, single channel conductance was determined to be 345 pS, which agrees well with conventional bilayer and patch-clamp experiments [25].

5.4 Gel/Gel Interface Bilayers [26]

In the third method, we describe how to incorporate channel proteins immobilized on a gel directly into bilayer membranes. In this section, we describe how to incorporate channels on a large gel bead. Using a thin gel layer instead of a large gel bead, one can visualize single fluorophores in the membrane as we did in Fig. 5.2.

Figure 5.6a shows a schematic drawing of the experimental apparatus. Gel beads with channel proteins were set on the dish. Channel proteins tagged with polyhistidine were immobilized to Co^{2+} affinity gel beads via the histidine tag. The horizontal lipid layer formed across the hole (100–200 μm in diameter) was moved downward until gently contacting a gel bead by using a micromanipulator. After the lipid layer contacted the bead, it became thinner, resulting in a rapidly formed lipid bilayer membrane (Fig. 5.6b). After several minutes, channel currents were

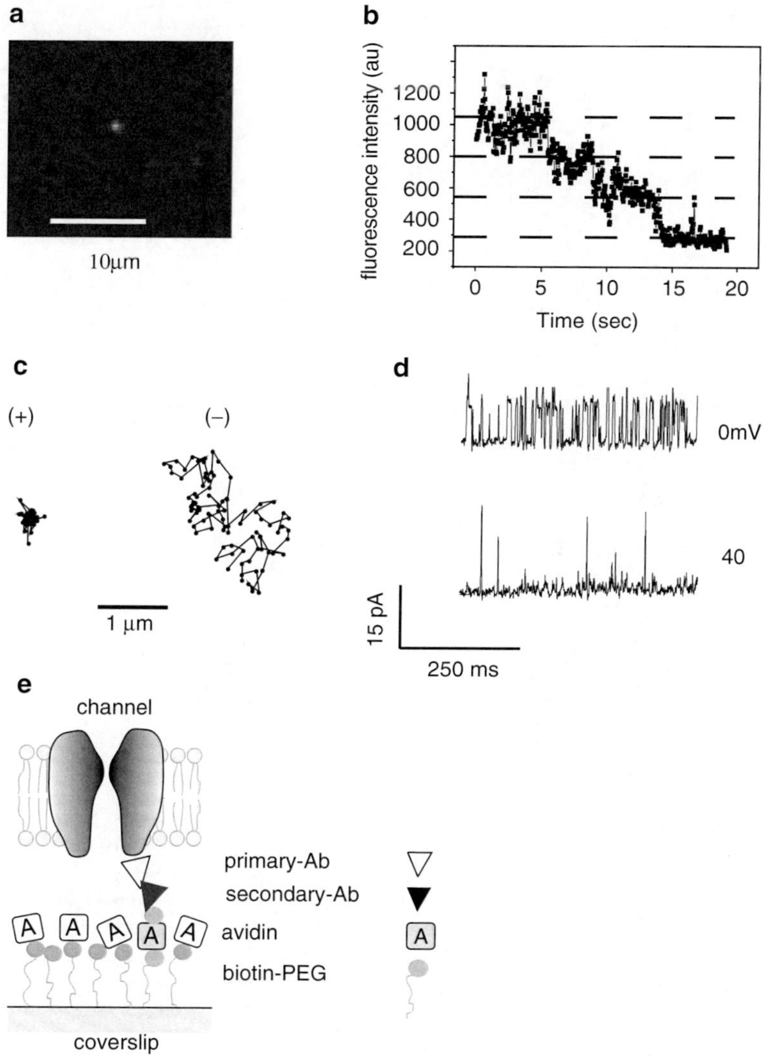

Fig. 5.5 Lateral diffusion of a single BK-channel. (**a**) Fluorescence image of a cy5 labeled uterine BK-channel incorporated into a bilayer membrane. (**b**) Stepwise decrease in fluorescence intensity of the spot shown in (a). (**c**) Trajectories of a single BK-channel protein observed in the absence (−) and in the presence (+) of antibodies. (**d**) Single channel records taken from the membrane shown in (c) (+). (**e**) A schematic diagram of a channel protein immobilized on the PEG-coated glass.

measured. As shown in Fig. 5.6c, channels could be incorporated even when the bilayer had already formed in solution. Channel proteins could avoid their denaturation because they were kept in and washed with detergent containing solutions during purification and when immobilized onto the beads, respectively. Therefore,

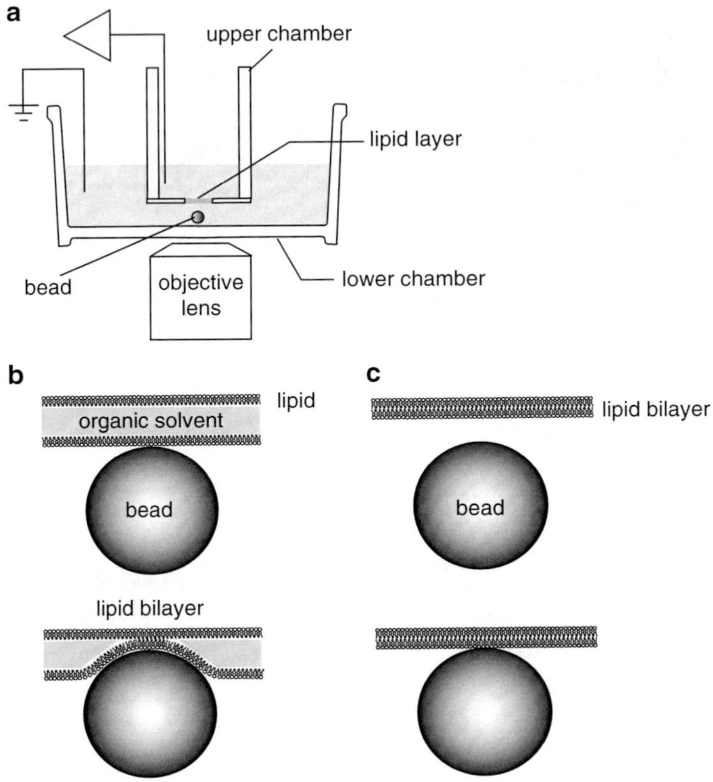

Fig. 5.6 Methods to form a lipid bilayer membrane on an agarose bead. (**a**) The bilayer apparatus [5, 9]. This apparatus consisted of two chambers. The upper chamber (100 ~ 200 μl) was manipulated by a micromanipulator. On the bottom of the upper chamber, a thin (0.1 ~ 0.2 mm thick) plastic film was attached. This film had a small pore (100 ~ 200 μm in diameter) in the center across which the bilayers formed. (**b**) A thick lipid layer formed in solution is pushed onto a Co^{2+} affinity gel bead causing the thick layer to spontaneously thin, leading to a bead-bilayer interface. The channel proteins on the beads were covered with surfactant molecules before incorporating them into the membranes. (**c**) A bilayer formed in solution is moved downward to contact a bead.

the hydrophobic surface of the proteins was covered with detergents before being incorporated into bilayers even in a recording solution without detergent (Fig. 5.6b, c). Lipid bilayer formation was observed with a normal transmission microscope. It appears that organic solvent was present in the thick lipid layer and that the two lipid monolayers contacted each other to form the bilayer.

Figure 5.7 shows the results of single channel recordings. Figure 5.7a (left) shows the current traces recorded from a bilayer formed on a gel bead. We incorporated KcsA channels into the bilayer and the current across the bilayer was recorded. The KcsA is a bacterial potassium channel whose crystal structure was determined. KcsA was open more frequently at positive voltages, which has been previously reported when the cytoplasmic side was held at virtual ground [27]. Here, the bead side was held at virtual ground indicating the cytoplasmic domain, which has the histidine tag, was at the bead side. We found that the KcsA channels were always incorporated in this direction. Single channel current amplitudes were plotted against membrane voltage in Fig. 5.7a (right). From this I–V relationship, the single channel chord conductance at 100 mV was about 175 pS, which is consistent with the conductance measured by the conventional planar bilayer method [28]. Figure 5.7b shows the results of single MthK channel currents. The MthK is another

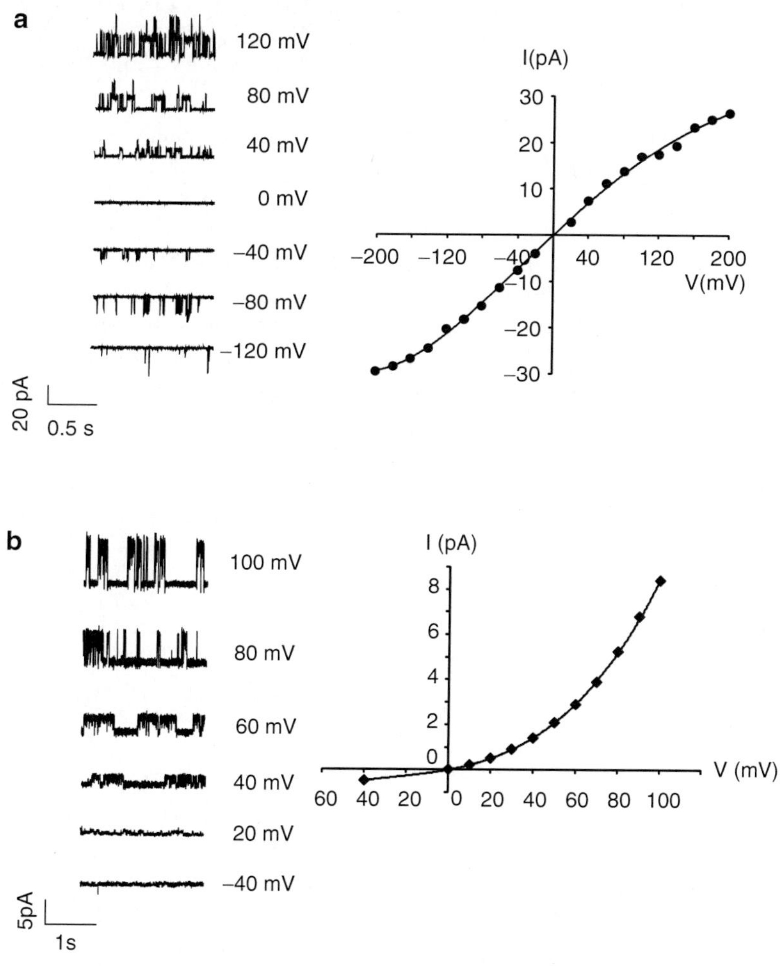

Fig. 5.7 Channels immobilized on a gel bead were directly incorporated into a bilayer. (**a**) KcsA channel current. (*left*) Trace of a single KcsA channel current. (*right*) Single KcsA channel current amplitude vs membrane voltage. (**b**) MthK channel current. (*left*) Single channel current at the indicated voltages. (*right*) I–V relationship. Single channel current amplitudes obtained in (*left*) were plotted against membrane voltage.

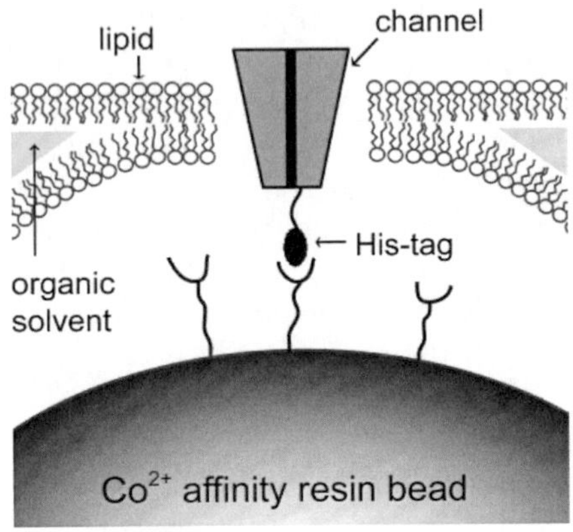

Fig. 5.8 Channel incorporation scheme. Channel proteins bound to a gel bead through a histidine-tag directly incorporate into the artificial lipid bilayer.

potassium channel which is activated by intracellular calcium. Channels were incorporated similarly to KcsA channels except for the high pH and 10 mM $CaCl_2$ added after the bilayer formed since high pH and millimolar Ca^{2+} in the cytoplasmic side is needed to activate MthK [29, 30]. Figure 5.7b (left) shows single channel current traces of MthK at the indicated membrane voltages. MthK reconstituted into the bilayer by our method showed rectification. Single channel current amplitudes plotted against membrane voltage showed the channel was inwardly rectifying (Fig. 5.7b, right), which agrees with conventional method results [31]. This rectification indicates that the channel was incorporated into the bilayer such that the cytoplasmic side of the channel was directed towards the bead. Addition of $CaCl_2$ in the upper side of the membrane failed to activate this channel. These results also show that the cytoplasmic domain, which has the histidine tag, exists at the bead side. We found that the channels were always incorporated in this direction.

From these results, we concluded that channel proteins bound to Co^{2+} affinity gel beads via the histidine tag directly incorporated into the bilayer membrane (Fig. 5.8).

5.5 Conclusion

We describe three different methods for immobilizing channel proteins in an aritificial bilayer. All these have been developed for future single molecular observation of ligand-channel interactions. Using TIRF, we can visualize single ligand bindings since fluorescently labeled ligands are visualized as a bright spot only when they bind and immobilize to a channel. When choosing an immobilization method, one needs to consider the purpose of the experiment and the properties of the channel. For example, while the annexin method is simplest, it is not applicable to channel proteins that are strongly affected by calcium or acidic lipids, which are required for annexin binding to a bilayer. Another example is that some channels change their activities when anchoring to a solid support through a PEG molecule.

References

1. Yanagida T, Ishii Y (2008) Single molecule dynamics in life science. Wiley-VCH
2. Sako Y, Yanagida T (2003) Single-molecule visualization in cell biology. Nat Rev Mol Cell Biol Suppl:SS1–SS5
3. Neher E, Sakmann B (1976) Single-channel currents recorded from membrane of denervated frog muscle fibres. Nature 260:799–802
4. Mueller P, Rudin DO (1968) Resting and action potentials in experimental bimolecular lipid membranes. J Theor Biol 18:222–258
5. Ide T, Yanagida T (1999) An artificial lipid bilayer formed on an agarose-coated glass for simultaneous electrical and optical measurement of single ion channels. Biochem Biophys Res Commun 265:595–599
6. Favre I, Sun YM, Moczydlowski E (1999) Reconstitution of native and cloned channels into planar bilayers. Method Enzymol 294:287–304
7. Schmidt T, Schutz GJ, Baumgartner W, Gruber HJ, Schindler H (1996) Imaging of single molecule diffusion. Proc Natl Acad Sci USA 93:2926–2929
8. Schutz GJ, Schindler H, Schmidt T (1997) Single-molecule microscopy on model membranes reveals anomalous diffusion. Biophys J 73:1073–1080
9. Ide T, Takeuchi Y, Yanagida T (2002) Development of an experimental apparatus for simultaneous observation of optical and electrical signals from single 1on channels. Single Molecules 3:33–42
10. Ide T, Takeuchi Y, Aoki T, Yanagida T (2002) Simultaneous optical and electrical recording of a single ion-channel. Jpn J Physiol 52:429–434
11. Ichikawa T, Aoki T, Takeuchi Y, Yanagida T, Ide T (2006) Immobilizing single lipid and channel molecules in artificial lipid bilayers with annexin A5. Langmuir 22:6302–6307
12. Gerke V, Moss SE (2002) Annexins: from structure to function. Physiol Rev 82:331–371
13. Rescher U, Gerke V (2004) Annexins – unique membrane binding proteins with diverse functions. J Cell Sci 117:2631–2639
14. Swairjo MA, Seaton BA (1994) Annexin structure and membrane interactions: a molecular perspective. Annu Rev Biophys Biomol Struct 23:193–213
15. Saurel O, Cezanne L, Milon A, Tocanne JF, Demange P (1998) Influence of annexin V on the structure and dynamics of phosphatidylcholine/phosphatidylserine bilayers: a fluorescence and NMR study. Biochemistry 37:1403–1410
16. Cezanne L, Lopez A, Loste F, Parnaud G, Saurel O, Demange P, Tocanne JF (1999) Organization and dynamics of the proteolipid complexes formed by annexin V and lipids in planar supported lipid bilayers. Biochemistry 38:2779–2786
17. Oling F, Bergsma-Schutter W, Brisson A (2001) Trimers, dimers of trimers, and trimers of trimers are common building blocks of annexin a5 two-dimensional crystals. J Struct Biol 133:55–63
18. Mo Y, Campos B, Mealy TR, Commodore L, Head JF, Dedman JR, Seaton BA (2003) Interfacial basic cluster in annexin V couples phospholipid binding and trimer formation on membrane surfaces. J Biol Chem 278:2437–2443
19. Richter RP, Him JL, Tessier B, Tessier C, Brisson AR (2005) On the kinetics of adsorption and two-dimensional self-assembly of annexin A5 on supported lipid bilayers. Biophys J 89:3372–3385
20. Peng S, Publicover NG, Airey JA, Hall JE, Haigler HT, Jiang D, Chen SR, Sutko JL (2004) Diffusion of single cardiac ryanodine receptors in lipid bilayers is decreased by annexin 12. Biophys J 86:145–151
21. Tokunaga M, Kitamura K, Saito K, Iwane AH, Yanagida T (1997) Single molecule imaging of fluorophores and enzymatic reactions achieved by objective-type total internal reflection fluorescence microscopy. Biochem Biophys Res Commun 235:47–53
22. Wazawa T, Ueda M (2005) Total internal reflection fluorescence microscopy in single molecule nanobioscience. Adv Biochem Eng Biotechnol 95:77–106

23. Isas JM, Cartailler JP, Sokolov Y, Patel DR, Langen R, Luecke H, Hall JE, Haigler HT (2000) Annexins V and XII insert into bilayers at mildly acidic pH and form ion channels. Biochemistry 39:3015–3022
24. Ide T, Takeuchi Y, Noji H, Tabata KV (2009) Simultaneous optical and electrical single channel recordings on a PEG glass. Langmuir 26(11):8540–8543
25. Magleby KL (2003) Gating mechanism of BK (Slo1) channels: so near, yet so far. J Gen Physiol 121:81–96
26. Hirano M, Takeuchi Y, Aoki T, Yanagida T, Ide T (2009) Current recordings of ion channel proteins immobilized on resin beads. Anal Chem 81:3151–3154
27. Cordero-Morales JF, Cuello LG, Perozo E (2006) Voltage-dependent gating at the KcsA selectivity filter. Nat Struct Mol Biol 13:319–322
28. LeMasurier M, Heginbotham L, Miller C (2001) KcsA: it's a potassium channel. J Gen Physiol 118:303–314
29. Jiang Y, Lee A, Chen J, Cadene M, Chait BT, MacKinnon R (2002) Crystal structure and mechanism of a calcium-gated potassium channel. Nature 417:515–522
30. Zadek B, Nimigean CM (2006) Calcium-dependent gating of MthK, a prokaryotic potassium channel. J Gen Physiol 127:673–685
31. Li Y, Berke I, Chen L, Jiang Y (2007) Gating and inward rectifying properties of the MthK K + channel with and without the gating ring. J Gen Physiol 129:109–120

Chapter 6
Single-Protein Dynamics and the Regulation of the Plasma-Membrane Ca^{2+} Pump

Carey K. Johnson, Mangala R. Liyanage, Kenneth D. Osborn, and Asma Zaidi

Abstract The plasma-membrane Ca^{2+} pump helps to maintain proper Ca^{2+} levels in the cell by using ATP to drive Ca^{2+} transport across the cell membrane. In this chapter, we describe single-molecule experiments in our laboratory to investigate the mechanisms of regulation of the plasma-membrane Ca^{2+} pump by the Ca^{2+} signaling protein calmodulin, which binds to a C-terminal autoinhibitory domain of the pump. Single-molecule fluorescence methods add new insights into the regulatory mechanisms by resolving heterogeneous populations and dynamics. We have used single-molecule polarization modulation and single-pair resonance energy transfer experiments to probe conformations and dynamics of the autoinhibitory domain. The results show that the mechanisms of regulation by the autoinhibitory domain are more subtle and complex than previously realized.

Keywords ATPase · Autoinhibitory domain · Burst measurements · Calmodulin (CaM) · FRET · Maximum likelihood · Modulation depth · Oxidative modification · Plasma membrane Ca^{2+}-ATPase (PMCA) · Polarization modulation

C.K. Johnson (✉) and M.R. Liyanage
Department of Chemistry, University of Kansas, Lawrence, KS 66044
e-mails: ckjohnson@ku.edu; mangala@ku.edu

K.D. Osborn,
Department of Chemistry, University of Kansas, Lawrence, KS 66045
Present address: Department of Math and Science, Fort Scott Community College, Fort Scott, KS 66701, USA
e-mail: kennetho@fortscott.edu

A. Zaidi
Department of Biochemistry, Kansas City University of Medicine and Biosciences, Kansas City, MO 64119
e-mail: Azaidi@kcumb.edu

Y. Sako and M. Ueda (eds.), *Cell Signaling Reactions: Single-Molecular Kinetic Analysis*, DOI 10.1007/978-90-481-9864-1_6, © Springer Science+Business Media B.V. 2011

6.1 Introduction: The Plasma-Membrane Calcium Pump

In this chapter we discuss the application of single-molecule methods to the plasma-membrane Ca^{2+} pump and its regulation by the Ca^{2+} signaling protein calmodulin (CaM). The full range of conformations and dynamics operative in PMCA regulation are difficult to unravel by conventional kinetic and biochemical methods. In the past several years we have developed single molecule strategies that have allowed us to study the interaction of the plasma-membrane Ca^{2+} pump with CaM at the molecular level. We have chosen specific examples from our studies of the Ca^{2+} pump in order to illustrate a range of single-molecule spectroscopic techniques and the information derived from them with particular relevance to their application to Ca^{2+} signaling interactions.

The plasma membrane Ca^{2+}-ATPase (PMCA) is a high-affinity Ca^{2+} pump that plays an important role in the maintenance of precise levels of intracellular Ca^{2+} in all eukaryotic cells (reviewed in [1–3]). PMCA uses the metabolic energy of ATP hydrolysis to transport Ca^{2+} from the cytosol to the extracellular environment, the stoichiometry of transport being 1:1. The protein structure is illustrated schematically in Fig. 6.1. PMCA is a ~138 kDa integral membrane protein with 10 transmembrane helices and several functional domains in the cytosolic portion of the protein [4]. The first cytosolic loop between transmembrane domains 2 and 3 is considered the

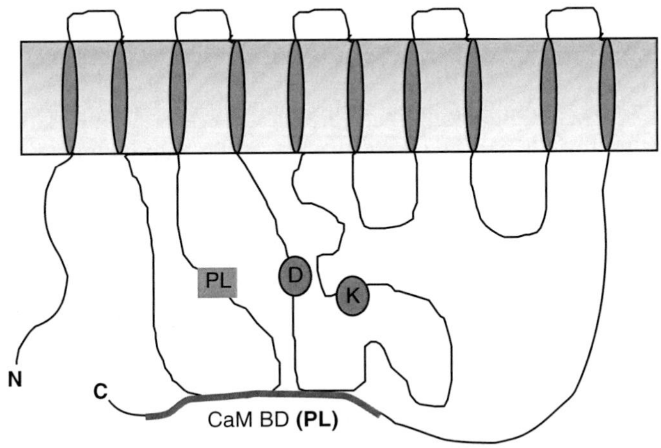

Fig. 6.1 Schematic of the structure of PMCA showing extensive cytoplasmic domains. These domains include an Asp residue (D) that is phosphorylated during the enzymatic cycle, the nucleotide binding site (K), and the CaM binding domain (CaM BD). PL denotes phospholipid binding domains. The C-terminal autoinhibitory domain interacts with residues 206–271 and 527–554, causing autoinhibition.

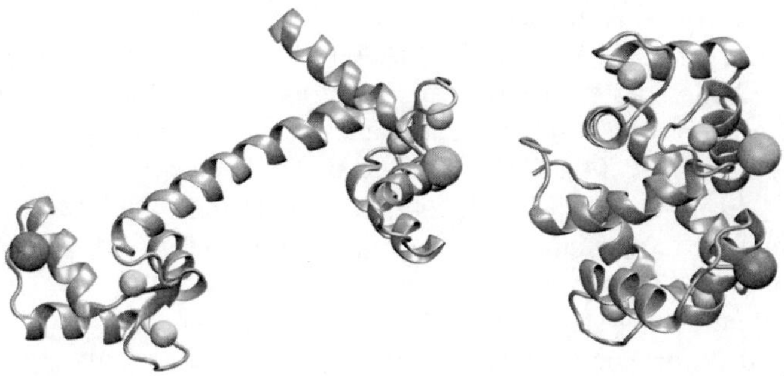

Fig. 6.2 Two crystal structures of Ca^{2+}-CaM showing an extended conformation (*left*) and a compact conformation (*right*). The four small spheres show Ca^{2+} binding sites. Also shown are sites 34 and 110 (*large gray spheres*) where we have attached fluorescent dyes. The crystal structures shown are available in the protein data bank (pdb 3cln [14], *left*, and pdb 1prw [15], *right*). Figures were generated in VMD for WIN32, version 1.8.4 [16]. Adapted from ref. [17].

"transducer domain." A large intracellular loop located between transmembrane 4 and 5 contains the ATP binding site and a conserved aspartate reside that is phosphorylated during the reaction cycle, typical of the P type class of ion motive ATPases [5, 6].

Biological signals are frequently transmitted by Ca^{2+} ions. CaM binds Ca^{2+} to function as a Ca^{2+}-activated molecular switch [7, 8]. CaM is a small (16.7 kD), acidic protein with two globular domains connected by a central linker (see Fig. 6.2). Each domain contains two EF-hand sites that bind Ca^{2+} ions when the Ca^{2+} concentration rises above resting Ca^{2+} concentrations of around 100 nM [9]. Upon binding Ca^{2+}, the EF-hand domains open to expose hydrophobic domains that can interact with a host of target proteins [10–12], including PMCA. The interactions of CaM with target proteins regulate many important cellular functions such as neurotransmission, ion transport, smooth muscle contraction, and neuronal plasticity [13]. How these responses occur and are relayed to target enzymes depends on the interplay of protein conformational changes and protein–protein interactions.

PMCA was one of the first enzymes shown to be regulated by the Ca^{2+} sensor protein CaM [18, 19], and a CaM-binding region was identified in the C-terminal domain [20]. Since this discovery much effort has been devoted to investigating the mechanisms underlying the stimulation of PMCA by CaM. A series of biochemical, biophysical, and analytical approaches have shown that the enzyme is autoinhibited in the absence of CaM. Binding of Ca^{2+}-CaM to its C terminal binding domain produces conformational changes in PMCA that displace the autoinhibitory domain away from the catalytic domain, thus relieving autoinhibition and resulting in several-fold stimulation [21]. However, the precise mechanisms of regulation and the interactions of the autoinhibitory domain with the catalytic core are not known at the molecular level. The structural changes induced in the protein during the enzymatic cycle and the heterogeneity and dynamics of the various conformations of PMCA are difficult to unravel by conventional kinetic and biochemical methods. Single molecule strategies have provided us with tools that can begin to address the interaction of PMCA with CaM at the molecular level.

Mammalian PMCA is encoded by four different genes that give rise to four isoforms, PMCA1 through PMCA4 [22]. Additionally, each gene transcript is alternatively spliced to generate several variants. Recent evidence has shown that the various PMCA isoforms and splice variants are specialized to regulate the spatial and temporal characteristics of the Ca^{2+} signal depending on the needs of individual cell types [23, 24]. PMCA1 and PMCA4, considered house-keeping forms for maintaining low levels of intracellular Ca^{2+}, are present in most cell types, while the expression of PMCA isoforms 2 and 3 is targeted more specifically to the central and peripheral nervous systems [24–26], where they play a crucial role in fast Ca^{2+} clearance needed for neuronal Ca^{2+} signaling [25]. Among the four isoforms, PMCA2b has the highest basal activity in the absence of CaM [25, 27] and a higher activation rate by CaM [28, 29] and by Ca^{2+} [25, 30]. PMCA2b also displays a longer memory than PMCA4b of a previous Ca^{2+} spike [30]. The mechanisms underlying the differential stimulation of the various isoforms by CaM remain largely unknown.

In addition to its stimulation by CaM, the functional versatility of PMCA is also mediated by the binding of acidic phospholipids, phosphorylation by kinases, limited proteolysis that cleaves off the autoinhibitory domain, and by an incompletely understood process of dimerization. More recently, PMCA has been shown to be localized in lipid rafts, cholesterol-enriched microdomains in the plasma membranes [31, 32]. PMCA activity is responsive to cholesterol depletion, suggesting an additional mechanism for the modulation of its function. The mechanistic aspects of stimulation of PMCA by the above mentioned CaM-independent regulatory agents are far from being understood. Lack of such knowledge prevents understanding of the molecular mechanisms underlying the maintenance of precise Ca^{2+} levels in diverse cell types and the mechanisms by which modification of PMCA leads to pathological conditions.

The role of the autoinhibitory domain in the regulation of PMCA is intriguing and worthy of detailed analysis since it may shed light onto the mechanism underlying the activation of PMCA and the origin of the different responses and multiple roles of the PMCA isoforms in various cell types. Although significant information about the structure–function relationship of P-type ATPases has been gleaned from the structurally homologous sarcoplasmic reticulum Ca^{2+}-ATPase (SERCA) whose crystal structure is now available [33], the lack of an autoinhibitory domain in SERCA keeps it from serving as an accurate model for the regulation of PMCA. In this chapter, we will show through several examples that single-molecule detection of PMCA–CaM interaction can add important new insights into the regulatory mechanism, conformations, and dynamics of PMCA.

6.2 Detection of Single PMCA Molecules

Many of the unanswered questions about the mechanisms of PMCA stimulation and regulation are difficult to answer by conventional ensemble measurements. This is a result of conformational heterogeneity (different molecules are in different conformational states) or temporal heterogeneity (the dynamics of different protein molecules in ensemble samples are not synchronized in time). In such situations, the single-molecule strategy is to resolve different states or kinetic events by probing individual members of the ensemble one at a time. For this strategy to be successful, however, the first and foremost challenge is to detect signals originating from single molecules of the protein. In this section we discuss various single-molecule detection strategies for PMCA.

To detect single protein molecules, a number of technical problems must be solved. These include three practical issues: (1) labeling the protein with an appropriate fluorophore; (2) spatially locating and isolating signals from single molecules; and (3) applying spectroscopic methods that yield interesting information at the single-molecule level. Each of these is addressed in this chapter in relation to the plasma-membrane Ca^{2+} pump.

6.2.1 Fluorescence Probes for PMCA

To detect single molecules of proteins, a bright fluorophore is needed, usually excited in the visible region of the spectrum. Most proteins are not naturally highly fluorescent, so a fluorophore must be introduced by fluorescence labeling of the protein molecule of interest. Fortunately, a number of methods exist for labeling proteins with fluorescence dyes [34], and the development of new techniques for fluorescence labeling is an emerging area of active research. For PMCA, one possible approach would be direct labeling of the protein. Typically, fluorescence labeling of proteins is accomplished by linking the fluorophore to the sulfhydryl group of a cysteine residue. We have avoided this method because PMCA has numerous Cys residues. As a result it would be challenging to select the location of the labeling site or even to control the number of sites labeled per molecule. We have rather used two alternative strategies to allow single-molecule detection of PMCA.

The first approach involves fluorescein isothiocynate (FITC), which is well known to bind preferably at Lys-591, located at the nucleotide binding site of PMCA (see Fig. 6.1) [35]. With FITC-labeled PMCA, single PMCA molecules can be detected, as we show below. However, this approach has the disadvantage that the presence of FITC in the nucleotide binding site interferes with the capacity of PMCA to bind its physiological substrate, ATP [36]. In the second method we have used for single-molecule detection of PMCA, a fluorescence label is attached to CaM, the Ca^{2+} sensor protein that binds to the PMCA and stimulates its activity, and the interaction of CaM with PMCA is detected. CaM has no native cysteine residues. It is therefore straightforward to insert one or two Cys residues by site-directed mutagenesis and then label these sites with maleimide derivatives of the fluorophore [37, 38]. PMCA–CaM complexes can then be detected by single-molecule spectroscopy so that the interactions with CaM that regulate PMCA can be probed.

6.2.2 *Detecting Signals from Single Molecules*

Single-molecule studies of biomolecules allow us to obtain a vast array of information pertaining to their dynamic behavior. However, the remarkable potential of this technique depends on finding ways to observe a given molecule for an extended period of time in order to track the full range of its dynamic behavior. Single-molecule fluorescence is detected with a fluorescence microscope. A schematic diagram of a confocal fluorescence microscope is shown in Fig. 6.3. Given a fluorescence-labeled protein, there are two general strategies for detecting signals from single molecules, differing according to whether the protein molecules are immobilized or allowed to diffuse freely. First, signals may be detected from immobilized molecules, sparsely distributed on a surface or in a medium. Alternatively, signals may be detected from molecules present at very low concentrations in solution as they diffuse through the detection region of a fluorescence microscope.

For many of the experiments reported in this chapter it was desirable to interrogate single PMCA molecules for relatively long periods of time, necessitating translational restriction of the molecule within the focal volume of the microscope for time periods ranging from seconds to minutes. In these situations, we have immobilized PMCA, either by trapping it in an agarose gel or by tethering it to a surface. PMCA reconstituted in detergent micelles can be immobilized in agarose gel (~2% by weight) [38, 39] where PMCA gets trapped in the gel pores. The smaller CaM (16.7 kDa) is highly mobile in agarose gels, so a fluorescently labeled

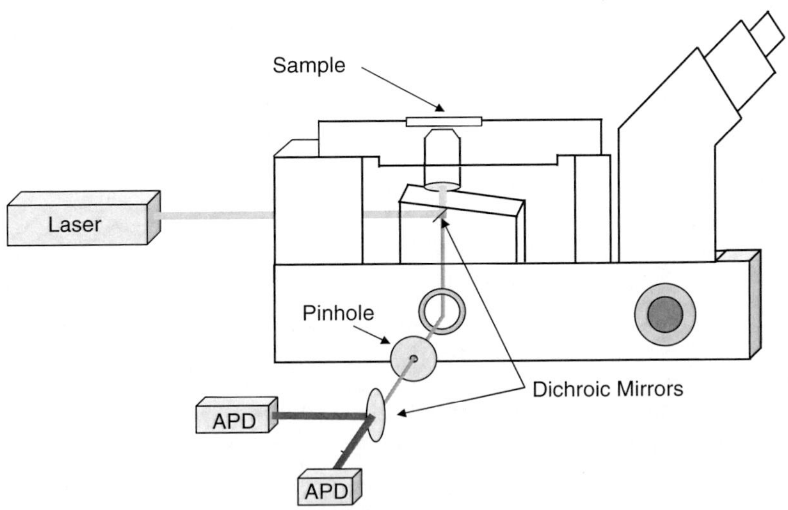

Fig. 6.3 Schematic diagram of the inverted confocal fluorescence microscope used in burst measurements. The excitation beam is directed to the objective lens by a dichroic mirror. Fluorescence is collected by the objective lens and directed through a confocal pinhole to a dichroic mirror, which separates green and red fluorescence. Signals are detected by avalanche photodiode (APD) detectors.

CaM will remain in the focal volume long enough for interrogation only if it is bound to PMCA. This discrimination between the bound and unbound CaM permits the detection of PMCA–CaM complexes and allows one to perform single-molecule experiments to probe the interaction between the two proteins.

Single-Molecule Burst Measurements of PMCA During fluorescence burst measurements, fluorescently labeled molecules diffuse through the ~2 femtoliter excitation volume of a laser beam focused by the objective lens of a microscope and emit short bursts of fluorescence that can be detected above the background noise level (see Fig. 6.4). Although burst analysis is a simple technique experimentally, the stochastic nature of burst data demands sophisticated methods of data analysis. Burst data are dominated by shot noise since bursts often consist of less than 100 photons, and each molecule can take random paths through the excitation/collection volumes. This results in a range of burst widths and intensities. Such transient signals are often analyzed by integrating spectroscopic signals over the duration of the burst [40]. From the integrated signals, various properties such as fluorescence lifetime or FRET efficiency can be calculated [41, 42].

PMCA in Ghost Membranes PMCA was first isolated from erythrocyte cell membranes [43] (also called "ghost membranes"), and erythrocyte cells continue to offer several advantages for the study of this protein. First, erythrocytes do not have any organelles such as nucleus, ribosomes, mitochondria etc. and so the membranes isolated from these cells are pure plasma membrane without any contamination coming from organellar membranes. In other cell types, known CaM-binding proteins such as phospholamban have been shown to reside in other membranes,

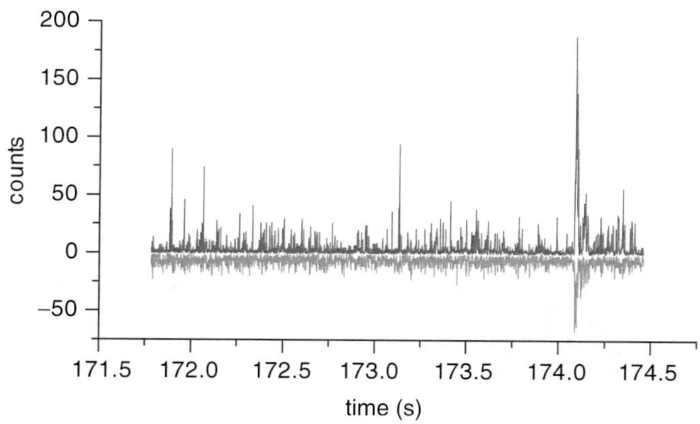

Fig. 6.4 A burst trajectory showing fluorescence bursts from PMCA-FITC and CaM labeled with Texas Red (TR). Bursts are visible from single molecules passing through the focal volume of the excitation beam. The signal in the green channel (FITC fluorescence) is plotted as positive-going bursts and the signal in the red channel (TR fluorescence) as negative-going bursts. In the segment of a burst trajectory shown, several short bursts are visible in the green channel, possibly originating from PMCA-FITC with no CaM bound or from free FITC. A longer, intense burst in both the green and red channels at ~174.1 s shows FRET in a PMCA-FITC-CaM-TR complex. Signal counts were binned by 1 ms for this analysis. Analysis of FRET from bursts for PMCA–CaM complexes is discussed further in Section 6.5 of this chapter.

complicating the plasma membrane isolation procedure. Only one other CaM-binding protein, namely spectrin, is known to exist in red blood cells [44]. This protein binds CaM only weakly and is removed with other cytosolic proteins in the preparation of the ghost membranes. This effectively eliminates any effect that spectrin may have on studies of CaM bound to PMCA. Ghost membranes are readily immobilized at low agarose gel concentrations, whereas CaM is highly mobile in these gels. This difference allows for the distinction between freely diffusing CaM and CaM bound to PMCA present in the immobilized ghost membranes at the 2% agarose gel concentration used in these experiments.

The overall concentration of PMCA within a typical 5 μm diameter ghost membrane was estimated to be many thousands of molecules. This density is too high to resolve single molecules of CaM-TMR bound one-to-one with PMCA. However, by adjusting the CaM concentration or by titrating labeled with unlabeled CaM, it was possible to reach levels where single molecules could be detected [45]. Figure 6.5a shows a ghost membrane saturated with CaM-TMR while Fig. 6.5b shows a single CaM-TMR molecule bound to the ghost membrane. To our knowledge, this was the first single-molecule detection of CaM bound to a target in a native membrane. However, the erythrocyte ghost membranes themselves were found to be moderately fluorescent over a wide range of excitation wavelengths. As a result, the use of ghost membranes resulted in high background levels from native fluorescence (autofluorescence), even with excitation in the red region of the spectrum. Hence, to carry out single-molecule studies in native membranes, approaches will be needed either to reduce autofluorescent background and/or to increase the brightness of the single-molecule probe. Therefore, in the work described in the rest of this chapter we have used PMCA extracted from the native membranes and purified in lipid micelles [46].

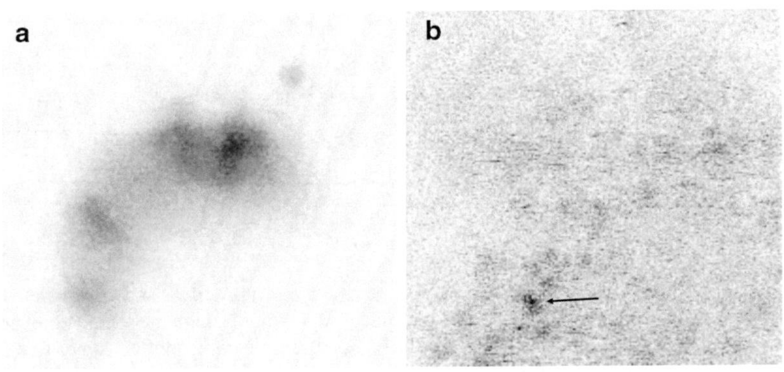

Fig. 6.5 (**a**) Fluorescence image of a ghost membrane with a saturating CaM-TMR concentration. (**b**) Fluorescence image of ghost membrane with CaM-TMR at a reduced concentration. The arrow points to a single PMCA–CaM–TMR complex. Note the high autofluorescent background. The scanned region is 10 × 10 μm. Adapted from ref. [45].

6.3 Calmodulin Binding Dynamics

One application of single-molecule spectroscopy is in the detection of protein–protein signaling interactions. As we show here, binding events can be detected in single molecules. As a result, single-molecule detection opens new experimental possibilities to investigate the interactions of signaling molecules with their substrate or regulatory molecules. One example is the possibility to monitor the kinetics of binding and dissociation at the single-molecule level. Measurement of kinetics by single-molecule detection can be useful for several reasons. First, with single-molecule detection, kinetics can be measured *at equilibrium*. At the practical level of experimental methodology, this means that a kinetic trigger (e.g. temperature jump, stop-flow, photolytic flash, etc.) is not necessary because the state of the molecule can be tracked by observing the equilibrium fluctuations in the state of the molecule. This is often a distinct advantage, not least because a much smaller sample size is required compared to conventional ensemble methods. More importantly, it may be possible to isolate individual steps in a complex kinetic mechanism. We show an example in Section 6.6 below in the detection of the rate of autoinhibitory domain interchange in PMCA.

In this section, we describe experiments to monitor binding of a single CaM molecule to a CaM-binding peptide called C28W, a 28-residue segment of the CaM-binding domain of PMCA [20, 47]. We previously found that fluorescence from CaM labeled with Alexa Fluor® 488 (CaM-AF488) is quenched upon binding to peptides like C28W that contain a single Trp residue in the binding domain [48]. This quenching can be exploited to detect binding events at the single molecule level. For these experiments, it was necessary first to immobilize CaM-AF488 so that single CaM molecules could be observed for sufficiently long periods of time. As pointed out above, CaM itself cannot be trapped in an agarose gel [37], so for the purpose of these studies we constructed a fusion protein that contained maltose binding protein (MBP) fused to the N-terminus of CaM by a flexible linker (MBP-T34C-CaM). MBP-CaM was labeled with AF488 at the 34 position of CaM. This fusion protein is readily immobilized in agarose [48], suggesting that MBP serves as an anchor to restrict the translational mobility of CaM in the agarose gel. Thus unlike other experiments described in this chapter where PMCA is trapped in a gel and CaM diffuses freely, here CaM is trapped in the gel by anchoring to MBP, and the CaM-binding domain C28W diffuses freely.

The dissociation of C28W from PMCA occurs on a time scale of tens of seconds. At the excitation rate required to detect single-molecule signals, AF488 photobleaches on the time scale of seconds. It was therefore necessary to prolong the observation time of single immobilized molecules for tens of seconds without sacrificing signal to background. Therefore, to follow the single molecules over longer time periods, an electronic shutter was opened for a brief period, (generally tens to hundreds of milliseconds) and then closed for a period of a second or more. This cycle was repeated. An example of a trajectory obtained in this way is shown in Fig. 6.6a with periods of quenched fluorescence. It was very rare to observe

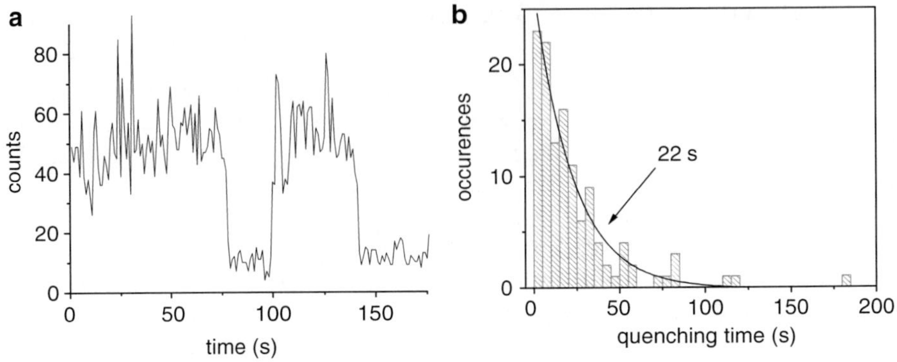

Fig. 6.6 (**a**) Example of a single-molecule trajectory of CaM-AF488 binding to C28W. (**b**) Histogram of "off" (quenching) times showing the persistence of peptide binding. The black solid line shows a fit to a single exponential with a time constant of 22 s. The MBP-CaM-AF488 fusion protein was immobilized in ~2.0% agarose gel, pH 7.4, 100 mM KCl, 100 μM $CaCl_2$, and 1.0 mM $MgCl_2$ as described in [37]. The MBP-CaM-AF488 concentration in the gel was 5 to 10 nM. The C28W concentration was 100 nM to 2.0 μM. Adapted from ref. [50].

prolonged periods of quenched fluorescence in the absence of peptide. Figure 6.6b shows a histogram of the duration of fluorescence quenching events for over 100 molecules that exhibited this behavior in the presence of C28W. Quenching reveals binding of the peptide C28W to CaM. Therefore, the histogram of quenching durations is indicative of the duration of CaM-C28W binding, and this can be analyzed to determine the C28W off-rate. The dissociation time constant (22 s) corresponds to an off rate of 0.045 s^{-1}, consistent with an intermediate component of the dissociation observed in previously reported kinetic measurements for the C28W binding domain of PMCA1b [49]. (A long component of the dissociation, ~270 s [49], would not have been observable in the typical trajectory lengths, and a shorter component of ~3 s would not be resolved given the binning time for the data in Fig. 6.6). Much larger data sets would be required to resolve multiple relaxation components. Single-molecule measurements such as the ones in Fig. 6.6 monitor binding and dissociation kinetics at equilibrium and require far smaller sample quantities than ensemble methods such as stop-flow experiments.

6.4 Conformational States of PMCA–CaM Complexes: Single-Molecule Polarization Modulation Measurements

The ability to detect PMCA–CaM complexes in agarose gels as described above paves the way for experiments to probe the conformations of these complexes. The objective is to relate conformations and conformational dynamics to the regulatory function of CaM bound to PMCA. To obtain conformational information, it is clearly not sufficient merely to detect single molecules. Spectroscopic methods sensitive to conformation need to be applied at the single-molecule level.

For experiments on PMCA we have used two such methods, one using modulation of the polarization of the exciting light and the other using resonance energy transfer. We will discuss each of these in turn, polarization modulation measurements in this section and energy transfer in Section 6.5.

Previous studies of PMCA had shown that the activity of the enzyme is related to conformations of the C-terminal autoinhibitory domain of the enzyme [21]. Due to the inherent heterogeneity of biological systems, it was possible that significant information pertaining to the regulatory mechanism had been hidden in bulk studies as a result of averaging over many molecules in the population. In fact, as we showed, a heterogeneous distribution of autoinhibitory domain conformations may exist with differential interactions with the catalytic core of the enzyme. Under these circumstances, the power of single molecule spectroscopy to reveal subpopulations of a heterogeneous distribution can play a major role in resolving these intermediate conformations.

6.4.1 Orientational Mobility Measurements

The first experimental method we used to probe conformations of the autoinhibitory domain involves modulation of the polarization of light. Polarization provides a link between the laboratory axis system and the molecular axis system. Light is polarized in a defined direction in a lab-fixed axis system (e.g. vertical). The link with molecular orientation arises because the probability of light absorption depends on the angle between the transition dipole of the dye and the polarization direction. When these directions are parallel, the excitation rate is a maximum, and when they are perpendicular the excitation rate is zero. If the molecule reorients freely and rapidly, then the polarization direction does not matter because the dye rapidly samples all orientations with respect to the polarization. On the other hand, if the molecule is fixed, then the excitation rate depends on the relative orientation of the dye transition dipole and the polarization. We exploited this dependence by rotating the excitation polarization, typically at 25 revolutions per second (see Fig. 6.7). If the molecule is highly orientationally mobile, then the excitation rate is changed little by rotating the polarization. However, for a molecule that is relatively orientationally immobile, the excitation rate, and therefore the fluorescence signal, is modulated by the polarization direction. The orientational mobility of molecules can therefore be measured by tracking the single-molecule fluorescence to determine if it is modulated by a rotating excitation polarization. We applied this method to CaM labeled with the fluorophore tetramethylrhodamine (CaM-TMR). Labeled CaM was allowed to bind to PMCA. Time-resolved fluorescence anisotropy measurements in our lab (J.R. Unruh et al., unpublished data) showed that TMR is orientationally coupled to the protein with nearly 2/3 of the anisotropy decay amplitude corresponding to reorientation of CaM rather than independent reorientation of the fluorophore alone. Thus the mobility of TMR reports directly on the mobility of the CaM to which it is attached. Because of

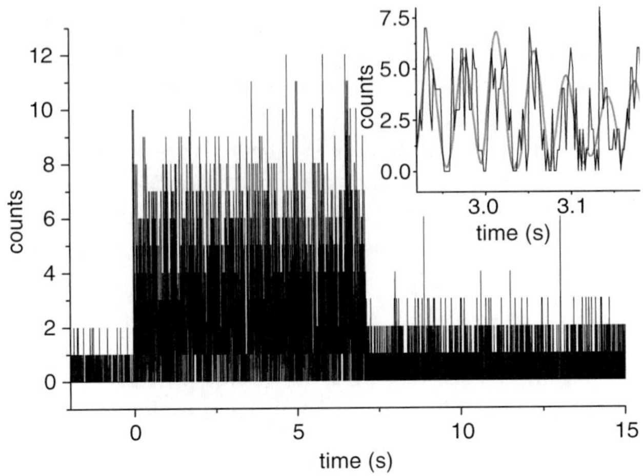

Fig. 6.7 Typical polarization modulation trajectory of CaM-TMR bound to porcine PMCA immobilized in an agarose gel. The excitation was turned on at 0 s, and photobleaching occurred at approximately 7 s. The inset shows modulated fluorescence due to polarization rotation at 25 Hz. The solid gray line is a fit period by period to the modulated signal based on a maximum likelihood estimator [38]. TMR was excited with a green He-Ne laser at 543 nm.

these factors, polarization modulation probes the orientational motion CaM bound to the autoinhibitory domain of PMCA. A maximum likelihood algorithm [38] was used to determine the extent of modulation (or the *modulation depth*) of the fluorescence. We define modulation depth as the ratio of the amplitude of modulation (above or below the average amplitude) to the average signal [38]. Thus a modulation depth of 1 represents maximum modulation of the signal. Figure 6.7 shows an example of a polarization modulation trajectory. The modulated signal can be fit to yield the modulation depth for each modulation period. The inset in Fig. 6.7 shows a fit to modulations with a high modulation depth.

Porcine PMCA PMCA purified from fresh porcine blood was studied by single-molecule polarization modulation to determine the Ca^{2+} dependence of the autoinhibitory domain mobility. The mobilities of individual molecules were combined in a histogram for all of the data taken at a minimally activating Ca^{2+} concentration of 1 μM. These data are shown in Fig. 6.8a. A bimodal distribution was observed with the dominant population centered at a modulation depth of 0.35 and the minor population centered at 0.48. Based on the unpolarized background in these experiments and the depolarization due to the high numerical aperture objective, the PMCA–CaM population centered at a modulation depth of 0.3–0.4 can be described as "mobile", and the population at ~0.5 can accurately be described as "immobile."

To see if the relative populations of these high and low mobility populations were sensitive to stimuli relevant to PMCA function, we tested the Ca^{2+} dependence of the modulation depth distribution. Holding all other conditions the same, the Ca^{2+} concentration was increased to a saturating Ca^{2+} concentration of 100 μM

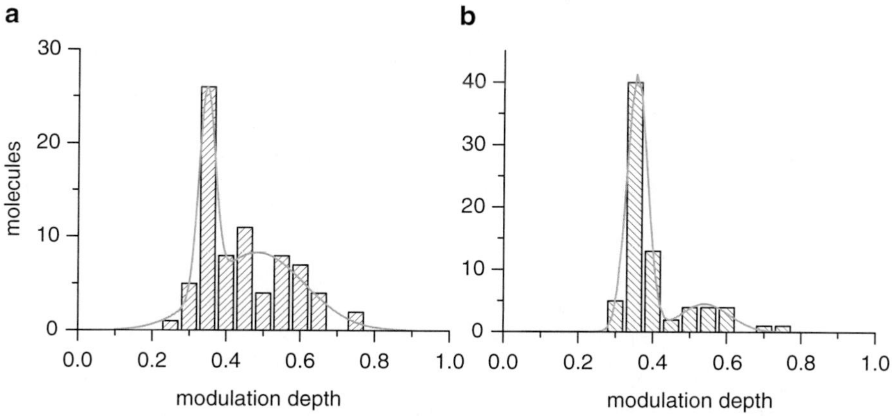

Fig. 6.8 Modulation depth histograms for CaM-TMR bound to porcine PMCA immobilized in agarose gel. The Ca^{2+} concentration was 1 μM (*panel A*) and 100 μM (*panel B*). The lines show fits to a double Gaussian function with peaks at 0.35 and 0.48 (**a**) and 0.36 and 0.54 (**b**).

(see Fig. 6.8b). A dramatic shift in mobilities of CaM-TMR ensued, with a decrease in the population having a high modulation depth (≥0.5). The single-molecule polarization modulation experiments thus show that there are Ca^{2+}-dependent orientational mobility populations of CaM bound to PMCA.

How can we understand these results? The orientational mobility of CaM bound to the autoinhibitory domain could relate to (a) conformations of CaM itself, with Ca^{2+}-dependent changes in the orientational mobility of CaM, or (b) conformations of the autoinhibitory domain of PMCA to which CaM is bound. Experiments described below strongly support the second possibility, correlating the orientational mobility of CaM-TMR bound to PMCA with conformations of the autoinhibitory domain. This suggests a Ca^{2+}-dependent shift in the conformation of the autoinhibitory domain with bound CaM. As discussed below, this interpretation, if correct, is inconsistent with the view of the autoinhibitory domain as a two-state, on-off switch with the pump in the "off" (autoinhibited) state in the absence of CaM and in the "on" (activated) state when CaM is bound to the autoinhibitory domain. To investigate the nature of these populations further, we turned to PMCA purified from human ghost membranes.

Human PMCA The body of work on human PMCA has been far more extensive than that on porcine PMCA. We therefore also carried out polarization modulation experiments on PMCA purified from human red blood cells. The modulation depth histograms are shown in Fig. 6.9. At a minimally activating Ca^{2+} concentration of 0.15 μM (Fig. 6.9a), the modulation depth histogram for human PMCA, like that of porcine PMCA, shows a high-mobility population (modulation depth ~0.4) and a low-mobility population (modulation depth ~0.6). At a high Ca^{2+} concentration (25 μM Ca^{2+}) a dramatic shift in the populations at high and low modulation depth was observed. As for porcine PMCA, the population of the high-modulation state nearly disappeared at the higher Ca^{2+} concentration (Fig. 6.9b).

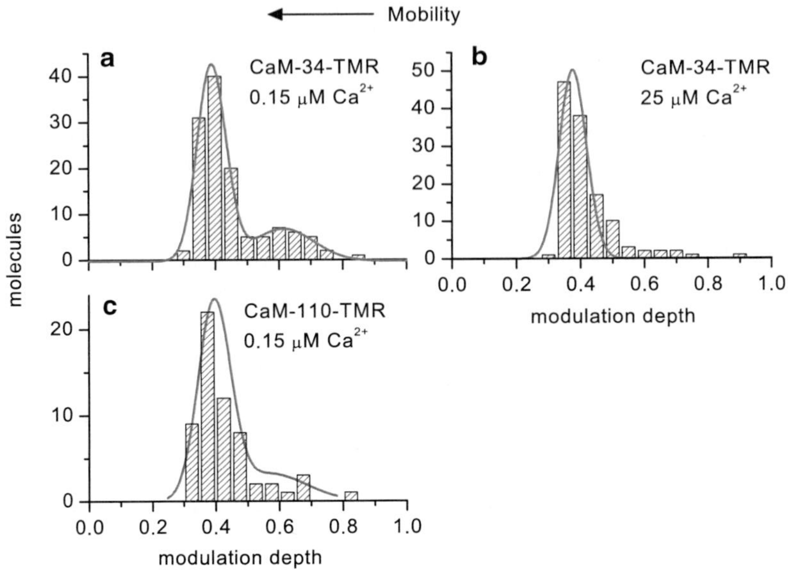

Fig. 6.9 Modulation-depth histograms for human PMCA. The modulation depths were averaged over modulation periods for individual molecules. The line shows fits to Gaussian distributions. Modulation depth distributions are shown for PMCA at 0.15 μM Ca^{2+} (*panel A*) and 25 μM Ca^{2+} (*panel B*) for CaM labeled on the N-terminal domain (CaM-34-TMR). *Panel C* shows the modulation depth distribution for CaM labeled on the C-terminal domain (CaM-110-TMR). (**a**) and (**b**) are adapted from ref. [39].

6.4.2 Model for PMCA–CaM Orientational Mobility States

Work by Squier and coworkers demonstrated that fluorescently labeled CaM bound to the autoinhibitory domain of PMCA exhibits a rotational correlation time much shorter than expected for the whole PMCA enzyme, consistent with segmental motion of the autoinhibitory domain [51]. Thus, we expected that a dissociated autoinhibitory domain would exhibit high orientational mobility and therefore low modulation consistent with the low modulation depth observed at a high Ca^{2+} concentration. The increase in population of the immobile (high modulation depth) state, in combination with the known decrease in activity upon decrease in Ca^{2+} concentration, suggests that the immobile population state is involved with self-inhibition of PMCA through binding of the autoinhibitory domain to the enzymatic core. If the orientationally mobile state corresponds to CaM bound to a dissociated autoinhibitory domain, as suggested above, we proposed that the less orientationally mobile state corresponds to molecules where the CaM-bound autoinhibitory domain of PMCA is not released from the enzymatic core [39]. This model is depicted in Fig. 6.10. Binding of CaM to PMCA with the autoinhibitory domain in both closed (non-dissociated autoinhibitory domain) and open (associated autoinhibitory domain) forms was also proposed by Caride and co-workers in

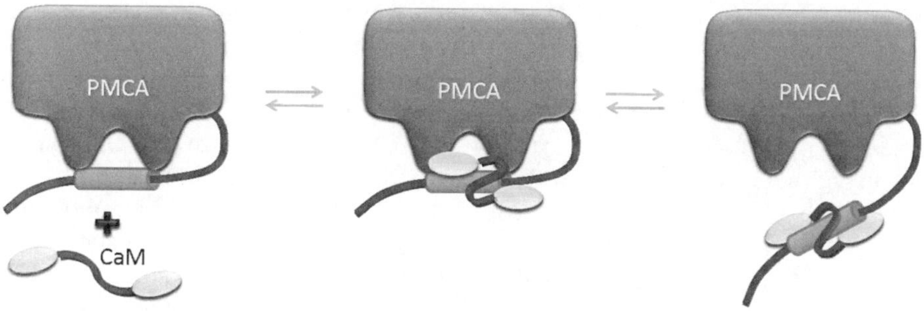

Fig. 6.10 Model for CaM activation of PMCA. The *left panel* shows PMCA with the autoinhibitory/CaM-binding domain bound near the enzymatic core. The *central panel* shows the state proposed to correspond to the low-mobility population, with CaM bound but the autoinhibitory domain not dissociated. In the *right panel*, CaM is bound and the autoinhibitory domain is dissociated, representing the population with high orientational mobility. Adapted from ref. [39].

a kinetic model for CaM binding to PMCA [52]. The presence of both high and low-mobility populations suggests that some PMCA molecules experience self-inhibition by a non-dissociated autoinhibitory domain, even while CaM is bound.

The results described give new insight into the mechanism of regulation of PMCA by CaM. The observation of two populations with different orientational mobilities is inconsistent with a two-state model (i.e., the left and right panels but without the center panel in Fig. 6.10). It is the dependence of the relative population on the Ca^{2+} concentration that demonstrates the inadequacy of such a model for PMCA activation by CaM. As the Ca^{2+} level was decreased, the fraction of immobile state increased. If the two-state model were correct, then a decrease in Ca^{2+} concentration would simply lead to a decreased number of CaM molecules bound to PMCA, but the conformations of PMCA with CaM bound would not be altered. Since we only detected PMCA with CaM bound, the nature of the distributions would not be affected. In contrast, the data shown in Figs. 6.8 and 6.9 show a shift in population towards lower mobilities at decreased Ca^{2+} concentration. The Ca^{2+}-dependent change in the orientational mobility distribution thus suggests a more complex picture, with at least two conformations of the autoinhibitory domain when CaM is bound, one with high orientational mobility favored at a high Ca^{2+} concentration and one with low orientational mobility that becomes more populated at a lower, more physiologically relevant Ca^{2+} concentration. This suggests that the PMCA regulatory state model be modified to include a state with CaM bound to PMCA with a non-dissociated autoinhibitory domain (Fig. 6.10, center panel). A state with CaM bound to PMCA but without dissociation of the autoinhibitory domain may function to fine-tune the regulation of PMCA by CaM. This state with the autoinhibitory domain bound would likely not be active beyond the basal activity of the pump and would be well situated to respond rapidly to Ca^{2+} influx into the cell, thus playing an important regulatory role.

An interesting question is whether the C-terminal and N-terminal domains of CaM experience the same orientational mobility when bound to PMCA. The N-terminal domain of CaM has a lower Ca^{2+} affinity than the C-terminal domain in the presence of a synthetic peptide corresponding to the CaM-binding region of the autoinhibitory domain [53]. We therefore carried out polarization modulation experiments with CaM labeled at the 110 site (CaM-110-TMR) for comparison with the results described above for CaM labeled at the 34 site (CaM-34-TMR). The Ca^{2+} concentration in both cases was 0.10–0.15 μM. Figure 6.9c shows the result of these measurements with CaM-110-TMR. As with CaM-34-TMR, there is a high-mobility population with modulation depths of 0.3–0.5, and a low-mobility population with modulation depths of 0.6–0.8. The distributions for CaM-34-TMR and CaM-110-TMR are not significantly different. This result suggests that the low-mobility population consists of CaM bound to the nondissociated autoinhibitory domain by both its C-terminal and N-terminal binding domains, as depicted in Fig. 6.10.

6.4.3 ATP-Dependent Changes of the PMCA–CaM Complex

Binding and hydrolysis of ATP by PMCA are coupled to the transport of Ca^{2+} across the plasma membrane [54]. To study a functional enzyme, it is therefore necessary to study the PMCA–CaM complex in the presence of its substrate ATP. As for other P-type ATPases, binding and hydrolysis of ATP in the nucleotide-binding site of PMCA drive structural motions of the enzyme that result in the transfer of phosphate to a highly conserved Asp residue in the phosphorylation site of PMCA. These events are coupled to motions of the transmembrane domains of the pump that alternatively expose the PMCA transmembrane Ca^{2+}-binding site to the interior and the exterior of the cell, with the result that Ca^{2+} is transported from the cytoplasmic side of the membrane and released outside the cell. The motions involved have been characterized by structural studies of the homologous SERCA Ca^{2+} pump [33].

This raises the question of how these structural changes are related to the interaction of the enzymatic core with the CaM-bound PMCA autoinhibitory domain. To address this question, we applied single-molecule polarization modulation to the mobility of the autoinhibitory domain of the PMCA–CaM complex in the presence of ATP. Figure 6.11 shows the average orientational mobility of over 200 molecules at both 0.15 μM (Fig. 6.11a) and 25 μM Ca^{2+} (Fig. 6.11b) in the presence of ATP. Unlike the results for PMCA in the absence of ATP, only the high-mobility population appears in the presence of ATP, and there is little or no Ca^{2+} dependence in the modulation depth population. At a high Ca^{2+} concentration the distribution is only slightly different from that observed in the absence of ATP, the most notable difference being the suppression of any low mobility population (modulation depth >0.5). At a more physiological Ca^{2+} concentration of 0.15 μM, however, there is no increase in the immobile fraction of PMCA–CaM complexes at reduced Ca^{2+} (Fig. 6.11a), unlike PMCA–CaM in the absence of ATP.

These results show that the presence of ATP changes the coupling between the autoinhibitory domain and the catalytic regions of the enzyme. Is this a consequence

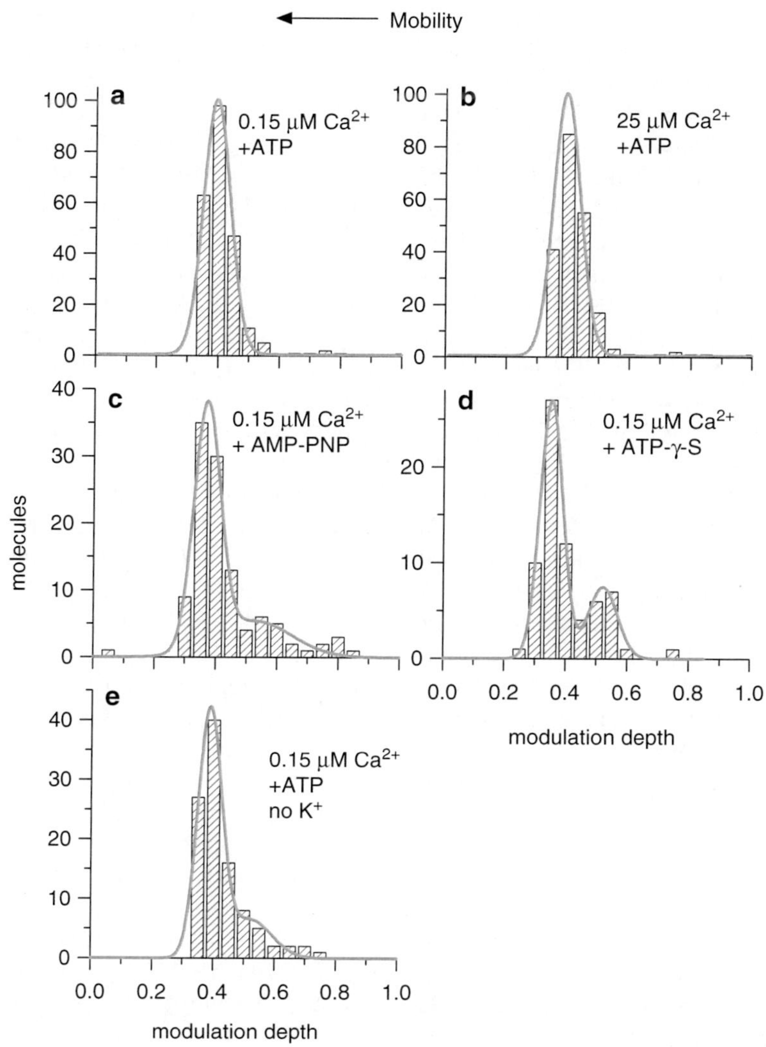

Fig. 6.11 *Top panels*: modulation depth histograms for PMCA with ATP (1 mM). The Ca^{2+} concentration was 0.15 μM (**a**) and 25 μM (**b**). The *middle panels* show modulation depth distributions measured with the non-hydrolyzable ATP analogs AMP-PNP (**c**) and ATP-γ-S (**d**), both 1 mM. The *bottom panel* (**e**) compares the modulation depth distribution in the absence of K$^+$ with 1 mM ATP, 0.15 μM Ca^{2+}, and 0.3 mM Mg^{2+}. *Panels A–C* are adapted from ref. [39]. *Panel E* is adapted from ref. [45].

of ATP binding itself or of subsequent enzymatic events, for example phosphorylation of an Asp residue in the P domain? To answer this question we measured single-molecule rotational mobilities of PMCA–CaM in the presence of non-hydrolyzable ATP analogs. These analogs bind to PMCA but do not support completion of the enzymatic cycle. Figure 6.11 shows the orientational mobility of PMCA–CaM in the presence of the non-hydrolyzable nucleotides ATP-PNP and ATP-γ-S. Both AMP-PNP and ATP-γ-S are widely used ATP analogs [55]. Unlike the effect of ATP, where only the high-mobility population was observed (Fig. 6.11, top panels), when the non-hydrolyzable ATP analog AMP-PNP was substituted for ATP a low-mobility state appeared at the reduced Ca^{2+} concentration (Fig. 6.11c,d) [39], as it did in the absence of ATP. It could be that the bent geometry of AMP-PNP renders it a poor analog of ATP. In AMP-PNP, the O atom between the β and γ phosphates of ATP is replaced by an N atom, which has a different coordination geometry. Therefore, we also measured single-molecule polarization modulation distributions for PMCA in the presence of 1 mM ATP-γ-S. Unlike AMP-PNP, in ATP-γ-S the β,γ-O is maintained. The results for ATP-γ-S, also shown in Fig. 6.11, are very similar to the modulation-depth distribution obtained with AMP-PNP [39], confirming that occupation of the nucleotide binding site alone does not explain the contrast in orientational mobility distributions that were observed for PMCA–CaM complexes in the presence versus the absence of ATP.

Binding and utilization of ATP by PMCA requires the presence of other ions such as Mg^{2+} [56]. Previous work by Adamo and co-workers [57] revealed that both K^+ and Mg^{2+} affect the activity of PMCA. Present at physiological levels near 100 mM, K^+ is necessary to achieve native enzymatic rates in the PMCA–CaM regulatory system. The presence of physiological amounts of K^+ concentrations was found to increase the population of phosphorylated PMCA [57]. By removing K^+ and limiting the Mg^{2+} concentrations in the buffer conditions containing 1 mM ATP, a partial recovery of the immobile state of PMCA was observed at 0.15 μM Ca^{2+} (Fig. 6.11e). These data suggest an increase in the presence of an inactive but CaM-bound state of PMCA in the absence of K^+. Under these conditions, the autoinhibitory domain reassociates with the nucleotide-binding site more readily leading to increased self-inhibition of the pump. This result thus implies a structural coupling between K^+ and the binding and/or hydrolysis of ATP. The recovery of the immobile state in the absence of K^+ suggests an important role for K^+ in the interaction between the autoinhibitory domain and the catalytic region of PMCA.

The role of ATP binding and hydrolysis in driving PMCA in active Ca^{2+} transport has been well documented [54]. The ATP-dependence of the autoinhibitory domain mobility and thus its likely role in the PMCA self-inhibition mechanism had not previously been reported and suggests new insights into the functional relationships of the various PMCA–CaM ligands on the activity of the pump. Typically the role of the autoinhibitory domain has been described in terms of preventing ATP binding and/or hydrolysis. Our results show that the there is also a reciprocal effect of ATP binding on the autoinhibitory domain conformation, that the coupling between the autoinhibitory domain and the catalytic site appears to be altered in the cycling enzyme. The reduction in the population of the immobile state when ATP is present suggests that conformational changes in the cycling enzyme limit association of the autoinhibitory domain with the nucleotide-binding site. The return of the immobile state in the presence of non-hydrolyzable ATP analogs further suggests that these enzymatic steps are blocked by binding of AMP-PNP or ATP-γ-S. Self-inhibition by the autoinhibitory domain requires the interaction of the PMCA autoinhibitory domain with two sites in the N domain near the nucleotide-binding site of the pump [58, 59] and at least one more site that may be in the "actuator" (A) domain of PMCA. The modulation depth distribution in the presence of ATP suggests that in the presence of ATP, PMCA adopts conformations that are not conducive to binding of the autoinhibitory domain to the enzymatic core, perhaps as a result of conformational changes in these domains.

6.4.4 *Effect of Oxidative Modification*

Oxidative modification of either PMCA [60] or CaM [61, 62] is known to result in reduced PMCA activity. Previous work has shown that PMCA is extremely susceptible to oxidative stress as observed in a variety of experimental paradigms [60, 63, 64]. Exposure of synaptic plasma membranes to oxidants of physiological relevance such as H_2O_2, peroxynitrite and azo initiators (peroxyl radical generating agents) resulted in rapid loss of the Ca^{2+}-activated ATPase activity of PMCA [63]. Decreased activity was due to a lowered V_{max} with no significant change in K_{act} or affinity for Ca^{2+}. To determine the effect of oxidative stress on PMCA present in its native cellular environment with its array of antioxidant defense mechanisms, recent experiments showed that exposure of primary cortical neurons to the superoxide generating agent paraquat inhibited PMCA activity [64]. The marked loss of PMCA activity coincided with the formation of PMCA aggregates apparently resulting from oxidation of cysteine residues and formation of intermolecular disulfide bonds. These results suggest that disulfide mediated PMCA aggregation is one mechanism for PMCA inactivation. More interestingly, the paraquat-treated cells showed evidence of PMCA proteolysis as evidenced by the appearance of PMCA fragments in immunoblots. Overall, this series of studies shows that PMCA undergoes functional and structural changes under relatively mild conditions of oxidative stress. These observations could be relevant in the lowered PMCA activity observed in the aging brain, a condition associated with elevated levels of oxidative stress [65, 66]. For CaM, exposure to oxidants such as peroxide results in oxidation of methionine residues [67]. Work in the Squier lab has shown specifically that it is oxidation of Met-144 and Met-145 that blocks activation of PMCA by CaM [68].

To investigate the mechanistic causes of activity loss, we carried out experiments to probe the effect of oxidative damage to CaM [69] or PMCA [70] on the distribution of orientational mobilities of PMCA–CaM complexes. In the absence of ATP, the orientational mobility populations for PMCA and oxidatively modified PMCA (PMCAox) were nearly the same, indicating that coupling of the autoinhibitory domain with the enzymatic core is unaltered. However, a different picture emerged in the presence of ATP. Results for PMCAox are shown in Fig. 6.12. The most salient finding was the appearance of a low-mobility (high modulation depth) distribution in the presence of ATP for PMCAox, even at high Ca^{2+} (25 μM), in contrast to what we had observed for native PMCA (Fig. 6.9). Thus, the structural

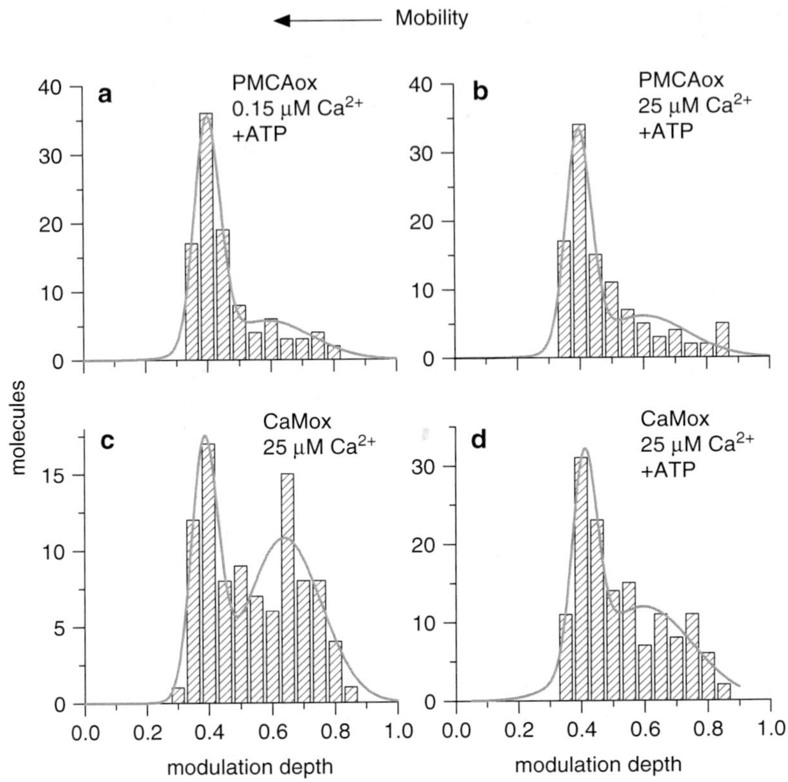

Fig. 6.12 *Top panels*: Modulation depth histograms of CaM-TMR bound to PMCAox in the presence of 1 mM ATP at 0.15 μM (**a**) and 25 μM Ca^{2+} (**b**). The increased low-mobility population correlates with the decreased activity of PMCAox. *Bottom panels*: Modulation depth histograms of CaMox-TMR with 25 μM Ca^{2+} in the absence (**c**) and presence (**d**) of 1 mM ATP. The increase in the population of the immobile state shown for CaMox-TMR with ATP present compared to PMCA bound to unoxidized CaM-TMR corresponds well with the observed decrease in bulk activity for CaMox [69].

coupling between ATP utilization and autoinhibitory domain dissociation appears to be disrupted in PMCAox. These findings suggest that the loss of activity resulting from oxidative damage is associated with a reduced dissociation of the autoinhibitory domain even when ATP is present.

Other experiments in our laboratory probed the effect of oxidative damage to CaM on the distribution of orientational mobilities of PMCA–CaM complexes [69]. Oxidation of CaM (CaMox) leads to reduced stimulation of PMCA, but the binding affinity of CaMox for PMCA is nearly the same as that for unoxidized CaM [71]. Our single-molecule polarization modulation experiments show that oxidation produces a significantly higher population of PMCA–CaMox complexes with low orientational mobility (Fig. 6.12c, d) compared to the results for unoxidized CaM (Figs. 6.9 and 6.11). Even in the presence of ATP, a significant fraction of PMCA–CaMox complexes populate a low-mobility state, suggesting that oxidative modification of CaM results in a loss of its ability to induce dissociation of the autoinhibitory domain [69]. Thus, the mechanism underlying the loss of potency for CaMox to activate PMCA is not a loss in binding affinity, but rather a loss in the effectiveness of binding. Squier and co-workers showed that oxidation of methionines 144 and 145 is responsible for activity loss [68]. Given that these residues are involved in binding of CaM to PMCA, their oxidation may result in alterations in binding interactions that are crucial for inducing conformational change in PMCA.

6.5 Single-Molecule FRET Measurements of PMCA

The polarization modulation measurements described above revealed populations of conformational states of PMCA–CaM complexes. However, detailed information about the nature of the conformational states is lacking from orientational mobility measurements. For structural information, we turn to Förster resonance energy transfer (FRET), a spectroscopic method that can provide detailed structural information [72–74]. FRET is a widely used to measure the efficiency of energy transfer from an excited fluorophore (the "donor") to a second fluorophore (the "acceptor"). It can be used effectively (with some caveats [73]) to measure distances on the scale of tens of Ångstroms. At the single-molecule level, single-pair FRET (spFRET) [75] has emerged as a powerful method for tracking dynamics and mechanisms in biomolecules [17, 76–83].

The approach we used for our PMCA FRET study is based on fluorescence burst detection. Burst measurements are more convenient and faster than methods based on immobilized molecules because single-molecule signals are obtained from molecules as they pass through the focal region of the microscope. Single-molecule resolution is achieved by use of a small confocal excitation volume defined by the microscopic objective and confocal pinhole, together with a low enough concentration (a few hundred picomolar) so that the probability of finding two molecules in the focal volume simultaneously is extremely low.

There are some practical caveats that must be heeded in single molecule burst measurements of autoinhibitory domain conformations. First, the loss of protein at very low concentrations due to adsorption onto the surface of the cover slip can be significant and would have a detrimental effect on the measurement. Surface adsorption not only disturbs the equilibrium between PMCA and CaM but also reduces the number of bursts that can be collected for a given sample. In order to prevent protein adsorption, the cover slips were treated with BSA to form a hydrophobic coating on the glass surface. Another limitation is that mixing of PMCA and CaM at a picomolar level is not conducive to formation of a complex since the dissociation constant of PMCA–CaM complex is several orders or magnitude higher (7 nM [71]). Since the concentration of each species is well below the dissociation constant, molecules in complex form at equilibrium would be very rare. However, taking the advantage of the long off rate of CaM from the CaM-PMCA complex (~0.015 s^{-1} for PMCA 4b [30, 84]), we devised a strategy to mix

CaM and PMCA at higher concentrations (~10 nM) by incubating the mixture for a period of time sufficient for complex formation, followed by a rapid dilution to the desired picomolar level immediately before taking burst measurements. Since burst collection requires only a few minutes to collect fluorescence from hundreds of molecules, the probability of finding CaM-PMCA complex molecules in the focal volume is still high because the off rate is longer than the data collection time. Therefore we were able to collect single molecule bursts from single CaM-PMCA molecules as shown in Section 6.2 (Fig. 6.4).

We labeled PMCA with FITC as described in Section 6.2 and then measured the efficiency of FRET between FITC (the donor) bound at the active site of PMCA and CaM labeled with Texas Red (TR, the acceptor) bound to the autoinhibitory domain. As described above (Section 6.2.1), FITC preferentially labels Lys-591 in the nucleotide binding site of PMCA. With single-molecule detection the method reports the distribution of single-molecule distances between the autoinhibitory and ATP-binding domains rather than merely the average distance over the entire sample. Single-molecule FRET results are shown in Fig. 6.13. These results demonstrate that single-molecule FRET is a sensitive probe of the distribution of conformations between CaM bound to the autoinhibitory domain and the nucleotide binding site of PMCA. These methods can therefore be used to probe the interaction between the CaM-bound autoinhibitory domain and the catalytic core of the enzyme to elucidate the mechanisms of regulation of PMCA.

The FRET results bolster our interpretation of the polarization modulation results presented above while adding important details. The results of the single molecule burst integrated FRET measurements we obtained for PMCA-FITC bound to CaM-TR reinforce the three-state model for regulatory mechanism of PMCA. At a Ca^{2+} concentration of 0.15 μM (Fig. 6.13a) there is a large population of complexes showing high FRET (FRET efficiency greater than 0.8) and therefore close proximity between the autoinhibitory domain and the nucleotide-binding domain. The large population of high FRET efficiencies is consistent with the high probability of an associated autoinhibitory domain detected in our previous polarization modulation experiments at submicromolar Ca^{2+} levels (Fig. 6.9a). There is also an increased population with interactions at intermediate FRET efficiency (0.4–0.8). The range of FRET efficiencies observed further suggests that at this Ca^{2+} concentration a heterogeneous population is present characterized by a rather wide range of separations between the nucleotide binding site and CaM, with apparent FRET efficiencies ranging from 0.4 to >0.9. Very interestingly at the low Ca^{2+} level, the appearance of states with energy transfer efficiency of ~0.4 may indicate conformations of the autoinhibitory domain that are only loosely bound or partially dissociated yet still in the proximity of the nucleotide binding site. Comparison with the high-Ca^{2+} result (upper right panel) suggests that this intermediate FRET state is favored at a physiological Ca^{2+} concentration but not at high Ca^{2+}. The measurements presented here do not discriminate between PMCA–CaM complexes with low FRET efficiency and PMCA without bound CaM. Thus, the population with a FRET efficiency $E < 0.4$ may comprise (1) FITC-PMCA with no bound CaM, (2) free FITC remaining in the sample; (3) conformations of

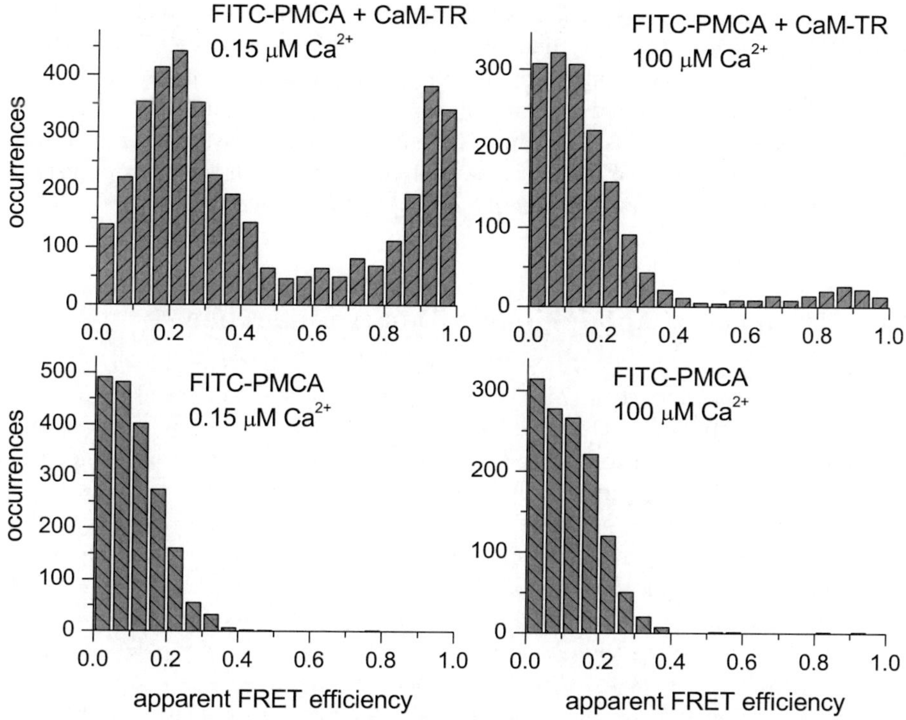

Fig. 6.13 *Top panels*: Single-pair FRET in PMCA-FITC-CaM-TR complexes. In the absence of CaM (*lower panels*) only PMCA-FITC was detected. Occurrences of apparent FRET efficiencies greater than 0 in the absence of CaM (*bottom panels*) result from donor cross talk and background counts in the acceptor channel. In contrast, the *top panels* show FRET in the presence of CaM. FRET efficiencies were measured in 1-ms bins in fluorescence bursts detected from freely diffusing complexes. The Förster radius for this dye pair is 52 Å. PMCA was labeled with FITC in the membrane and then purified following published procedures [35]. FITC-PMCA was excited at 488 nm. The signal from direct excitation of Texas Red at 488 nm was negligible. The apparent FRET efficiencies were not corrected for background or cross-talk between channels.

PMCA–CaM complexes with low FRET efficiency. PMCA–CaM complexes that have dissociated contribute to the donor-only FRET signal.

The single molecule FRET studies display a Ca^{2+} concentration dependence consistent with single molecule polarization modulation studies. At high Ca^{2+} concentration (upper right panel) there is only a small population of PMCA–CaM complexes with FRET efficiencies above 0.4, indicating low energy transfer consistent with a separation of at least tens of Ångstroms between CaM bound to the autoinhibitory domain and FITC at the nucleotide binding site. The absence of a larger population of high FRET states at elevated Ca^{2+} adds structural information about the dissociated autoinhibitory domain. The FRET results thus support our interpretation of the high-mobility occurrences as complexes where the autoinhibitory domain is dissociated and therefore not in close proximity to the nucleotide binding site. A comparison with the results from PMCA-FITC alone in the absence of CaM shown in the lower right panel suggests a subtle difference that may result from PMCA–CaM complexes that have a low but non-zero FRET efficiency between FITC and TR. A small population of complexes also appears with $E > 0.4$, indicating conformations where the autoinhibitory domain and nucleotide binding domains are in close proximity (compare control measurements in lower panels), consistent with the minor fraction of complexes with a high polarization modulation ratio in polarization modulation experiments (Fig. 6.9b).

6.6 Dynamics of Autoinhibitory Domain Interchange

The biological function of proteins is not explained by an understanding of conformations and structures alone, as important as those are. Knowledge of the *dynamics* of proteins is also required. To understand protein function we need to know the time scales, amplitudes, and nature of the protein motions. This section describes single molecule polarization modulation studies used to determine the time scale of interchange between high-mobility and low-mobility populations.

The detection of conformations of PMCA–CaM complexes with associated and dissociated autoinhibitory domains as described above raises the question of dynamics: on what time scale do these conformations interchange? The answer to this question is pertinent to the response of cells to Ca^{2+} signals, because the enzyme's response to a change in Ca^{2+} includes the rates of association or dissociation of the autoinhibitory domain. In line with our interpretation of the nature of these two populations, the interchange time yields the time scale of association and dissociation of the autoinhibitory domain with the catalytic core of the enzyme PMCA. Single-molecule measurements allow the autoinhibitory interchange step itself to be isolated from other kinetic steps in PMCA activation.

In the single-molecule polarization modulation experiments described above (Section 6.4), the single-molecule modulation depth distributions at a physiologically relevant Ca^{2+} concentration of ~0.15 μM appeared in most cases to reflect fluctuations within a single orientational mobility population, but in about 20–30% of the molecules the modulation depth distributions seem to explore both low mobility and high mobility populations [39]. Since the typical fluorescence trajectory length is about 5–10 s, this suggested that orientational mobility populations may interchange at a time scale longer than 10 s. Therefore to track interchange between orientational mobility states, it was necessary to probe single PMCA–CaM complexes over a time scale of tens of seconds. To extend interrogation times in single-molecule trajectories to tens of seconds, we devised an experimental method to track the orientational mobility of PMCA–CaM complexes and applied it to detect conformational changes of the autoinhibitory domain [85]. We used an electronic shutter to take a brief "snap shot" of the orientational mobility periodically. In this way, the exposure time was reduced and thus photobleaching was limited. Since the modulation depth is readily determined from a time period of 0.5–1 s, this method ensures that there is no loss in precision of the measurement of

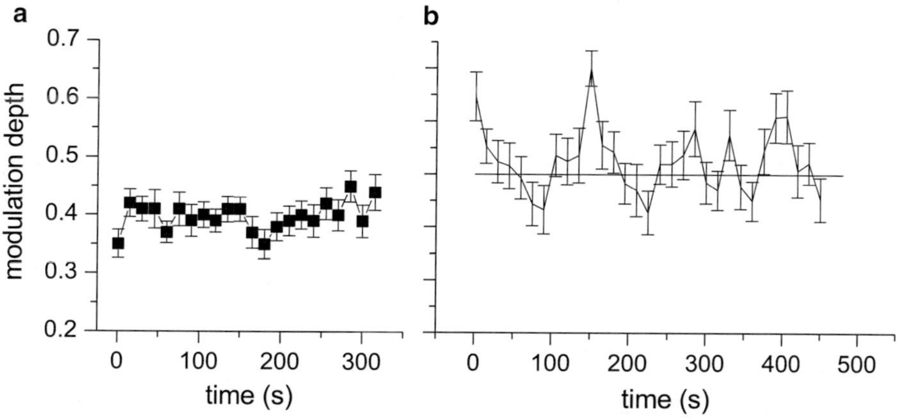

Fig. 6.14 Modulation depth trajectories for CaM-TMR bound to PMCA at 25 μM (**a**) and 0.15 μM (**b**). Persistence times above and below a threshold with modulation depth 0.5 (*horizontal line*) were tabulated for kinetic measurements. Modulation depths were measured every 15 s by opening the shutter for a period of 1 s. Each exposure period was fit by a maximum-likelihood estimator [38] to determine values of the modulation depth. Note the extended trajectory lengths before photobleaching resulting from limited exposure to the excitation beam. Adapted with permission from ref. [85].

modulation depth while allowing polarization modulation trajectories to be collected over 100s of seconds. Figure 6.14 shows shutter control polarization modulation trajectories of CaM bound to fresh PMCA molecule immobilized in agarose gel.

At a high Ca^{2+} concentration (25 μM) the modulation depths in most trajectories showed no transitions away from the low modulation depth population (see Fig. 6.14a), consistent with most PMCA–CaM complexes remaining in the high mobility population. In contrast, at a subsaturating Ca^{2+} concentration (0.15 μM) the modulation depth distributions displayed frequent interconversion of modulation depths for single CaM-TMR molecules bound to a fresh PMCA. To analyze these trajectories, we plotted the distribution of "lag times" or "persistence times" in states with modulation depth above or below a threshold of 0.5, a value intermediate between the low modulation depth and high modulation depth populations (see Fig. 6.9). The duration of high and low modulation depths were counted for all single molecules. Histograms of lag times are shown in Fig. 6.15. As in the FRET experiments described above, these measurements take advantage of the slow off-rate of CaM from PMCA 4b (~0.015 s^{-1} [30]). Dissociation of PMCA–-CaM complexes would lead to termination of a single-molecule trajectory in a manner indistinguishable from photobleaching in these experiments.

The lag time distributions were fit to single-exponential curves to determine the average lag times and from them the rates of association and dissociation of the autoinhibitory domain of PMCA were determined. The lag times in high and low-mobility states were fit to simple exponential decays, yielding a time constant of 24 s in the high modulation depth states, corresponding to an associated autoinhibitory

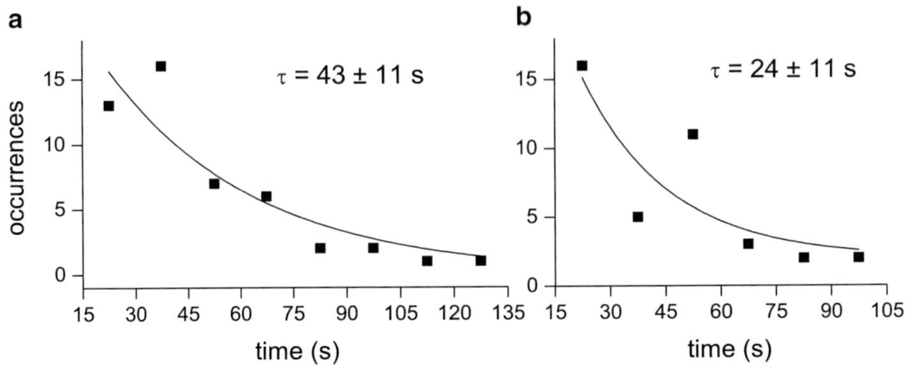

Fig. 6.15 Persistence time histograms for CaM-TMR bound to PMCA in the high-mobility state (*left*) and low-mobility state (*right*) at a Ca^{2+} concentration of 0.15 μM. The solid lines show fits to exponential functions with the decay times shown. The persistence time histograms thus yield the kinetics of autoinhibitory domain interchange. Taken with permission from ref. [85].

domain, and a time constant is of 44 s in the low modulation depth states, corresponding to a dissociated autoinhibitory domain. The corresponding rate constants at a Ca^{2+} concentration of 0.15 μM are $0.042 \pm 0.011\ s^{-1}$ for autoinhibitory domain dissociation and $0.023 \pm 0.006\ s^{-1}$ for association. The ratio of these rate constants is consistent with the relative populations of high and low-mobility states (Fig. 6.9).

6.7 Conclusions

Taken together, our results show that the response of PMCA to CaM, Ca^{2+}, and ATP is more complex and mechanistically intricate than previously realized. A simplistic picture of PMCA regulation in which the autoinhibitory domain blocks access to the ATP binding site in the absence of CaM and is released in the presence of CaM now appears incomplete. Results from single-molecule spectroscopy [39] and enzyme kinetics [52] indicate the existence of an intermediate state of PMCA in which CaM is bound but the autoinhibitory domain is still associated with the catalytic domain of the enzyme (see Fig. 6.9). Single-molecule studies also suggested that the relative population of this intermediate is not directly linked to nucleotide binding but to subsequent steps in the enzymatic cycle that occur during or after ATP hydrolysis [39]. Thus the autoinhibitory domain is more than a simple on-off switch.

In addition to its dependence on the rate of CaM binding to the autoinhibitory domain, the response time of PMCA to changes in Ca^{2+} concentration also depends on the rate of interchange of autoinhibitory domain conformations. Penniston and coworkers showed that after a Ca^{2+} spike the response of PMCA activity to a subsequent Ca^{2+} spike is enhanced for a period of tens of seconds for PMCA4b and for up to a minute for PMCA2b [30]. They attributed this Ca^{2+} "memory" to the slow dissociation of CaM from PMCA. Because the interchange of autoinhibitory domain conformations that we observed occurs on a similar time scale as CaM dissociation, we expect that dynamics of the autoinhibitory domain may also play a role in the "memory" of previous Ca^{2+} spikes. This phenomenon would be particularly useful in excitable cells such as neurons that need to respond quickly to transient increases in intracellular calcium.

As illustrated in this chapter, single-molecule methods offer new insights due to the possibility to resolve distributions of properties (such as polarization modulation depths or FRET efficiencies), thus resolving multiple conformational states. Time-domain analysis of trajectories further allows interchange rates between conformational states to be measured at equilibrium. The result of these approaches is identification of conformations and dynamics that are difficult or impossible to isolate by conventional methods, leading to new insights into signaling mechanisms.

Acknowledgments We acknowledge the many insights and ideas gained from discussions with Tom Squier, Ramona Bieber-Urbauer, and Jeff Urbauer, who were indispensible to the work described here. We further acknowledge present and former graduate students and postdoctoral associates who contributed to the work described in this chapter: Michael Allen, Abhijit Mandal, Manoj Singh, and Brian Slaughter. This research was supported by the American Heart Association (grant no. 0455487Z) and the National Institutes of Health (GM-58715).

References

1. Brini M (2009) Plasma membrane Ca^{2+}-ATPase: from a housekeeping function to a versatile signaling role. Pflugers Arch 457:657–664
2. Di Leva F, Domi T, Fedrizzi L, Lim D, Carafoli E (2008) The plasma membrane Ca^{2+} ATPase of animal cells: structure, function and regulation. Arch Biochem Biophys 476(1):65–74
3. Strehler EE, Caride AJ, Filoteo AG, Xiong Y, Penniston JT, Enyedi A (2007) Plasma membrane Ca^{2+} ATPases as dynamic regulators of cellular calcium handling. Ann NY Acad Sci 1099:226–236
4. Verma AK, Filoteo AG, Stanford DR, Wieben ED, Penniston JT, Strehler EE, Fischer R, Heim R, Vogel G, Mathews S, Strehler-Page M-A, James P, Vorherr T, Krebs J, Carafoli E (1988) Complete primary structure of a human plasma membrane Ca^{2+} pump. J Biol Chem 263:14152–14159
5. James P, Zvaritch EI, Shakhparonov MI, Penniston JT, Carafoli E (1987) The amino acid sequence of the phosphorylation domain of the erythrocyte Ca^{2+} ATPase. Biochem Biophys Res Commun 149:7–12
6. Apell HJ (2004) How do P-type ATPases transport ions? Bioelectrochemistry 63:149–156
7. Chin D, Means AR (2000) Calmodulin: a prototypical calcium sensor. Trends Cell Biol 10:322–328
8. Yap KL, Ikura M (2004) Calmodulin. Handbook of metalloproteins 3:447–458
9. Means AR (2004) Calmodulin-mediated signaling. Handbook Cell Signal 2:83–85
10. Hoeflich KP, Ikura M (2002) Calmodulin in action: diversity in target recognition and activation mechanisms. Cell 108:739–742

11. Nelson MR, Chazin WJ (1998) An interaction-based analysis of calcium-induced conformational changes in Ca^{2+} sensor proteins. Protein Sci 7:270–282
12. Zhang M, Tanaka T, Ikura M (1995) Calcium-induced conformational transition revealed by the solution structure of apo calmodulin. Nat Struct Biol 2:758–767
13. Carafoli E (2002) Calcium signaling: a tale for all seasons. Proc Natl Acad Sci USA 99:1115–1122
14. Babu YS, Bugg CE, Cook WJ (1988) Structure of calmodulin refined at 2.2 Å resolution. J Mol Biol 204:191–204
15. Fallon JL, Quiocho FA (2003) A closed compact structure of native Ca^{2+}-calmodulin. Structure 11:1303–1307
16. Humphrey W, Dalke A, Schulten K (1996) VMD: visual molecular dynamics. J Mol Graphics 14:33–38
17. Johnson CK (2006) Calmodulin, conformational states, and calcium signaling. A single-molecule perspective. Biochemistry 45:14233–14246
18. Gopinath RM, Vincenzi FF (1977) Phosphodiesterase protein activator mimics red blood cell cytoplasmic activator of (Ca^{2+}-Mg^{2+})ATPase. Biochem Biophys Res Commun 77:1203–1209
19. Jarrett HW, Penniston JT (1977) Partial purification of the Ca^{2+}-Mg^{2+} ATPase activator from human erythrocytes: its similarity to the activator of 3′:5′-cyclic nucleotide phosphodiesterase. Biochem Biophys Res Commun 77:1210–1216
20. James P, Maeda M, Fischer R, Verma AK, Krebs J, Penniston JT, Carafoli E (1988) Identification and primary structure of a calmodulin binding domain of the Ca^{2+} pump of human erythrocytes. J Biol Chem 263:2905–2910
21. Enyedi A, Vorherr T, James P, McCormick DJ, Filoteo AG, Carafoli E, Penniston JT (1989) The calmodulin binding domain of the plasma membrane Ca^{2+} pump interacts both with calmodulin and with another part of the pump. J Biol Chem 264:12313–12321
22. Strehler EE, Zacharias DA (2001) Role of alternative splicing in generating isoform diversity among plasma membrane calcium pumps. Physiol Rev 81:21–50
23. Domi T, Di Leva F, Fedrizzi L, Rimessi A, Brini M (2007) Functional specificity of PMCA isoforms? Ann NY Acad Sci 1099:237–246
24. Strehler EE, Filoteo AG, Penniston JT, Caride AJ (2007) Plasma-membrane Ca^{2+} pumps: structural diversity as the basis for functional versatility. Biochem Soc Trans 35:919–922
25. Caride AJ, Filoteo AG, Penheiter AR, Paszty K, Enyedi A, Penniston JT (2001) Delayed activation of the plasma membrane calcium pump by a sudden increase in Ca^{2+}: fast pumps reside in fast cells. Cell Calcium 30:49–57
26. Stauffer TP, Guerini D, Carafoli E (1995) Tissue distribution of the four gene products of the plasma membrane Ca^{2+} pump. J Biol Chem 270:12184–12190
27. Elwess NL, Filoteo AG, Enyedi A, Penniston JT (1997) Plasma membrane Ca^{2+} pump isoforms 2a and 2b are unusually responsive to calmodulin and Ca^{2+}. J Biol Chem 272:17981–17986
28. Ba-Thein W, Caride AJ, Enyedi A, Paszty K, Croy CL, Filoteo AG, Penniston JT (2001) Chimaeras reveal the role of the catalytic core in the activation of the plasma membrane Ca^{2+} pump. Biochem J 356:241–245
29. Caride AJ, Elwess NL, Verma AK, Filoteo AG, Enyedi A, Bajzer Z, Penniston JT (1999) The rate of activation by calmodulin of isoform 4 of the plasma membrane Ca^{2+} pump is slow and is changed by alternative splicing. J Biol Chem 274:35227–35232
30. Caride AJ, Penheiter AR, Filoteo AG, Bajzer Z, Enyedi A, Penniston JT (2001) The plasma membrane calcium pump displays memory of past calcium spikes. Differences between isoforms 2b and 4b. J Biol Chem 276:39797–39804
31. Jiang L, Fernandes D, Mehta N, Bean JL, Michaelis ML, Zaidi A (2007) Partitioning of the plasma membrane Ca^{2+}-ATPase into lipid rafts in primary neurons: effects of cholesterol depletion. J Neurochem 102:378–388

32. Sepulveda MR, Berrocal-Carrillo M, Gasset M, Mata AM (2006) The plasma membrane Ca^{2+}-ATPase isoform 4 is localized in lipid rafts of cerebellum synaptic plasma membranes. J Biol Chem 281:447–453
33. Toyoshima C (2008) Structural aspects of ion pumping by Ca^{2+}-ATPase of sarcoplasmic reticulum. Arch Biochem Biophys 476:3–11
34. Haugland RP (2006) Molecular probes handbook – a guide to fluorescent probes and labeling technologies, 10th edn. Invitrogen, Eugene, OR
35. Filoteo AG, Gorski JP, Penniston JT (1987) The ATP-binding site of the erythrocyte membrane Ca^{2+} pump. J Biol Chem 262:6526–6530
36. Adamo HP, Filoteo AG, Penniston JT (1996) The plasma membranc Ca^{2+} pump mutant lysine591→ arginine retains some activity, but is still inactivated by fluorescein isothiocyanate. Biochem J 317:41–44
37. Allen MW, Urbauer RJB, Zaidi A, Williams TD, Urbauer JL, Johnson CK (2004) Fluorescence labeling, purification and immobilization of a double cysteine mutant calmodulin fusion protein for single-molecule experiments. Anal Biochem 325:273–284
38. Osborn KD, Singh MK, Urbauer RJB, Johnson CK (2003) Maximum likelihood approach to single-molecule polarization modulation analysis. ChemPhysChem 4:1005–1011
39. Osborn KD, Zaidi A, Mandal A, Urbauer RJB, Johnson CK (2004) Single-molecule dynamics of the calcium-dependent activation of plasma-membrane Ca^{2+}-ATPase by calmodulin. Biophys J 87:1892–1899
40. Fries JR, Brand L, Eggeling C, Kollner M, Seidel CAM (1998) Quantitative identification of different single molecules by selective time-resolved confocal fluorescence spectroscopy. J Phys Chem A 102:6601–6613
41. Eggeling C, Berger S, Brand L, Fries JR, Schaffer J, Volkmer A, Seidel CAM (2001) Data registration and selective analysis for single-molecule fluorescence spectroscopy using BIFL. J Biotechnol 86:163–180
42. Deniz AA, Laurence TA, Dahan M, Chemla DS, Schultz PG, Weiss S (2001) Ratiometric single-molecule studies of freely diffusing biomolecules. Annu Rev Phys Chem 52:233–253
43. Niggli V, Penniston JT, Carafoli E (1979) Purification of the (Ca^{2+}-Mg^{2+})-ATPase from human erythrocyte membranes using a calmodulin affinity column. J Biol Chem 254:9955–9958
44. Sobue K, Muramoto Y, Fujita M, Kakiuchi S (1981) Calmodulin-binding protein of erythrocyte cytoskeleton. Biochem Biophys Res Commun 100:1063–1070
45. Osborn KD (2003) Single-molecule studies of the interaction of calmodulin with the plasma membrane Ca^{2+}-ATPase Chemistry. University of Kansas, Lawrence, KS, p 150
46. Niggli V, Adunyah ES, Penniston JT, Carafoli E (1981) Purified (Ca^{2+}-Mg^{2+})-ATPase of the erythrocyte membrane. Reconstitution and effect of calmodulin and phospholipids. J Biol Chem 256:395–401
47. Verma AK, Enyedi A, Filoteo AG, Penniston JT (1994) Regulatory region of plasma membrane Ca^{2+} pump. 28 residues suffice to bind calmodulin but more are needed for full auto-inhibition of the activity. J Biol Chem 269:1687–1691
48. Allen MW, Bieber Urbauer RJ, Johnson CK (2004) Single-molecule assays of calmodulin target binding detected with a calmodulin energy-transfer construct. Anal Chem 76:3630–3637
49. Slaughter BD, Urbauer RJ, Urbauer JL, Johnson CK (2007) Mechanism of calmodulin recognition of the binding domain of isoform 1b of the plasma membrane Ca^{2+}-ATPase: kinetic pathway and effects of methionine oxidation. Biochemistry 46:4045–4054
50. Slaughter BD (2005) Single-molecule studies of calmodulin conformation, dynamics, and folding Department of Chemistry. University of Kansas, Lawrence, KS, p 233
51. Yao Y, Gao J, Squier TC (1996) Dynamic structure of the calmodulin-binding domain of the plasma membrane Ca-ATPase in native erythrocyte ghost membranes. Biochemistry 35:12015–12028

52. Penheiter AR, Bajzer Z, Filoteo AG, Thorogate R, Török K, Caride AJ (2003) A model for the activation of plasma membrane calcium pump isoform 4b by calmodulin. Biochemistry 42:12115–12124
53. Yazawa M, Vorherr T, James P, Carafoli E, Yagi K (1992) Binding of calcium by calmodulin: influence of the calmodulin binding domain of the plasma membrane calcium pump. Biochemistry 31:3171–3176
54. Graf E, Verma AK, Gorski JP, Lopaschuk G, Niggli V, Zurini M, Carafoli E, Penniston JT (1982) Molecular properties of calcium-pumping ATPase from human erythrocytes. Biochemistry 21:4511–4516
55. Bagshaw C (2001) ATP analogues at a glance. J Cell Sci 114:459–460
56. Adamo HP, Rega AF, Garrahan PJ (1990) Magnesium-ions accelerate the formation of the phosphoenzyme of the (Ca^{2+} + Mg^{2+})-activated ATPase from plasma membranes by acting on the phosphorylation reaction. Biochem Biophys Res Commun 169:700–705
57. Herscher CJ, Rega AF, Adamo HP (1996) Pre-steady-state kinetic study of the effects of K^+ on the partial reactions of the catalytic cycle of the plasma membrane Ca^{2+}-ATPase. Biochem J 315:673–677
58. Falchetto R, Vorherr T, Brunner J, Carafoli E (1991) The plasma membrane Ca^{2+} pump contains a site that interacts with its calmodulin-binding domain. J Biol Chem 266:2930–2936
59. Falchetto R, Vorherr T, Carafoli E (1992) The calmodulin-binding site of the plasma membrane Ca^{2+} pump interacts with the transduction domain of the enzyme. Protein Sci 1:1613–1621
60. Zaidi A, Barron L, Sharov VS, Schoneich C, Michaelis EK, Michaelis ML (2003) Oxidative inactivation of purified plasma membrane Ca^{2+}-ATPase by hydrogen peroxide and protection by calmodulin. Biochemistry 42:12001–12010
61. Gao J, Yao Y, Squier TC (2001) Oxidatively modified calmodulin binds to the plasma membrane Ca-ATPase in a nonproductive and conformationally disordered complex. Biophys J 80:1791–1801
62. Bigelow DJ, Squier TC (2005) Redox modulation of cellular signaling and metabolism through reversible oxidation of methionine sensors in calcium regulatory proteins. Biochim Biophys Acta 1703:121–134
63. Zaidi A, Michaelis ML (1999) Effects of reactive oxygen species on brain synamptic plasma membrane Ca^{2+}-ATPase. Free Radical Biol Med 27:810–821
64. Zaidi A, Fernandes D, Bean JL, Michaelis ML (2009) Effects of paraquat-induced oxidative stress on the neuronal plasma membrane Ca^{2+}-ATPase. Free Radic Biol Med 47:1507–1514
65. Zaidi A, Gao J, Squier TC, Michaelis ML (1998) Age-related decrease in brain synaptic membrane Ca^{2+}-ATPase in F344/BNF1 rats. Neurobiol Aging 19:487–495
66. Qin Z, Zaidi A, Gao J, Krainev AG, Michaelis ML, Squier TC, Bigelow DJ (1998) Decrease in Ca-ATPase activity in aged synaptosomal membranes is not associated with changes in fatty acyl chain dynamics. Mech Ageing Dev 105:291–300
67. Anbanandam A, Bieber Urbauer RJ, Bartlett RK, Smallwood HS, Squier TC, Urbauer JL (2005) Mediating molecular recognition by methionine oxidation: conformational switching by oxidation of methionine in the carboxyl-terminal domain of calmodulin. Biochemistry 44:9486–9496
68. Bartlett RK, Bieber Urbauer RJ, Anbanandam A, Smallwood HS, Urbauer JL, Squier TC (2003) Oxidation of Met144 and Met145 in calmodulin blocks calmodulin dependent activation of the plasma membrane Ca-ATPase. Biochemistry 42:3231–3238
69. Osborn KD, Bartlett RK, Mandal A, Zaidi A, Urbauer RJB, Urbauer JL, Galeva N, Williams TD, Johnson CK (2004) Single-molecule dynamics reveal an altered conformation for the autoinhibitory domain of plasma-membrane Ca^{2+}-ATPase bound to oxidatively modified calmodulin. Biochemistry 43:12937–12944
70. Osborn KD, Zaidi A, Urbauer RJB, Michaelis ML, Johnson CK (2005) Single-molecule characterization of the dynamics of calmodulin bound to oxidatively modified plasma-membrane Ca^{2+}-ATPase. Biochemistry 44:11074–11081

71. Liyanage MR, Zaidi A, Johnson CK (2009) Fluorescence polarization assay for calmodulin binding to plasma membrane Ca^{2+}-ATPase: dependence on enzyme and Ca^{2+} concentrations. Anal Biochem 385:1–6
72. Stryer L, Haugland RP (1967) Energy transfer: a spectroscopic ruler. Proc Natl Acad Sci USA 58:716–726
73. van der Meer BW, Coker G, Chen S-Y (1994) Resonance energy transfer: theory and data. VCH, New York
74. Selvin PR (1995) Fluorescence resonance energy transfer. Methods Enzymol 246:300–334
75. Ha T, Enderle T, Ogletree DF, Chemla DS, Selvin PR, Weiss S (1996) Probing the interaction between two single molecules: fluorescence resonance energy transfer between a single donor and a single acceptor. Proc Natl Acad Sci USA 93:6264–6268
76. Deniz AA, Dahan M, Grunwell JR, Ha T, Faulhaber AE, Chemla DS, Weiss S, Schultz PG (1999) Single-pair fluorescence resonance energy transfer on freely diffusing molecules: observation of Förster distance dependence and subpopulations. Proc Natl Acad Sci USA 96:3670–3675
77. Weiss S (2000) Measuring conformational dynamics of biomolecules by single molecule fluorescence spectroscopy. Nature Struct Biol 7:724–729
78. Schuler B, Lipman EA, Eaton WA (2002) Probing the free-energy surface for protein folding with single-molecule fluorescence spectroscopy. Nature 419:743–747
79. Ha T (2001) Single-molecule fluorescence resonance energy transfer. Methods 25:78–86
80. Margittai M, Widengren J, Schweinberger E, Schroeder GF, Felekyan S, Haustein E, Koenig M, Fasshauer D, Grubmueller H, Jahn R, Seidel CAM (2003) Single-molecule fluorescence resonance energy transfer reveals a dynamic equilibrium between closed and open conformations of syntaxin 1. Proc Natl Acad Sci USA 100:15516–15521
81. Slaughter BD, Allen MW, Unruh JR, Urbauer RJB, Johnson CK (2004) Single-molecule resonance energy transfer and fluorescence correlation spectroscopy of calmodulin in solution. J Phys Chem B 108:10388–10397
82. Schuler B, Lipman EA, Steinbach PJ, Kumke M, Eaton WA (2005) Polyproline and the "Spectroscopic ruler" Revisited with single-molecule fluorescence. Proc Natl Acad Sci USA 102:2754–2759
83. Kapanidis AN, Laurence TA, Lee NK, Margeat E, Kong X, Weiss S (2005) Alternating-laser excitation of single molecules. Acc Chem Res 38:523–533
84. Penheiter AR, Caride AJ, Enyedi A, Penniston JT (2002) Tryptophan 1093 is largely responsible for the slow off rate of calmodulin from plasma membrane Ca^{2+} pump 4b. J Biol Chem 277:17728–17732
85. Mandal A, Liyanage MR, Zaidi A, Johnson CK (2008) Interchange of autoinhibitory domain conformations in plasma-membrane Ca^{2+}-ATPase-calmodulin complexes. Protein Sci 17:555–562

Chapter 7
Single-Molecule Analysis of Cell-Virus Binding Interactions

Terrence M. Dobrowsky and Denis Wirtz

Abstract Adhesion assays based on single molecule interactions are a useful option when discerning between avidity and affinity in complex systems. This is especially true for viral adhesion to living cells which typically involves a complex system of proteins working together to lead to productive infection. Here, we discuss assays that have been used to quantitatively study the adhesion of viral and cellular receptors including surface plasmon resonance, real time fusion assays involving viral fluorescent tags and single molecule force spectroscopy (SMFS). We highlight the advantages of SMFS over other methods, including its specificity, versatility and application to studying the adhesion of HIV-1 to human cells. We discuss how using SMFS with infectious virus and living cells allow us to distinguish the adhesion of HIV-1 surface protein, gp120, to its primary cellular receptor, CD4, from the adhesion of gp120 to its secondary co-receptor, CCR5.

Keywords Single molecule force spectroscopy · Virus · Cell · Adhesion · HIV · gp120 · CD4 · CCR5 · Avidity · Affinity · Surface plasmon resonance · Viral fusion

---

T.M. Dobrowsky (✉)
Department of Chemical and Biomolecular Engineering, The Johns Hopkins University, 3400 North Charles Street, Baltimore, MD 21218, USA
e-mail: terrence@jhu.edu

D. Wirtz
Department of Chemical and Biomolecular Engineering, The Johns Hopkins University, 3400 North Charles Street, Baltimore, MD 21218, USA
and
Physical Science Oncology Center, The Johns Hopkins University, 3400 North Charles Street, Baltimore, MD 21218, USA
e-mail: wirtz@jhu.edu

Y. Sako and M. Ueda (eds.), *Cell Signaling Reactions: Single-Molecular Kinetic Analysis*, DOI 10.1007/978-90-481-9864-1_7,

7.1 Introduction

Viral particles are complex macromolecular machines built for the purpose of infecting target cells. The productive infection of a viral particle is a multi-step process that includes the initial binding and subsequent stable adhesion of the viral particle to the host cell surface, overcoming the plasma membrane of the cell through either fusion of their respective membranes or direct transport through a membrane pore, the directed transport of the viral genetic material through the cytoplasm, its transfer to the nucleus, the transcription of the viral genetic material, the assembly of viral particles, and their ultimate dispersion in the extracellular environment [1].

Viral species form either enveloped or non-enveloped particles. Enveloped particles are coated by a lipid bilayer generated during the budding of the particles from their host cell, while non-enveloped particles do not contain a lipid bilayer coat. Enveloped particles directly form at, and bud off from the plasma membrane or form by budding within the cell from internal membranes including the endoplasmic reticulum. In contrast, non-enveloped particles are typically released from their host cells through cell lysis. To enter target cells, enveloped particles fuse with the plasma membrane or the membrane of endosomes, while non-enveloped particles are thought to enter through membrane pores. For this review, the viral particle can be simplified as a macromolecular structure containing genetic cargo and surrounded by viral receptors organized on the particle surface. These viral surface proteins are the first critical protagonists for effective viral infection. The binding of viral surface proteins promotes the clustering of cellular receptors in the interfacial space just underneath the virus and induce cellular responses via surface protein signaling prior to the virion overcoming the cellular membrane. While much more is known about enveloped viral particle entry into the cell compared to non-enveloped particle entry, the initial adhesion step is thought to be similar for both viral species [2].

Here we use the enveloped human immunodeficiency virus type 1 (HIV-1) as a model system for reviewing the complex and dynamic nature of viral adhesion to the host cell surface. HIV is an enveloped lentivirus that infects and degrades the host's immune system by drastically reducing the number of CD4 positive T cells, eventually resulting in acquired immune deficiency syndrome (AIDS)[2]. According to the World Health Organization, 33 million people were infected by HIV worldwide by the end of 2007 [3]. That year, 2.7 million of these infected individuals newly contracted the disease, outnumbering the 2 million who died from AIDS-related causes [3]. HIV is one of the most studied infectious diseases: over 18,000 articles reporting on HIV/AIDS research were published in 2003 alone [4]. The application of highly active anti-retroviral therapy (HAART) has significantly improved the treatment of HIV by using drug cocktails targeting multiple points of the viral replication cycle, concentrating heavily on the replication machinery of HIV and indeed, the treatment can often achieve immeasurable viral loads within a patient [5]. However, recent work indicates that a small population of memory

T cells infected with HIV do not actively produce virus in their dormant state [6]. Sporadically activated over time, this subset of memory T cells can reintroduce high viral loads in a patient with a flagging commitment to HAART [6]. Therefore, a promising new target for antiretroviral therapy is the fusion of HIV with host cellular membrane, preventing the entry of the virus into the cells altogether. Most fusion-inhibiting drugs (e.g. BMS-806, AMD3100 or Enfuvirtide) prevent viral fusion by binding specific conserved regions of viral surface receptors, which either prevent receptor adhesion or halt the protein conformation changes required for fusion [5]. The biophysical mechanism by which a virus enters its host is not well understood.

HIV adhesion and fusion are governed by the dynamic interactions of two types of protein on the viral surface and two types of receptors on the cell surface. The envelope glycoprotein (Env) gp120 mediates viral adhesion with host cells. Located on the viral surface, gp120 is non-covalently bound to the viral transmembrane glycoprotein gp41 [7–10]. The gp120/gp41 protein pair is organized on the viral surface in ~14 trimers [11]. Initially, gp120 binds its primary cellular receptor CD4, which induces a change in gp120 conformation [12–15], subsequently allowing gp120 to bind the cellular co-receptor, most commonly CCR5 or CXCR4 [16–21] (Fig. 7.1). The binding of the cellular co-receptor induces yet another conformation change, this time in gp41, in which six helical regions stacked three on three lengthwise (HR1 and HR2 per gp41), fold in onto themselves [22–24]. This conformation change forms a coiled-coil complex, which drives the fusion of the viral membrane with the cellular membrane [25] (Fig. 7.1).

For effective infection, viral particles must maintain a balance of specific adhesion that is neither too strong nor too labile so that the particle may rapidly probe a large area away from the original cell in the extracellular space. The escape

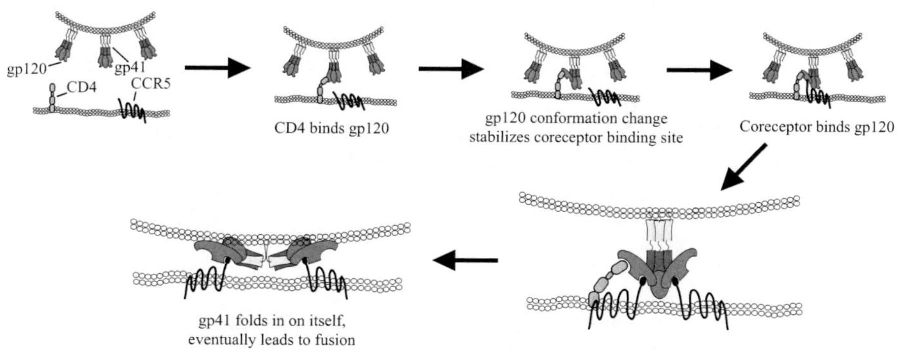

Fig. 7.1 Description of dynamic protein interactions leading to viral fusion. Cellular receptor CD4 and co-receptor CCR5 interact with viral surface protein gp120 which is noncovalently bound to gp41, the viral transmembrane protein which contains two helical regions, HR1 and HR2. Initially, CD4 binds gp120 which initiates a conformation change in gp120 stabilizing the gp120 co-receptor binding site. Co-receptor binding results in another conformation change in gp41 resulting in a coiled-coil complex. As a result of the conformation change, viral fusion ensues.

of HIV-1 from its cell of origin is aided by multiple adhesion-deterring factors. First, as described above, effective adhesion and fusion with a target cell is dependent on a cascade of binding events [1]. Second, HIV-1 infection of a host cell results in decreased expression of the major receptor CD4 on the cellular surface [26]. In addition to adhesion-based deterrents, viral particles mature in the extracellular space using internal proteases that cleave the structural Gag proteins converting an uninfectious particle into an infectious one [1]. Despite their importance in the design of future of viral entry inhibitors, the biochemical (i.e. rates of association and dissociation, stoichiometry of binding) and biophysical properties (i.e. bond strength, reactive compliance) remain ill defined.

7.2 Current Methods to Measure Cell-Virus Binding Interactions

The design of new therapies halting the entry of HIV into cells requires an improved, quantitative knowledge of the mechanism of binding of HIV to the host cell surface. The interaction between the viral surface protein gp120 and its primary cellular receptor CD4 has been quantitatively characterized using adhesion techniques, which probe either global avidity or single-receptor affinity including fluorescent viral membrane fusion and cell-cell binding assays. There have also been both indirect methods of analysis where downstream events are measured, including stoichiometric infection assays and assays monitoring binding directly like those that employ surface plasmon resonance. Methods that measure bonds directly are further divided into those that measure bonds as a step function with either bound or unbound states and finally those that measure the distribution of micromechanical properties of individual bonds.

Stoichiometric assays have recently been used to probe the HIV viral particle binding to its host [27, 28]. These assays measure the ratio of competent to incompetent fusion proteins expressed on the particles surface. This technique was used to determine the global requirements of fusion proteins on a single particle for productive fusion and subsequently infection. Briefly, viral particles with varying ratios of functional:nonfunctional gp120 proteins were made by transfecting cells to produce virus with three DNA vectors. The first, encodes the core structure of the HIV virion as well as GFP to be used as in infection tracer. The second and third vectors encode functional and nonfunctional envelope proteins. It was shown that resulting functional:nonfunctional gp120 expression correlated to the ratio of the corresponding DNA vectors used to make the virus [27, 28]. Results comparing the effectiveness of viral particles were assessed by the percentage of cells ultimately infected by that stock of virus. Results obtained using stoichiometric assays suggest that only a single gp120:gp41 trimer on the viral surface is required for infection and that only two of the three gp120:gp41 pairs within the trimer are required for infection [27, 28]. Hence, stoichiometric assays offer biochemical insight into the global interactions of a viral particle with a plasma membrane. However, these assays rely on events far downstream from adhesion and fusion - including transcription of viral DNA after cellular transport and nuclear incorporation - which are quantitative readouts of infection rather than of early interaction events. It has been established that a large fraction of a viral load is not capable of productive infection, therefore assays relying on complete infection will not account for viral particles that can productively adhere to host cells but not infect.

Another method for quantifying bimolecular adhesion is surface plasmon resonance spectroscopy [29]. This method has gained popularity due to its ease of use and the high throughput nature of the experimental setup. Indeed, this assay has been used in an attempt to quantify gp120-CD4 bond interactions [30–32]. However, severe limitations with the broad application of surface plasmon resonance spectroscopy are often overlooked. Briefly, receptors are covalently crosslinked to a glass surface of which the refractive index is measured in real time. Ligand molecules are washed over the receptor functionalized glass and binding of receptor-ligand pairs alters the refractive index of the glass to a measurable extent which is then followed by a washing cycle to remove bound ligands from the surface. The association and dissociation rates are then measured by the rate and extent of change of the refractive index during the ligand and wash cycles [29]. This assay measures the binding of receptor ligand pairs directly using surface plasmon resonance and offers a high throughput method for comparing receptor ligand pairs [29]. However, kinetic values obtained depend on purified proteins, ignoring the important modulating effects of cellular signaling, the cytoskeletal network, and the physical properties of the plasma membrane itself. Moreover, the use of purified proteins in such an assay instead of live cells does not guarantee proper post-translation modifications of the proteins, such as phosphorylation and glycosylation.

In an attempt to observe the real-time fusion of viral particles with host cellular membranes, assays have been developed to optically measure the fusion via multiple fluorescent tags associated with the viral particle [33, 34]. In this approach, the internal cargo of each viral particle contains a fluorescent tag which is incorporated into the viral particles during their assembly [33, 34]. Once these viral particles are collected and purified, they are adsorbed onto a glass surface and their viral lipid bilayer membranes are stained using a hydrophobic membrane fluorescent tag [33, 34]. The virus-coated glass is then seeded with infection-competent cells expressing CD4 and co-receptors at 4°C. Indeed, fusion can be halted in what is known as a temperature arrested state (TAS) [35]. Briefly, binding of gp120 to CD4 can occur at temperatures below physiological while co-receptor binding cannot. Therefore, by allowing the cells to contact the glass and form gp120-CD4 adhesion complexes at 4°C, the real-time fusion of viral particles with host cell membranes can be monitored, coordinating a collective zero time point for fusion by simply increasing the temperature to physiological. The fusion of the two membranes is monitored by following the diffusion of the viral membrane stain into the plasma membrane of the cell, indicating the formation of a fusion pore between the cell and virus [33, 34]. The subsequent dispersion of the viral cargo into the cytoplasm of the cell is monitored by following the diffusion of the internal viral stain, indicating successful delivery of viral proteins [33, 34]. This multiple fluorescent tag assay can monitor fusion of individual viral particles in real time and could be a useful tool to observe the efficiency of fusion-inhibiting drugs. However this assay is unable to offer insight into the adhesion complex between the virion and the cell prior to co-receptor engagement, such as the organization of cellular receptors and actin-mediated cellular response to initial gp120 binding.

In addition to virus-cell assays, cell-cell adhesion assays [36–41] have proved useful when observing the adhesive properties of viral surface proteins and target cellular receptors through the formation of a viralogical synapse [42]. So far, only cells grown in culture have been used to depict the viralogical synapse using various microscopy techniques. However, we suggest that cell-cell adhesion assays, which use either micropipette suction or atomic force microscopy (AFM) [43, 44], could provide useful insight into the growth kinetics of such a complex. Briefly, a micropipette cellular adhesion assay uses a micropipette capped on the tip by an antigen presenting cell. The cell is brought in contact with either a receptor functionalized surface or a receptor presenting cell and retracted after some varying contact time. By using cells with well characterized mechanical properties, such as red blood cells (RBC), and monitoring subtle changes in pressure within the micropipette during retraction, the force of bonds breaking on the cellular surface can be measured [36–38, 41]. In addition, changing the initial pressure difference in the micropipette allows for a tuning of force sensitivity [37]. Controlling the contact time allows this assay to use Poisson distribution statistics to ensure that only the rupture force of single bonds on the cellular surface is analyzed. This assay has been used to measure MHC-CD8 adhesion using T cells, suggesting it could be used to quantify other immunological adhesion events [38]. While this is the most physiological assay discussed so far, there are still major drawbacks to this approach. In particular, the positional resolution during retraction directly decreases the accuracy of the observed loading rates (the rate at which force is applied to the bonds formed).

Global adhesion between two biological surfaces is often mediated by multiple receptor protein subpopulations. The above assays cannot detect the presence of these subpopulations on cells nor can they distinguish how many individual bonds simultaneously adhere to produce effective adhesion between cell and virion. Next we discuss how single molecule force spectroscopy can distinguish between the contribution of individual bonds and the global adhesion event.

7.3 Principles of Single-Molecule Force Spectroscopy

SMFS characterizes the micromechanical properties of single bimolecular bonds. Similar in architecture to an atomic force microscope, a molecular force probe (MFP) utilizes a piezoelectric motor to move a flexible cantilever presenting a virion or antigen into contact with receptors expressed on the surface of a living cell [45, 46] (Fig. 7.2a). During the contact between the antigen and the cell, molecular bonds form, are subsequently stretched, and are finally ruptured upon cantilever

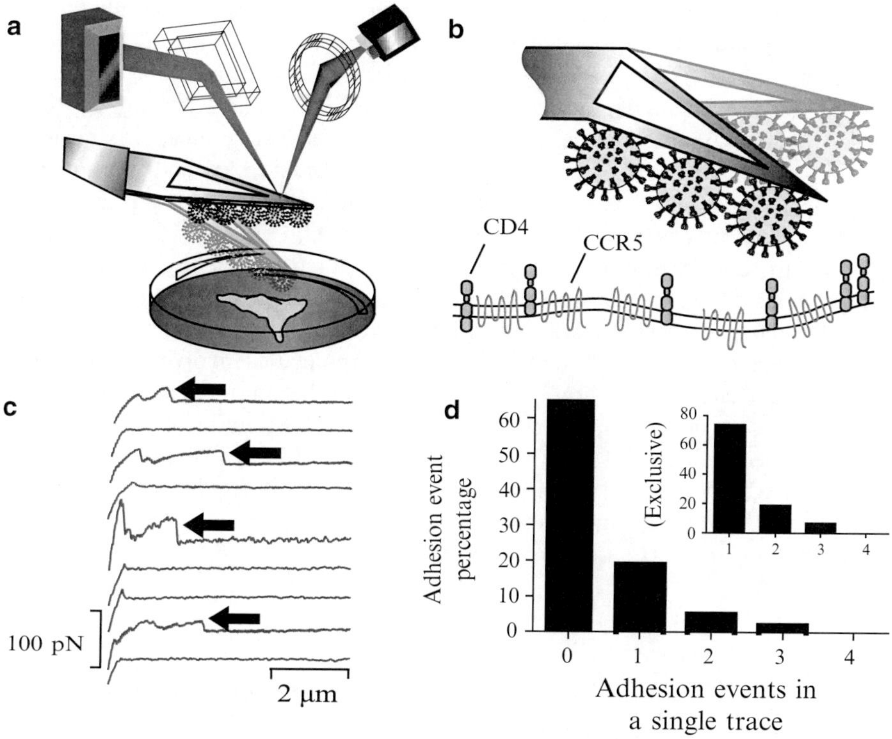

Fig. 7.2 Schematic of the instrument used to measure the micromechanics and kinetic properties of single molecular bonds between an infectious HIV-1 virion and individual cell receptors on a live host cell. (**a**) Schematic of the detection components of the molecular force probe (MFP) and the flexible cantilever placed just above a host cell. (**b**) Pseudovirus particles are cross-linked to a triangular cantilever, which is delicately brought into contact with a cell displaying either major receptor CD4, co-receptor CCR5 or both on its surface. (**c**) Typical force-deflection traces recorded during the retraction of the cantilever. Rupture of virion-cell bonds are marked by *arrows*. (**d**) Probability of formation of bonds between a virion and a host cell. The distribution displays Poisson characteristics (see more in text). (*Inset*) Probability of formation of bonds when analyzing only force-deflection traces displaying at least one bond adhesion. The time of contact between cell and virion was ~1 μs.

retraction. The exact position of the flexible cantilever is measured by laser reflection onto an electric photodiode (Fig. 7.2b, c). By using the spring constant of the cantilever, the exact displacement of the cantilever can be translated directly into a measured force via Hooke's law (Fig. 7.2c).

Unlike previously described assays, single molecule force spectroscopy (SMFS) can distinguish the difference between the affinity of an individual cellular receptor for its ligand and the avidity or global adhesion between multiple cellular receptors working in concert to bind a viral particle. Avidity may depend on multiple parameters, including the surface area of contact between cell and virion, cellular membrane stiffness, and receptor density. These parameters cannot be controlled or even measured during typical assays. Measuring single molecule affinity allows for the comparison of different cellular conditions without concern that uncontrolled parameters would affect the measurements. In addition, SMFS can detect the presence of subpopulations of receptors of different ligand binding capacity. While similar in principle to the cell-cell adhesion assays described earlier, SMFS does not require the use of cell lines with well characterized and reproducible mechanical properties. This could greatly reduce the versatility of receptor-antigen pairs that could be observed. In addition, while the cell-cell adhesion assay might be altered to address viral protein adhesion, it would not be able to do so on actual viral particles which are far too small for that experimental setup.

Traditional assays assume a linear correlation between global adhesion strength between cell and virus and the number of adhesion molecules involved, which is not necessarily the case. SMFS, however, measures the adhesion force of a single bond with piconewton force resolution and nanometer spatial resolution [43, 45–51]. The assay can directly distinguish how individual bonds contribute to the global avidity. In addition, SMFS does not require protein labeling and it is able to probe single-bond micromechanics using living cells. Hence, receptors have proper orientation and physiological post-translational modifications.

7.4 Single-Molecule Force Spectroscopy of Cell-Virion Binding Interactions

Using SMFS, we recently demonstrated a novel, sensitive, specific, and versatile approach to measure the adhesion between the surface proteins of infectious viral particles and their primary and secondary viral receptors located on the surface of living host cells [47]. Specifically, we monitored the adhesion forces between individual gp120 molecules and CD4 receptors, as well as gp120 and CCR5 to quantify the number and micromechanical properties of these bonds as a function of time of contact between cell and viral particle [47].

Infectious HIV-1 pseudovirus is crosslinked to flexible cantilevers (Fig. 7.2b). Crosslinking was verified to not affect the ability and efficiency of the virus to infect cells. In this assay, we utilized a cell line stably transfected to express either CD4 or CCR5. To probe individual gp120-CD4 bonds, these HIV-1 functionalized cantilevers were brought in contact with cells expressing only CD4. To probe individual gp120-CCR5 bonds, the conformation change in gp120 on the HIV-1 functionalized cantilevers was induced using soluble CD4 (sCD4). Cantilevers were then brought in contact with CCR5-expressing cells to measure gp120(sCD4)-CCR5 bonds. In either case, the presence of function-blocking antibodies against either CD4 or CCR5 reduced adhesion events to background levels, demonstrating the specificity of the binding interactions between the cantilever and the cell (Fig. 7.3a, b).

Quickly generating, while measuring, bond rupture events occur through the rapid repetition of a series of steps. First, the cantilever tip is moved toward the cell and bonds are allowed to form between cantilever viral particles and the cellular receptors. Second, the cantilever is retracted at a constant speed, eventually reaching the required applied force to break the existing bond. Third, immediately after bond rupture the cantilever quickly deflects back to the zero force position. This method quickly gathers many force versus distance curves and using Poisson distribution statistics, the breakage of single bimolecular bonds can be discerned [36](Fig. 7.2d). Briefly, the frequency with which a binding event occurs is dependent on the area of contact between the cantilever and the cell as well as the total time spent in contact with one another. By tuning the applied force on the cell (and thereby the contact area) as well as the dwell time of contact between cell and virus, the frequency of adhesion can be controlled. According to Poisson distribution statistics [36], an event frequency of ~30% out of all contacts between cell and cantilever corresponds to observing the results of mostly single events. Specifically, 80% of the events that occur when 30% of all cell-cantilever contacts result in binding are the rupture of single bonds. SMFS can measure subtle quantitative differences in the micromechanical properties of ligand-receptor pairs, including differences in bond strength with piconewton resolution, differences in bond lifetime with ms temporal resolution, and reactive compliance of the bonds with sub-angstrom resolution.

We use two separate models in the analysis of HIV receptor/co-receptor bond micromechanics. The well-established Bell model,

$$\langle F \rangle = \frac{k_b T}{x_\beta} ln\left(\frac{x_\beta r_f}{k_{off}^0 k_b T}\right),$$

can be used to compute an equilibrium dissociation rate constant, k_{off}^0, and reactive compliance, x_β, of the molecular bond from the logarithmic relationship between the measured bond loading rate, r_f, and mean measured rupture force, $\langle F \rangle$ [52]. This model has been extensively used to quantify properties of a large number of receptor-ligand pairs [43, 45–51]. However, as use of SMFS has been extended to more physiologically relevant experiments the large amount of data over a range of loading rates required for Bell model analysis has become cumbersome. Originally, multiple rupture probability distributions are recorded over a range of retraction velocities (e.g. 5, 10, 20, 30, 50 μm/s) resulting in a step-wise increase in loading rate. The rupture forces from these distributions are then binned together with their corresponding loading rates to produce a single Bell model data set. However, the second model that we have utilized was developed by Hummer et al. [53] and allowed us to quantitatively compare the micromechanical properties of individual bonds between cell and virus by analyzing the rupture probability distribution of a single loading rate. This model makes two basic assumptions. First it assumes that there is only one energy potential well, i.e. that there are no intermediate steps in the de-adhesion of the bond. This assumption can be tested by verifying that the Bell model only displays a simple linear dependence between the mean rupture force and the log of the loading rate (Fig. 7.3d), which is indicative of a single energy potential

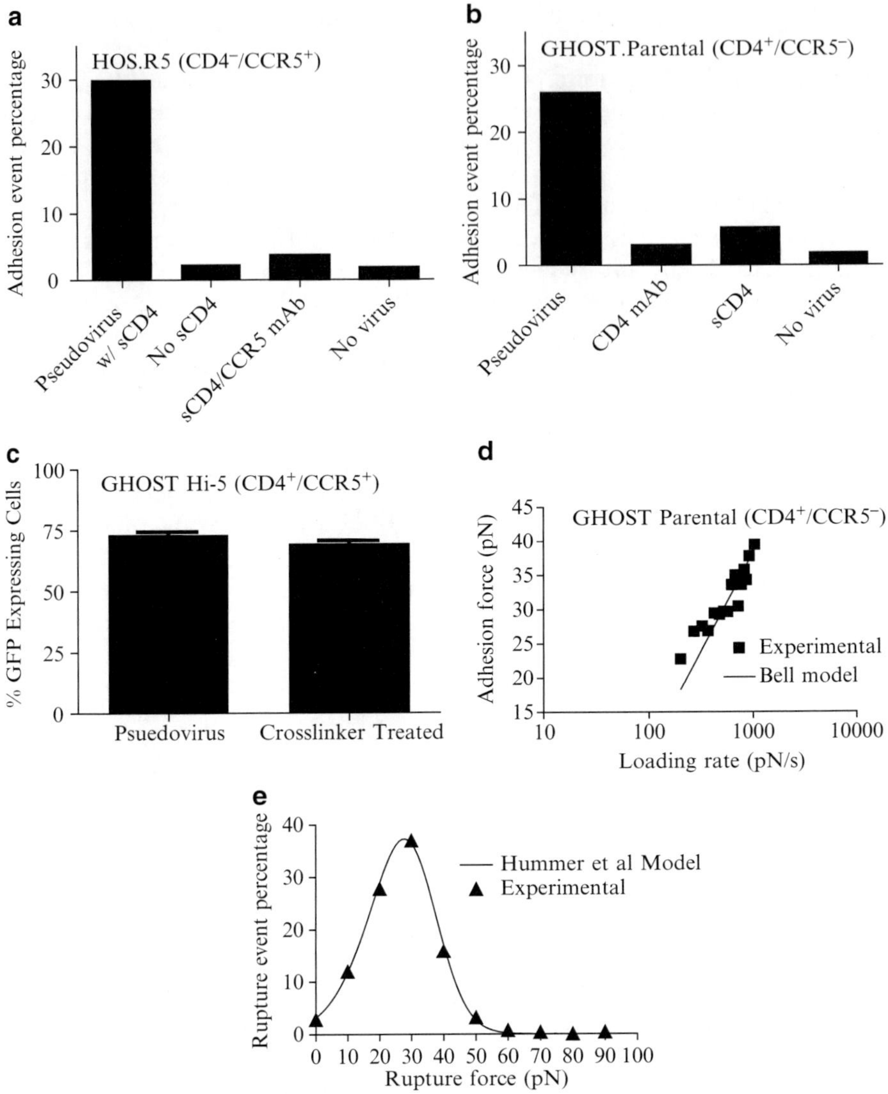

Fig. 7.3 Test of binding specificity and characterization of virion-receptor interactions at single-molecule resolution. (**a**) Test of specificity of molecular force probe (MFP) measurements and frequency of binding interactions between Env glycoprotein gp120, and $CD4^{-}CCR5^{+}$ living cells in the presence of soluble CD4 (sCD4), or in the absence of sCD4, or in the presence of a function-blocking antibody against CCR5, or in the absence of virions attached to the cantilever, respectively. (**b**) Test of specificity of MFP measurements and frequency of binding interactions between Env glycoproteins and $CD4^{+}CCR5^{-}$ living cells in the absence of added molecules, or in the presence of a function-blocking antibody against CD4 (B4), or in the presence of sCD4, or in the absence of virions, respectively. (**c**) Comparison of $CD4^{+}CCR5^{+}$ cells infected to express GFP with pseudotyped virus with and without the crosslinking treatment required for cantilever functionalization. (**d**) Mean adhesion force of the gp120-CD4 bond as a function of loading rate (pN/s) for $CD4^{+}CCR5^{-}$ parental cells. Fit of this curve using Bell's model yielded a bond dissociation constant, k°_{off}, of 3.73 s^{-1} and a bond reactive compliance, x_{d}, of 0.34 nm. (**e**) Example fit of Hummer et al. model to experimental probability distribution.

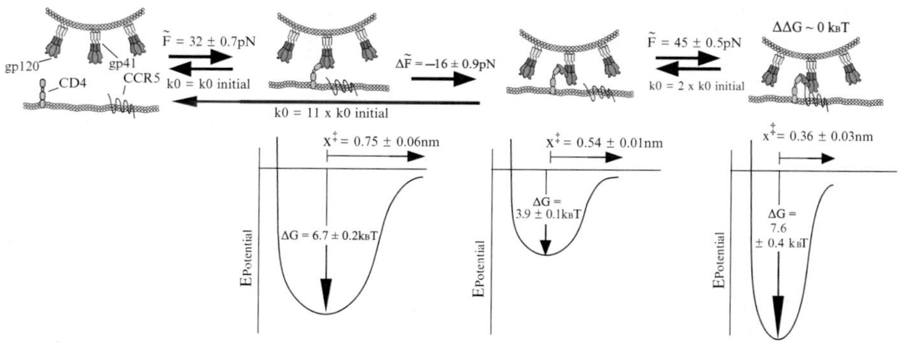

Fig. 7.4 Schematic of kinetic and mechanical parameters describing early fusion dynamics of HIV-1. Mean adhesion force, relative dissociation constant, change in free energy, and distance from the free energy minimum to bond rupture ($x^{\ddagger}$) for a multiple of binding events. The initial binding of CD4 to gp120, the conformation change of gp120 and finally gp120 binding CCR5 are illustrated here above their corresponding energy potentials.

well. Second, it is assumed that the reaction coordinate during the dissociation of the ligand from the receptor is one-dimensional which was deemed appropriate given the experimental setup. Briefly, together these models provide the basis for high-throughput analysis of SMFS data (Fig. 7.3d, e).

The initial Bell model fit to gp120-CD4 bond data resulted in an equilibrium (or unstressed) dissociation rate constant of 3.73 ± 0.45 s^{-1}, a reactive compliance of 3.4 ± 0.6 Å and a bond lifetime, $1/k^0_{off}$, of 0.27 ± 0.03 s. As discussed above, the data fit well to a single linear regime over a wide range of loading rates, which is indicative of a single energy potential well along the reaction coordinate. Hummer et al.'s model was used to analyze how the micromechanical properties of gp120-CD4 and gp120(sCD4)-CCR5 bonds changed with time. We demonstrated that the gp120-CD4 bond is a more flexible and less stable bond than the gp120(sCD4)-CCR5 bond with molecular spring constants of ~100 and ~500 pN/nm, and minimum potential energies of ~6.7 and ~7.5 k_bT, respectively. In addition, we observed that after 0.3s, the gp120-CD4 bond became more unstable when CCR5 was present on the cell surface along with CD4 resulting in dissociation rate constant roughly 11 times greater than at t = 0s. However, gp120(sCD4)-CCR5 bond micromechanical properties remained consistent over time. See Fig. 7.4 for a summary of how the micromechanical properties for HIV-1 bounds change during the course of viral adhesion.

This method, utilizing infectious viral particles and living cells expressing viral receptors is versatile and requires materials that are easily produced and purified. Viral particles were crosslinked to the cantilevers in a manner nonspecific to HIV-1 that did not affect viral infectivity (Fig. 7.3c). As such, this method can easily be reproduced using a variety of other viral strains both enveloped and non-enveloped.

Multiple assays exist to probe a variety of protein bond interactions each with their own advantages and disadvantages. However, when the specific micromechanics of single receptor-ligand bonds are the objective, SMFS is a more dependable tool for both physiological and mechanical reasons.

Acknowledgments The authors acknowledge partial support from the Howard Hughes Medical Institute and the National Cancer Institute (U54CA143868)

References

1. Fields BN, Knipe DM, Howley PM (2007) Fields virology, 5th edn. Wolters Kluwer Health/ Lippincott Williams & Wilkins, Philadelphia
2. Coffin JM, Hughes SH, Varmus H (1997) Retroviruses. Cold Spring Harbor Laboratory Press, Plainview, NY
3. 2008 Report on the Global AIDS Epidemic (2008) World Health Organization. UNAIDS Publication. ISBN 9291737119
4. Becker K, Hu Y, Biller-Andorno N (2006) Infectious diseases – a global challenge. Int J Med Microbiol 296(4–5):179–185
5. Barbaro G, Scozzafava A, Mastrolorenzo A, Supuran CT (2005) Highly active antiretroviral therapy: current state of the art, new agents and their pharmacological interactions useful for improving therapeutic outcome. Curr Pharm Des 11(14):1805–1843
6. Blankson JN, Persaud D, Siliciano RF (2002) The challenge of viral reservoirs in HIV-1 infection. Annu Rev Med 53:557–593
7. Earl PL, Doms RW, Moss B (1990) Oligomeric structure of the human immunodeficiency virus type 1 envelope glycoprotein. Proc Natl Acad Sci USA 87(2):648–652
8. Kwong PD, Wyatt R, Robinson J, Sweet RW, Sodroski J, Hendrickson WA (1998) Structure of an HIV gp120 envelope glycoprotein in complex with the CD4 receptor and a neutralizing human antibody. Nature 393(6686):648–659
9. Thomas DJ, Wall JS, Hainfeld JF, Kaczorek M, Booy FP, Trus BL et al (1991) gp160, the envelope glycoprotein of human immunodeficiency virus type 1, is a dimer of 125-kilodalton subunits stabilized through interactions between their gp41 domains. J Virol 65 (7):3797–3803
10. Wyatt R, Sodroski J (1998) The HIV-1 envelope glycoproteins: fusogens, antigens, and immunogens. Science 280(5371):1884–1888
11. Zhu P, Liu J, Bess J Jr, Chertova E, Lifson JD, Grise H et al (2006) Distribution and three-dimensional structure of AIDS virus envelope spikes. Nature 441(7095):847–852
12. Alkhatib G, Broder CC, Berger EA (1996) Cell type-specific fusion cofactors determine human immunodeficiency virus type 1 tropism for T-cell lines versus primary macrophages. J Virol 70(8):5487–5494
13. Arthos J, Deen KC, Chaikin MA, Fornwald JA, Sathe G, Sattentau QJ et al (1989) Identification of the residues in human CD4 critical for the binding of HIV. Cell 57(3):469–481
14. Deen KC, McDougal JS, Inacker R, Folena-Wasserman G, Arthos J, Rosenberg J et al (1988) A soluble form of CD4 (T4) protein inhibits AIDS virus infection. Nature 331(6151):82–84
15. Landau NR, Warton M, Littman DR (1988) The envelope glycoprotein of the human immunodeficiency virus binds to the immunoglobulin-like domain of CD4. Nature 334 (6178):159–162
16. Berger EA, Doms RW, Fenyo EM, Korber BT, Littman DR, Moore JP et al (1998) A new classification for HIV-1. Nature 391(6664):240
17. Kozak SL, Platt EJ, Madani N, Ferro FE Jr, Peden K, Kabat D (1997) CD4, CXCR-4, and CCR-5 dependencies for infections by primary patient and laboratory-adapted isolates of human immunodeficiency virus type 1. J Virol 71(2):873–882
18. Rizzuto C, Sodroski J (2000) Fine definition of a conserved CCR5-binding region on the human immunodeficiency virus type 1 glycoprotein 120. AIDS Res Hum Retroviruses 16(8):741–749

19. Sattentau QJ, Moore JP (1991) Conformational changes induced in the human immunodeficiency virus envelope glycoprotein by soluble CD4 binding. J Exp Med 174(2):407–415
20. Sattentau QJ, Moore JP, Vignaux F, Traincard F, Poignard P (1993) Conformational changes induced in the envelope glycoproteins of the human and simian immunodeficiency viruses by soluble receptor binding. J Virol 67(12):7383–7393
21. Zhang W, Canziani G, Plugariu C, Wyatt R, Sodroski J, Sweet R et al (1999) Conformational changes of gp120 in epitopes near the CCR5 binding site are induced by CD4 and a CD4 miniprotein mimetic. Biochemistry 38(29):9405–9416
22. Chan DC, Fass D, Berger JM, Kim PS (1997) Core structure of gp41 from the HIV envelope glycoprotein. Cell 89(2):263–273
23. Kilby JM, Hopkins S, Venetta TM, DiMassimo B, Cloud GA, Lee JY et al (1998) Potent suppression of HIV-1 replication in humans by T-20, a peptide inhibitor of gp41-mediated virus entry. Nat Med 4(11):1302–1307
24. Weissenhorn W, Dessen A, Harrison SC, Skehel JJ, Wiley DC (1997) Atomic structure of the ectodomain from HIV-1 gp41. Nature 387(6631):426–430
25. Melikyan GB, Markosyan RM, Hemmati H, Delmedico MK, Lambert DM, Cohen FS (2000) Evidence that the transition of HIV-1 gp41 into a six-helix bundle, not the bundle configuration, induces membrane fusion. J Cell Biol 151(2):413–423
26. Bandres JC, Shaw AS, Ratner L (1995) HIV-1 Nef protein downregulation of CD4 surface expression: relevance of the lck binding domain of CD4. Virology 207(1):338–341
27. Yang X, Kurteva S, Lee S, Sodroski J (2005) Stoichiometry of antibody neutralization of human immunodeficiency virus type 1. J Virol 79(6):3500–3508
28. Yang X, Kurteva S, Ren X, Lee S, Sodroski J (2006) Subunit stoichiometry of human immunodeficiency virus type 1 envelope glycoprotein trimers during virus entry into host cells. J Virol 80(9):4388–4395
29. Malmqvist M, Karlsson R (1997) Biomolecular interaction analysis: affinity biosensor technologies for functional analysis of proteins. Curr Opin Chem Biol 1(3):378–383
30. Brigham-Burke M, Edwards JR, O'Shannessy DJ (1992) Detection of receptor-ligand interactions using surface plasmon resonance: model studies employing the HIV-1 gp120/CD4 interaction. Anal Biochem 205(1):125–131
31. Navratilova I, Pancera M, Wyatt RT, Myszka DG (2006) A biosensor-based approach toward purification and crystallization of G protein-coupled receptors. Anal Biochem 353 (2):278–283
32. Sharma D, Balamurali MM, Chakraborty K, Kumaran S, Jeganathan S, Rashid U et al (2005) Protein minimization of the gp120 binding region of human CD4. Biochemistry 44 (49):16192–16202
33. Markosyan RM, Cohen FS, Melikyan GB (2005) Time-resolved imaging of HIV-1 Env-mediated lipid and content mixing between a single virion and cell membrane. Mol Biol Cell 16(12):5502–5513
34. Melikyan GB, Barnard RJ, Abrahamyan LG, Mothes W, Young JA (2005) Imaging individual retroviral fusion events: from hemifusion to pore formation and growth. Proc Natl Acad Sci USA 102(24):8728–8733
35. Henderson HI, Hope TJ (2006) The temperature arrested intermediate of virus-cell fusion is a functional step in HIV infection. Virol J 3:36
36. Chesla SE, Selvaraj P, Zhu C (1998) Measuring two-dimensional receptor-ligand binding kinetics by micropipette. Biophys J 75(3):1553–1572
37. Evans E, Ritchie K, Merkel R (1995) Sensitive force technique to probe molecular adhesion and structural linkages at biological interfaces. Biophys J 68(6):2580–2587
38. Huang J, Edwards LJ, Evavold BD, Zhu C (2007) Kinetics of MHC-CD8 interaction at the T cell membrane. J Immunol 179(11):7653–7662
39. Ounkomol C, Xie H, Dayton PA, Heinrich V (2009) Versatile horizontal force probe for mechanical tests on pipette-held cells, particles, and membrane capsules. Biophys J 96 (3):1218–1231

40. Pierres A, Eymeric P, Baloche E, Touchard D, Benoliel AM, Bongrand P (2003) Cell membrane alignment along adhesive surfaces: contribution of active and passive cell processes. Biophys J 84(3):2058–2070
41. Zarnitsyna VI, Huang J, Zhang F, Chien YH, Leckband D, Zhu C (2007) Memory in receptor-ligand-mediated cell adhesion. Proc Natl Acad Sci USA 104(46):18037–18042
42. Morita E, Sundquist WI (2004) Retrovirus budding. Annu Rev Cell Dev Biol 20:395–425
43. Bajpai S, Correia J, Feng Y, Figueiredo J, Sun SX, Longmore GD et al (2008) {alpha}-Catenin mediates initial E-cadherin-dependent cell-cell recognition and subsequent bond strengthening. Proc Natl Acad Sci USA 105(47):18331–18336
44. Panorchan P, Lee JS, Daniels BR, Kole TP, Tseng Y, Wirtz D (2007) Probing cellular mechanical responses to stimuli using ballistic intracellular nanorheology. Method Cell Biol 83:115–140
45. Bajpai S, Feng Y, Krishnamurthy R, Longmore GD, Wirtz D (2009) Loss of alpha-catenin decreases the strength of single E-cadherin bonds between human cancer cells. J Biol Chem 284(27):18252–18259
46. Chang MI, Panorchan P, Dobrowsky TM, Tseng Y, Wirtz D (2005) Single-molecule analysis of human immunodeficiency virus type 1 gp120-receptor interactions in living cells. J Virol 79 (23):14748–14755
47. Dobrowsky TM, Zhou Y, Sun SX, Siliciano RF, Wirtz D (2008) Monitoring early fusion dynamics of human immunodeficiency virus type 1 at single-molecule resolution. J Virol 82 (14):7022–7033
48. Hanley W, McCarty O, Jadhav S, Tseng Y, Wirtz D, Konstantopoulos K (2003) Single molecule characterization of P-selectin/ligand binding. J Biol Chem 278(12):10556–10561
49. Hanley WD, Wirtz D, Konstantopoulos K (2004) Distinct kinetic and mechanical properties govern selectin-leukocyte interactions. J Cell Sci 117(Pt 12):2503–2511
50. Panorchan P, George JP, Wirtz D (2006) Probing intercellular interactions between vascular endothelial cadherin pairs at single-molecule resolution and in living cells. J Mol Biol 358 (3):665–674
51. Panorchan P, Thompson MS, Davis KJ, Tseng Y, Konstantopoulos K, Wirtz D (2006) Single-molecule analysis of cadherin-mediated cell-cell adhesion. J Cell Sci 119(Pt 1):66–74
52. Bell GI (1978) Models for the specific adhesion of cells to cells. Science 200(4342):618–627
53. Hummer G, Szabo A (2001) Free energy reconstruction from nonequilibrium single-molecule pulling experiments. Proc Natl Acad Sci USA 98(7):3658–3661

Chapter 8
Visualization of the COPII Vesicle Formation Process Reconstituted on a Microscope

Kazuhito V. Tabata, Ken Sato, Toru Ide, and Hiroyuki Noji

Abstract Transport from the endoplasmic reticulum to the Golgi body is ensured by a protein complex called COPII. Because the COPII vesicles are covered by the COPII coat protein – which consists of the low molecular weight GTPase Sar1p, Sec23/24p, and Sec13/31p, the transported proteins are selectively incorporated into the COPII vesicles by binding directly to the COPII coat. In this study, we reconstituted the formation of COPII vesicles on artificial planar lipid bilayer membranes, and visualized the dynamics of fluorescent-labeled transported proteins at a single molecular level, using a Total Internal Reflection Fluorescence Microscope (TIRFM). Then, the clusters of cargo molecules were observed by addition of Sec13/31p, revealing that the cargo molecules were concentrated inside the clusters. In addition, it has been revealed that the non-cargo molecules were excluded from the clusters. In this communication, we discuss the dynamics of cargo molecule in the process of COPII vesicle formation.

Keywords COPII · Vesicular transport · Membrane traffic · Golgi body · Endoplasmic reticulum · Bet1p · Ufe1p · Sar1p · Sec23/24p · Sec13/31p · Minimum component · Reconstitution · Low molecular weight GTPase · Cargo concentration · TIRFM · Single molecule observation · Artificial lipid bilayer · Real time imaging

K.V. Tabata (✉) and H. Noji
The Institute of Scientific and Industrial Research, Osaka University, Ibaraki, Osaka 567-0047, Japan
e-mail: kazuhito@sanken.osaka-u.ac.jp; hnoji@sanken.osaka-u.ac.jp

K. Sato
Department of Life Sciences, Graduate School of Arts and Sciences, University of Tokyo, Meguro-ku, Tokyo 153-8902, Japan
e-mail: kensato@bio.c.u-tokyo.ac.jp

T. Ide
Graduate School of Frontier Biosciences, Osaka University, Suita, Osaka 565-0871, Japan
e-mail: ide@phys1.med.osaka-u.ac.jp

Y. Sako and M. Ueda (eds.), *Cell Signaling Reactions: Single-Molecular Kinetic Analysis*, DOI 10.1007/978-90-481-9864-1_8,

8.1 Vesicular Transport

Intracellular vesicular transport in eukaryotic cells is organized such that membrane-bound vesicles, 50–100 nm in diameter, loaded with select protein cargoes, bud from intracellular organelles, fuse with the membranes of recipient organelles, and thereby transfer their cargoes to their various destinations. These transport vesicles are coated with a protein complex called "coat protein." Their binding is regulated by low-molecular-weight GTPase. Inside the cell, there are several vesicular transport pathways; each one uses a different coat protein and low-molecular-weight GTPase [1]. Among these, COPII vesicles, which are coated with the Sec23/24p and Sec13/31p complexes (also known as coat protein complex II) undertake vesicular transport from the endoplasmic reticulum (ER) to the Golgi body (Fig. 8.1) [2]. The formation of COPII vesicles begins when Sar1p, a low-molecular-weight GTPase, is converted from a GDP (inactive) form to a GTP (active) form by the guanine nucleotide exchange factor (GEF), Sec12p, which is present on the ER [3–5]. As a result, the GTP form of Sar1p undergoes a structural change and binds to the membrane of the ER via its N-terminal region [6]. The Sar1p bound to the membrane of the ER binds to the Sec23/24p complex via Sec23p; Sec24p binds to cargo molecules on the ER membrane [7]. The Sar1p-Sec23/24p-cargo complex, formed on the ER membrane, is specifically referred to as the "prebudding complex." Sec13/31p complexes are thought to play a role in the assembly of prebudding complexes on the ER membrane by mediating attachment of the complexes to each other [8]. The assembled coat proteins change the curvature of the membranes and form COPII vesicles [9, 10].

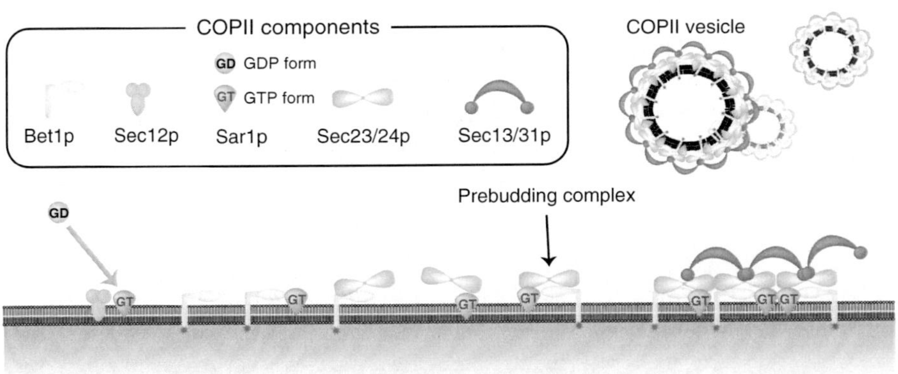

Fig. 8.1 Minimum machinery required for coat protein complex II (COPII) vesicle formation. The *illustration* shows the process of COPII vesicle formation (*model*). The prebudding complex constructed from Bet1p, Sar1p, and Sec23/24p. The *inset* shows COPII components. Bet1p; cargo protein, Sec12p; guanine nucleotide exchange factor (GEF), Sar1p; low molecular weight GTPase, Sec23/24p; cargo binding protein (Sec24) and Sar1p and Sec31 binding protein (Sec23), Sec13/31p; outer shell protein of COPII vesicle. Sec23/24p and Sec13/31p called coat protein complex II.

Experiments investigating the vesicle formation process, consisting of in vitro reconstitution of vesicle formation, have been carried out for more than 10 years [11]. However, because ER membranes that contained large amounts of proteins unrelated to vesicle formation were used, it was difficult to determine the vesicle formation process in detail. In recent years, however, a complete reconstitution system of the reactions involved in the formation of COPII vesicles has been established [12]. This system prepares proteoliposomes to reconstitute cargo molecules (Bet1p) and forms COPII vesicles by the addition of purified Sar1p, Sec23/24p, and Sec13/31p. These protein groups are called "minimum components" and are known to be the minimal requirements for the formation of COPII vesicles. This reconstitution system allows further understanding of the process of COPII vesicle formation. However, the elementary steps in the formation of COPII vesicles are still not clear. Therefore, we used a single-molecule observation technique to visualize the vesicle formation process under a microscope in real time [13]. This technique is known to be a powerful tool for studying reaction dynamics and the elementary steps in a process. In this communication, we report a case in which we reproduced the reactions involved in COPII vesicle formation under a microscope. We also discuss the dynamics of the proteins that are incorporated into the vesicles.

8.2 Microscopic Analysis of the Planar Lipid Bilayer

In case vesicle formation reactions were reconstituted under the microscope, there was need to form a lipid bilayer membrane on the microscope. Therefore, we utilized a microscope system developed by Ide et al. [14]. This system was developed for the purpose of observing the fluorescence emitted by dye-labeled ion channels at a single molecular level, as well as measuring electric current. Figure 8.2a shows an overview of this system. In this system, two chambers are set up on the objective lens: the lower and upper chambers. At the bottom surface of the upper chamber, there is an aperture with a diameter of 100–200 μm; lipid molecules dissolved in a solvent were applied around it. Because both chambers are full of buffer, the immersion of the upper chamber into the lower chamber causes the formation of a lipid bilayer membrane at the aperture of the upper chamber. This is known as "black lipid membrane" (Fig. 8.2b). The formation of a lipid bilayer membrane can be used to distinguish and monitor the capacitive current from the lipid membrane. Because the lipid membrane is a dielectric substance, sandwiching the lipid membranes between two conductors (buffer) allows this system to function as a capacitor. In fact, a decrease in the thickness of the lipid membrane leads to an increase in the capacitive current. Figure 8.2c shows the change in capacitive current during bilayer formation. Comparison of the capacitive current of the lipid membrane (Dotted line) and the capacitive current after formation of the lipid bilayer membrane (Solid line) shows that the

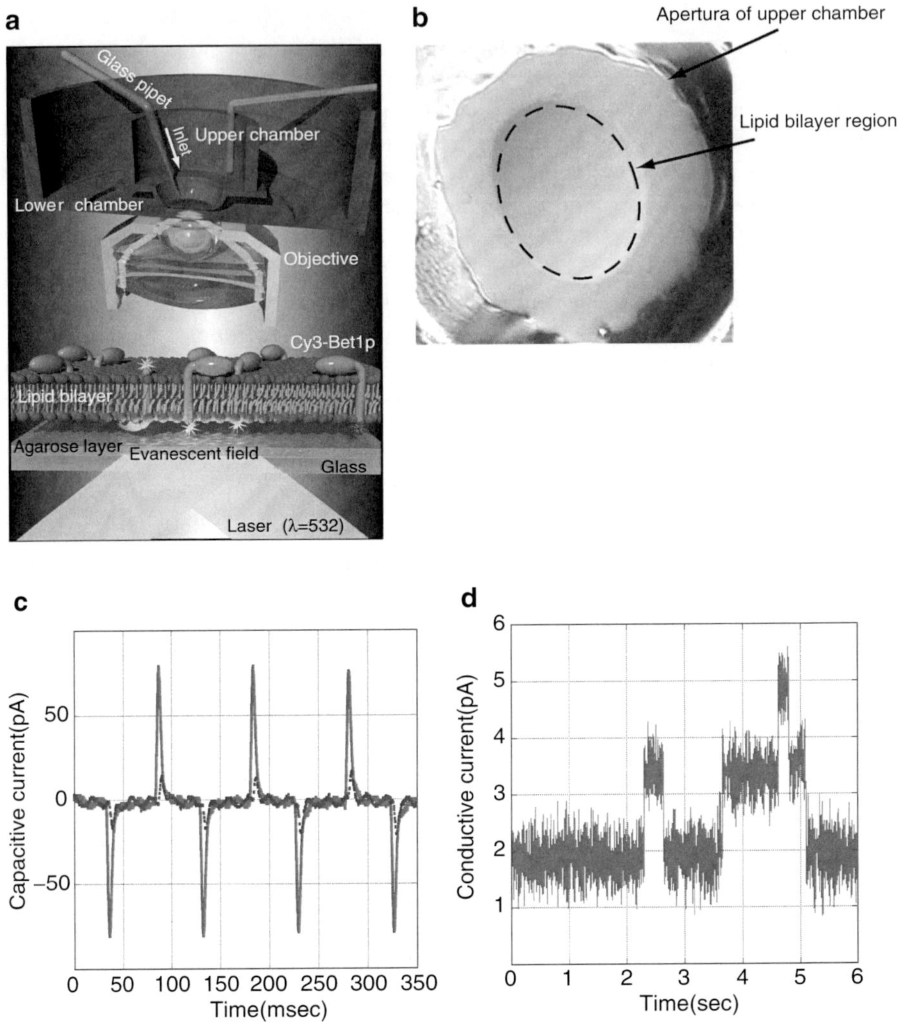

Fig. 8.2 Schematice illustration of the planar lipid bilayer microscope and character of planar lipid bilayer membrane. (**a**) Overview of the horizontal planar lipid bilayer formation apparatus. The apparatus consists of 2 chambers. A thin plastic film with a small hole (~100 μm) was attached to the bottom of the upper chamber, and an artificial planar lipid bilayer was formed horizontally across the hole on an agarose-coated coverslip. Planar lipid bilayers were formed horizontally by placing the major-minor lipid mix dissolved in squalene or *n*-decane at the hole of the upper chamber, as described previously [12, 14]. Upper chamber could be moved vertically using a micromanipulator. (**b**) Microscopic view of planar lipid bilayer. (**c**) Variations in the capacitive current during formation of the lipid bilayer membrane. *Dotted line* indicates the low capacitive current before planar lipid bilayer membrane formation. *Solid line* indicates the high capacitive current after lipid bilayer membrane formation. The value of the characteristic membrane capacitance is 0.45 $\mu F/cm^2$ in a lipid bilayer membrane formed using *n*-decane and 0.92 $\mu F/cm^2$ in a lipid bilayer membrane formed using squalene. (**d**) Channel current of gramicidin D reconstituted in the planar lipid bilayer. Gramicidin D (100 ng/ml) was introduced into the planar lipid bilayer membrane and +100 mV voltage was applied. Channel current is produced by coupling of the half channels of gramicidin D on the upper and lower leaflet of the lipid bilayer membrane.

lipid membrane was electrically thinner. Because the lipid bilayer membrane is of uniform thickness, the capacitive current indicates a constant value by forming the lipid bilayer membrane. Gramicidin D, shown in Fig. 8.2d, conducts an electrical current by inducing the formation of a dimer from the monomers present in each monolayer. To summarize, the channel current can be observed only when a lipid bilayer membrane is formed. The channel electric current was measured using the lipid bilayer that was formed (Fig. 8.2d). These results show that lipid bilayer membrane formation was observed under the microscope. Therefore, the required evanescent illumination was achieved under the microscope by creating a laser induction optical system using a commercial inverted microscope (IX-71, Olympus). Single-molecule observation of the fluorescent dye attached to proteins reconstituting the planar lipid bilayer membrane was performed by bringing the planar lipid bilayer membrane into contact with the glass surface of the lower chamber (which was coated with agarose), and by evanescent illumination. For the reconstitution of proteins inside the lipid bilayer, proteoliposomes, which incorporate membrane proteins, were prepared and sprayed onto the planar membrane through a glass pipette. Then, the proteoliposomes were autonomously fused with the planar lipid bilayer membrane. As a result, the membrane proteins were introduced into the planar lipid bilayer membrane.

8.3 Visualization of the COPII Vesicle Formation Process

8.3.1 Microscopic Observation of Vesicle Budding from the Planar Lipid Bilayer Membrane

First, we conducted experiments on the budding of COPII vesicles from the planar lipid bilayer membrane. The experimental system used was set up as shown in Fig. 8.3a. The planar lipid bilayer membrane was reconstituted in the upper chamber, which was located 10–20 μm above the glass surface of the lower chamber. Proteoliposomes were introduced through the upper part of the upper chamber. Bet1p-Cy3 (Bet1p labeled with a fluorescent dye) was reconstituted inside the planar lipid bilayer membrane. Subsequently, the components of COPII, Sar1p, Sec23/24p, and Sec13/31p, were introduced between the upper and lower chambers. The resulting COPII vesicles were isolated from the planar membrane; they reached by their diffusion into the surface of the lower chamber, which was illuminated with an evanescent light. COPII vesicles were detected as fluorescent spots. The results obtained from this experiment are shown in Fig. 8.3b. A difference in the rate of increase of fluorescent spots was found between the occurrence (GTP) and nonoccurrence (GDP) conditions of vesicle formation. In light of these results, the budding of COPII vesicles is thought to occur from the planar lipid bilayer membrane.

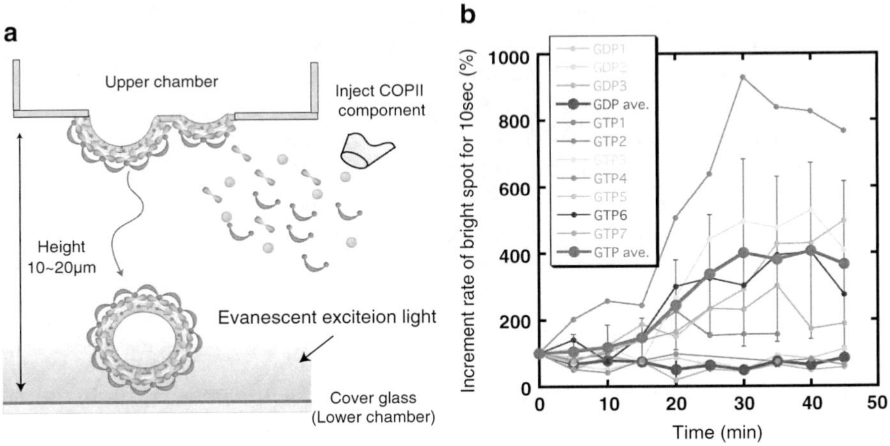

Fig. 8.3 COPII vesicle budding from the planar lipid bilayer membrane. (**a**) Schematic illustration of downward COPII vesicle budding. (**b**) Time-dependent increase in the number of fluorescent particles that appeared on the coverslip after the addition of the COPII components. Each thin trace corresponds to an individual experiment. The numbers of fluorescent particles were normalized so that the initial number of fluorescent particles on the coverslip was 100%. Thick traces with standard error deviations represent the averages of all thin traces. Sar1p (70 ng), Sec23/24p (410 ng), and Sec13/31p (1.4 μg) were added to the bilayer membrane reconstituted with Bet1p-Cy3 (95.6 molecules μm^{-2}) and Sec12Δlum (without luminal domain of Sec12p) at $t = 0$ s in the presence of GTP (0.1 mM) or GDP (0.1 mM) as illustrated in the (a).

8.3.2 *Single-Molecule Observation of the Cargo Protein (Bet1p-Cy3) Reconstituted in Planar Lipid Bilayer Membranes*

Because the budding of COPII vesicles from planar lipid bilayer membranes had been confirmed from the results described in the previous section, we observed the vesicle formation process. We reconstituted Bet1p-Cy3, at labeling rates of 90% or more, in planar lipid bilayer membranes and observed their behavior. Planar lipid bilayer membranes loaded with Bet1p-Cy3 were evanescent-illuminated, and fluorescent spots, laterally diffused inside the planar lipid bilayer membranes, were observed (Fig. 8.4a). The diffusion coefficient was calculated from the trajectory of the fluorescent spots found, and was found to be 4.5 ± 2.0 $\mu m^2/s$. This is consistent with the diffusion coefficients of commonly known membrane proteins in lipid bilayer membranes [15–17]. Additionally, under the same conditions, we suppressed the diffusion of Bet1p-Cy3 in the planar lipid bilayer membranes by using the membrane-binding protein annexin V, and analyzed the intensity of the resulting fluorescent spots [18]. It was apparent that the histogram, drawn on the basis of the intensity of fluorescence found in all fluorescent spots, had a single peak (Fig. 8.4b). Bet1p-Cy3 is present mostly as a single molecule. This is because analysis of the

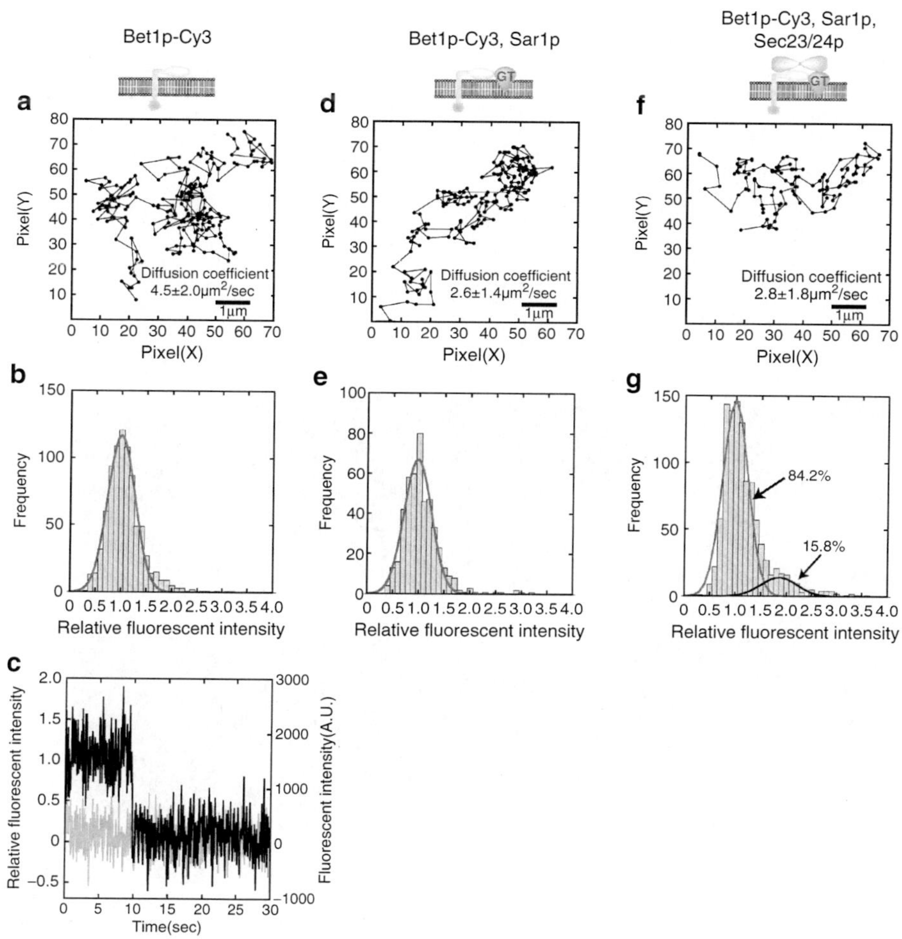

Fig. 8.4 Single-molecule analysis of reconstituted Bet1p-Cy3 in a planar lipid bilayer. (**a**, **d**, and **f**) Trajectory of a single Bet1p-Cy3 in the planar lipid bilayer membrane. We calculated the value of the diffusion coefficient (*inset*). The time interval between 2 consecutive observations was 5 ms. (**b**, **e**, and **g**) Distribution of the fluorescence intensity of Bet1p-Cy3 in the presence of the COPII component(s). We added Sar1p (70 ng) (**e**), Sar1p (70 ng), and Sec23/24p (410 ng) (**g**) to the bilayer membranes reconstituted with Bet1p-Cy3 (0.28 molecules μm^{-2}) and Sec12Δlum (**b**). (**c**) A typical fluorescent spot of Bet1p-Cy3 in the bilayer membrane was photobleached in a single step (*black*) to the background level (*gray*).

time course of the intensity of the fluorescent spots in the peak revealed that most spots exhibited a single-step bleaching (Fig. 8.4c). We also conducted experiments by adding Sar1p only or both Sar1p and Sec23/24p to the planar lipid bilayer membranes reconstituted with Bet1p-Cy3; their diffusion coefficients were $2.6 \pm 1.4\ \mu m^2/s$ and $2.8 \pm 1.8\ \mu m^2/s$, respectively. In either case, diffusion

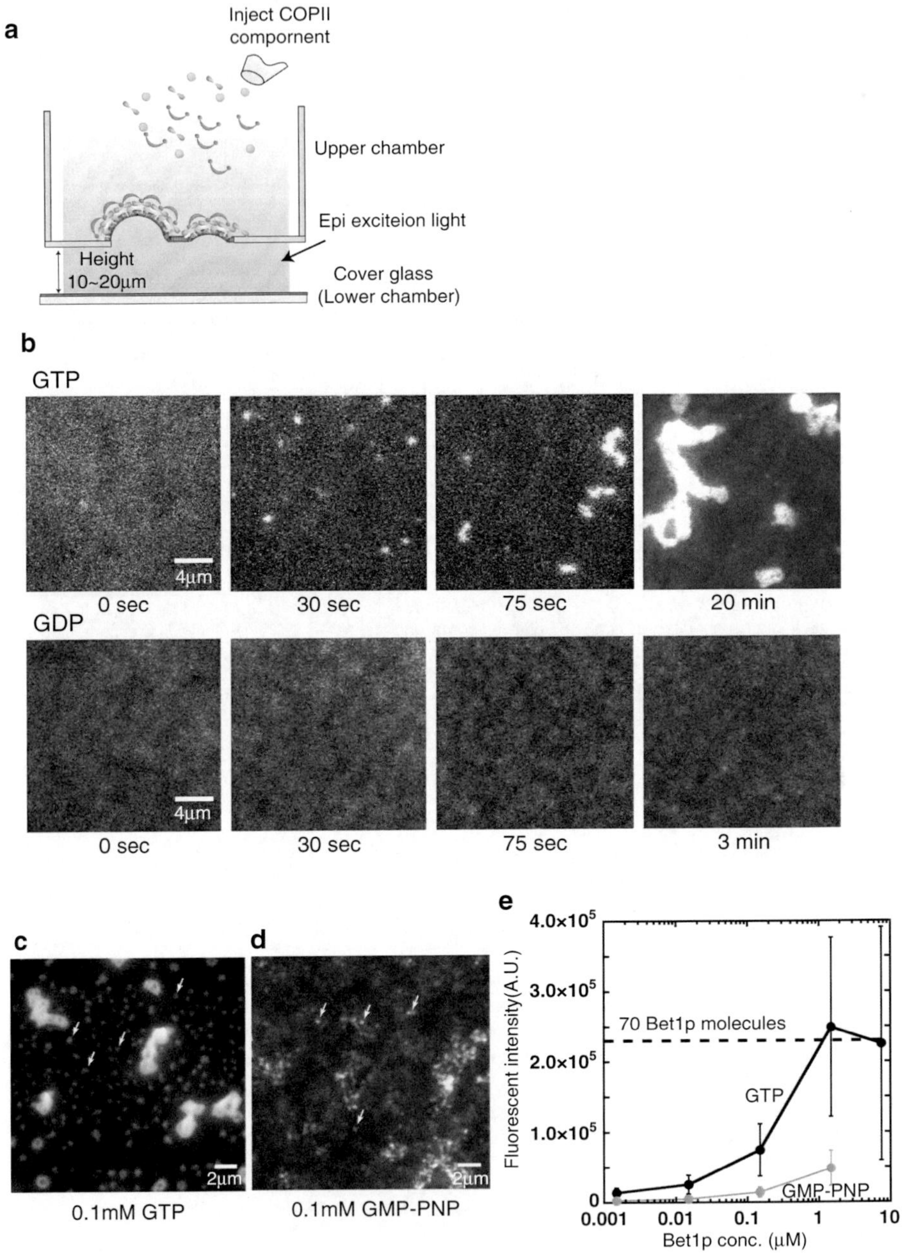

Fig. 8.5 Bet1p-Cy3 is concentrated in the cluster by GTP hydrolysis of Sar1p. (**a**) Schematic illustration of Bet1p-Cy3 molecules clustering experiment. (**b**) Sequential images of the Bet1p-Cy3 clustering in the lipid bilayer membrane. We added Sar1p (70 ng), Sec23/24p (410 ng), and Sec13/31p (1.4 mg) to the lipid bilayer membrane reconstituted with Bet1p-Cy3 (95.6 0molecules μm^{-2}) and Sec12Δlum at $t = 0$ s in the presence of GTP or GDP as illustrated in the *upper*

coefficients decreased by roughly 50% in comparison to those that were only reconstituted with Bet1p-Cy3 (Fig. 8.4d, f). These results suggest that Bet1p-Cy3, Sar1p, and Sec23/24p interact on the planar lipid bilayer membranes. In addition, analysis of the intensity of fluorescence revealed that the shape of the histogram after the addition of Sar1p was the same as in the case of planar lipid bilayer membranes containing Bet1p-Cy3 alone (Fig. 8.4e), whereas a slight peak for the dimer was observed when Sar1p and Sec23/24p were added simultaneously (Fig. 8.4g). These results also show that the prebudding complex (Bet1p-Cy3, Sar1p, and Sec23/24p complex) weakly forms a dimer on the planar lipid bilayer membranes.

8.3.3 Clusterization of the Prebudding Complex and Concentration of Bet1p-Cy3 by Sec13/31p

The behavior of prebudding complexes has been studied in previous experiments. In this study, we conducted experiments where Sec13/31p, which is responsible for the formation of the outermost layers of the COPII coat, was used. Using the same method described previously, we added COPII components to the planar lipid bilayer membranes reconstituted with Bet1p-Cy3 (Fig. 8.5a). After the addition, observation with epi-excitation was started. As shown in Fig. 8.5b, in the presence of GTP, fluorescent spots grow bigger with time. However, this phenomenon was not observed in the presence of GDP. Moreover, because this phenomenon was not observed in the above experiments, even after the addition of Sec23/24p and Sar1p, it shows that Sec13/31p induced the binding of prebudding complexes with each other, leading to the formation of clusters. Our results are the first to show real-time clustering induced by Sec13/31p. In addition, when diffusion of the formed cluster molecules was suppressed in the presence of GTP in order to analyze the clusters in detail, small clusters that were previously invisible because of their rapid diffusion on the membranes became visible (Fig. 8.5c). Plotting of fluorescence intensity of the clusters versus the concentration of Bet1p-Cy3 reconstituted in the membranes revealed saturation of the fluorescence intensity in the region with high concentrations of Bet1p-Cy3 (Fig. 8.5e). By dividing the saturation fluorescent intensity by

Fig. 8.5 (continued) panel. Epifluorescence images were acquired at frame rate of 66 ms/frame. (**c** and **d**) COPII components (70 ng, Sar1p; 410 ng, Sec23/24p; and 1.4 mg, Sec13/31p) were added from the upper chamber to the bilayer membrane (used *n*-decane as a solvent) formed in an aqueous environment and fluorescent spots were acquired in the presence of GTP (*left panel*) or GMP-PNP (*right panel*). The bilayer membrane was placed on a coverslip. Fluorescent images were taken under evanescent field illumination. Typical small cluster is indicated by arrowhead. (**e**) A correlation plot of the Bet1p-Cy3 concentration versus the mean fluorescent intensity of the clusters obtained from (**c** and **d**).

the fluorescent intensity of single molecules of Bet1p-Cy3, it could be presumed that 70 molecules of Bet1p-Cy3 were present inside the small clusters. This result might have indicated the number of cargo molecules incorporated into COPII vesicles. Previous studies have shown there are two known types of the COPII cage structure: cuboctahedral and icosidodecahedral structures [19, 20]. If the Bet1p and Sec23/24p contained in these structures bind at a ratio of 1:1, the number of Bet1p molecules contained per structure is thought to be 48 and 120, respectively [21]. Previous results also demonstrate that our results might reflect the number of cargo molecules inside the vesicles. Therefore, it is likely that these small clusters correspond to the COPII vesicle buds that are formed before the vesicles pinch off. An important question is whether the large clusters observed among the small clusters were also active in vesicle generation. However, this was not clear because of the difficulty involved in resolving the movement of individual COPII vesicles away from the membrane with our detection system. Next, we changed the conditions of the clusterization experiment. Among the proteins included in the components of COPII was Sar1p: a low-molecular-weight GTPase with hydrolytic effects on GTP. In addition, the nucleotide state of Sar1p is also known to control its binding to and dissociation from membranes. Thus, these experiments were performed by the addition of GTP. In this study, however, we performed experiments with the non-hydrolyzable GTP analog, GMP-PNP. Compared to the aforementioned results obtained with GTP, the clusters that were formed with GMP-PNP had a markedly different appearance – they had a speckled pattern (Fig. 8.5d). In addition, the fluorescence intensity of the clusters was at most, one-tenth that of the clusters formed using GTP (Fig. 8.5e). In previous reports, it was reported that the incorporated proteins in COPII vesicles were concentrated [22]. However, the dynamics of concentration were not reported. Our results provide the novel finding that GTP hydrolysis by Sar1p causes concentration of Bet1p-Cy3 inside the clusters.

8.3.4 Why Does Hydrolysis of GTP by Sar1p Require Concentration of Cargo Proteins Inside the Clusters?

Why does this concentration occur? This will be discussed on the basis of the previous results. As mentioned earlier, Sar1p controls its binding to and dissociation from membranes through its nucleotide state. In addition, the binding and dissociation of Sar1p is also accompanied by binding and dissociation of other COPII components [23, 24]. We speculate that in our experiments, only prebudding complexes selectively remain on the membrane – this is how concentration occurs (Fig. 8.6). Sec13/31p binds with Sec23/24p [25] and forms a cluster. Therefore, only the complex including Sec23/24p on the lipid bilayer membrane is introduced into the cluster. Because of this, in our experiment, the complexes that were incorporated into a cluster were prebudding complexes, Bet1p-Cy3-Sec23/24p complexes, and Sar1p-Sec23/24p complexes (Fig. 8.6a). However, under our

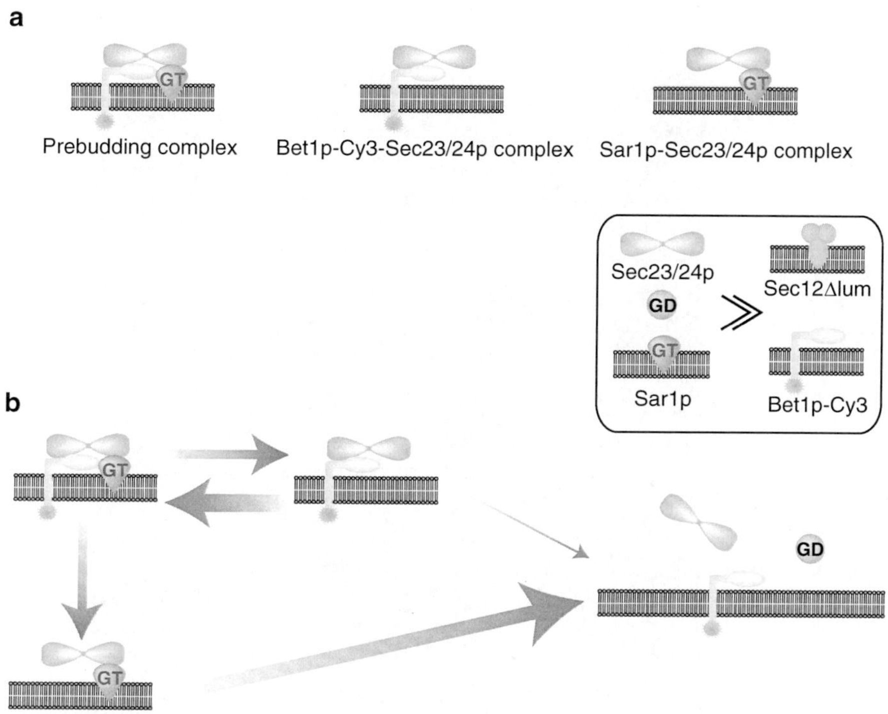

Fig. 8.6 Model of Bet1p-Cy3 concentration process. (**a**) These molecules incorporate into the cluster by Sec13/31p. (**b**) Prebudding complex concentration scheme on the lipid bilayer membrane. The *inset* shows the ratio of number of molecules under the experimental condition.

experimental conditions, Sar1p was largely in excess compared to Bet1p-Cy3. Consequently, the Bet1p-Cy3-Sec23/24p complex binds to Sar1p and becomes a prebudding complex. The dissociation of Sar1p from the membranes, which accompanies the hydrolysis of GTP, will cause the prebudding and Sar1p-Sec23/24p complexes to dissociate completely from the membranes. Sato et al. showed that the dissociation of Sec23/24p from prebudding complexes was more than several folds its dissociation from mutated Bet1p prebudding complexes in which Bet1p could not bind to Sec23/24p. Therefore, the dissociation of Sec23/24p from the prebudding complex revealed that the dissociation occurs delaying from the hydrolysis of GTP by Sar1p [24]. This shows that the dissociation of Sec23/24p occurs after dissociation of Sar1p from the membrane, and that the dissociation of Sec23/24p is expected to be slow because of the binding between Sec23/24p and Bet1p. In other words, the prebudding complex is thought to assume the state of the Bet1p-Cy3-Sec23/24p or Sar1p-Sec23/24p complex before it completely dissociates from the membrane. Then, through rebinding to Sar1p, the Bet1p-Cy3-Sec23/24p complex returns to the prebudding complex. Bet1p-Cy3 released from the complex binds with Sar1p and Sec23/24p and forms the prebudding complex again. Consequently, the prebudding complex becomes the predominantly present complex on the

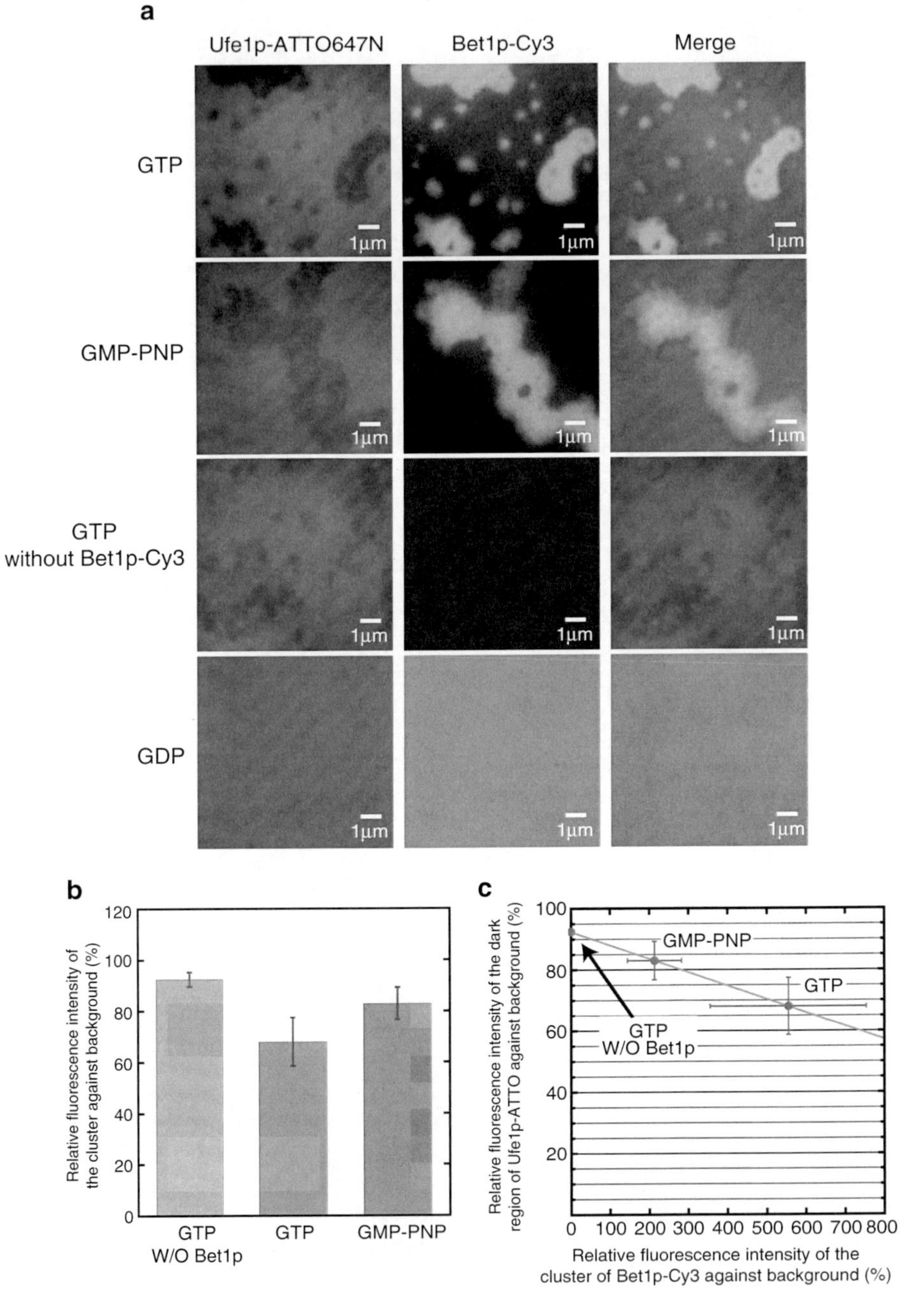

Fig. 8.7 Ufe1p-ATTO647N is excluded from Bet1p-Cy3 cluster in the presence of the COPII components. (**a**) Sar1p (70 ng), Sec23/24p (410 ng), and Sec13/31p (1.4 μg) were added from the upper chamber to the bilayer membrane (used *n*-decane as a solvent) reconstituted with Ufe1p-ATTO647N (41.3 molecules μm^{-2}), Bet1p-Cy3 (113.1 molecules μm^{-2}), and Sec12Δlum

membranes (Fig. 8.6b). Meanwhile, in case the non-hydrolyzable GMP-PNP is used up, Sar1p does not dissociate from the membrane. As a result, the prebudding complexes containing Bet1p-Cy3 and the Sar1p-Sec23/24p complexes lacking Bet1p-Cy3 remain incorporated inside the cluster. Because the Sar1p-Sec23/24p complexes lacking Bet1p-Cy3 are present in large amounts on the membrane, the clusters appear to be speckled. Therefore, it is thought that as a result of the binding and dissociation of Sar1p, which accompanies the hydrolysis of GTP, only prebudding complexes containing cargo proteins will accumulate on the membrane; consequently, concentration proceeds. Under our experimental conditions, Sar1p and Sec23/24p were added in excess to Bet1p-Cy3. However, there is no example of quantitative examination of whether the amounts of Sar1p and Sec23/24p are greater than the amount of Bet1p in the cell. There is no reason to believe that our model is incorrect because vesicle formation has been confirmed in our system, and there are several results that correspond to previous reports.

8.3.5 Exclusion of Non-cargo Proteins by the Minimum COPII Components

In COPII vesicular transport, cargo proteins recognized by COPII components are incorporated into the vesicles and transported to the Golgi body. On the other hand, non-cargo proteins are not incorporated into the vesicles; instead, they must remain where they are. However, there certainly are proteins that are transported by mistake. Such proteins are returned from the Golgi body to the ER by COPI transport vesicles [26]. This correction mechanism prevents mistakenly transported proteins from functioning elsewhere. Nonetheless, a large number of non-cargo proteins are present on the ER. Without a mechanism for the exclusion of non-cargo proteins from transport vesicles, the non-cargo proteins present on the membrane of the ER would be incorporated into vesicles in their existing concentrations, thereby lowering the efficiency of transportation. Thus, we reconstituted two proteins together inside planar lipid bilayer membranes: Ufe1p, which was not transported by a COPII vesicle, and Bet1p. We examined how non-cargo proteins would behave during cluster formation. According to previous studies, when Ufe1p is used as a non-cargo protein, it is known to be unrecognized by COPII components [22, 23, 27]. Ufe1p was labeled with a fluorescent dye, ATTO647N, and was observed together with Bet1p-Cy3, with two colors. The results are displayed in Fig. 8.7a. Dark contrasting

Fig. 8.7 (continued) molecules in the presence of GTP or GMP-PNP. Addition of Bet1p-Cy3 was omitted in some cases. The following procedures were same as those described in Fig. 8.5. ATTO647N (*left*) and Cy3 (*center*) fluorescence channels and merged image (*right*) are shown. ATTO647N and Cy3 fluorescence is shown in *left* and *center*, respectively. (**b**) We compared the relative density of Ufe1p-ATTO647N in the large clusters with that outside the cluster. (**c**) A correlation plot of the Ufe1p-ATTP647N exclusion rate from (b) versus the Bet1p-Cy3 concentration ratio calculated from the Bet1p-Cy3 channel.

regions were observed in the presence of GTP or GMP-PNP in the results of the Ufe1p-ATTO647N experiment. In addition, it was understood that the cluster shape, found in the Bet1p-Cy3 experiment, coincided with the dark contrasting regions (in a merged image). Furthermore, in the results of the Ufe1p-ATTO647N experiment, calculation of the exclusion rate expressed as the ratio of the fluorescence intensity of the dark region to the background of the dark region revealed that the exclusion rate of the GTP condition was the highest, and that its fluorescence intensity decreased by as much as 35% (Fig. 8.7b). These results show that Ufe1p-ATTO647N was excluded from the Bet1p-Cy3 clusters. Although it has previously been considered that a mechanism based on the recognition between ER exit signals and COPII components is less likely to underlie cargo exclusion events, our findings demonstrate, for the first time, that the exclusion of non-cargo proteins is at least partly driven by the minimal COPII machinery. Why is Ufe1p excluded outside the cluster despite the fact that it is not recognized by the COPII components? One possible model to explain this is that tight structure of the COPII vesicle may lead to physical exclusion of non-cargo proteins that are not already incorporated into the vesicle. Further analysis is necessary to make this discussion possible. It is worth noting that the results of the Ufe1p-ATTO647N experiment conducted without Bet1p-Cy3 showed that fluorescent intensity was uneven, and there was an overall decrease in fluorescent intensity. In addition, the calculated exclusion rate was several percent (Fig. 8.7b). This suggests that exclusion occurs only in COPII components. Moreover, using the results obtained with Bet1p-Cy3, we calculated the concentration ratio as the ratio of the fluorescent intensity of the background to that inside the clusters; the relationship with the exclusion rate was plotted (Fig. 8.7c). The exclusion rate and the concentration ratio were found to be correlated. From these results, it can be thought that the more Bet1p-Cy3 is incorporated inside the clusters, the more Ufe1p-ATTO647N is excluded. These results suggest the possibility that exclusion is associated with the internal structure of a cluster. Stagg et al. were able to create a cage structure using only Sec13/31p, and they reconstituted its 3D structure from transmission electron microscope images [10, 19, 20]. By hypothesizing the size of vesicles capable of entering the cage on the basis of their structure, and by allocating the number of Sec23/24p molecules corresponding to that of Sec13/31p, we probably coated 80% of the surface of the vesicles [21]. Their surface was almost entirely unexposed. In addition, because Bet1p and Sar1p can enter the cage, more empty spaces at the surface of the vesicles are lost. However, because it is unknown whether the internal structure of clusters is uniform, further studies will be required to resolve this issue.

8.4 Conclusion

In this study, we succeed in visualizing the process of formation of COPII vesicles from several proteins under the microscope, and we were able to clarify the behavior of the cargo proteins during the process of vesicle formation. In addition, we also revealed that in the process of vesicle formation, the non-cargo proteins are excluded. In the future, we plan to examine how the proteins involved in vesicle formation behave during the process. Additionally, we showed that planar bilayer microscopy is an effective tool for real-time imaging of phenomena which occur on the cell membrane and organelle membranes.

Acknowledgments We thank Toshio Yanagida and Takayuki Nishizaka for their assistance in facilitating our understanding of planar bilayer microscopy and total internal reflection fluorescence microscopy (TIRFM), respectively. We thank Akihiko Nakano for discussions.

References

1. Bonifacino JS, Glick BS (2004) The mechanisms of vesicle budding and fusion. Cell 116:153–166
2. Barlowe C, Orci L, Yeung T, Hosobuchi M, Hamamoto S et al (1994) COPII: A membrane coat formed by Sec proteins that drive vesicle budding from the endoplasmic reticulum. Cell 77:895–907
3. Oka T, Nishikawa S, Nakano A (1991) Reconstitution of Gtp-binding Sar1 protein function in ER to Golgi transport. J Cell Biol 114:671–679
4. Futai E, Hamamoto S, Orci L, Schekman R (2004) GTP/GDP exchange by Sec12p enables COPII vesicle bud formation on synthetic liposomes. EMBO J 23:4146–4155
5. Barlowe C, Denfert C, Schekman R (1993) Purification and characterization of Sar1p, a small Gtp-binding protein required for transport vesicle formation from the endoplasmic-reticulum. J Biol Chem 268:873–879
6. Huang M, Weissman JT, Beraud-Dufour S, Luan P, Wang C et al (2001) Crystal structure of Sar1-GDP at 1.7. A resolution and the role of the NH_2 terminus in ER export. J Cell Biol 155:937–948
7. Bi XP, Mancias JD, Goldberg J (2007) Insights into COPII coat nucleation from the structure of Sec23 center dot Sar1 complexed with the active fragment of sec31. Dev Cell 13:635–645
8. Antonny B, Madden D, Hamamoto S, Orci L, Schekman R (2001) Dynamics of the COPII coat with GTP and stable analogues. Nat Cell Biol 3:531–537
9. Lee MCS, Orci L, Hamamoto S, Futai E, Ravazzola M, Schekman R (2005) Sar1p N-terminal helix initiates membrane curvature and completes the fission of a COPII vesicle. Cell 122:605–617
10. Stagg SM, LaPointe P, Balch WE (2007) Structural design of cage and coat scaffolds that direct membrane traffic. Curr Opin Struct Biol 17:221–228
11. Salama NR, Yeung T, Schekman RW (1993) The Sec13p complex and reconstitution of vesicle budding from the ER with purified cytosolic proteins. EMBO J 12:4073–4082
12. Matsuoka K, Orci L, Amherdt M, Bednarek SY, Hamamoto S et al (1998) COPII-coated vesicle formation reconstituted with purified coat proteins and chemically defined liposomes. Cell 93:263–275
13. Tabata KV, Sato K, Ide T, Nishizaka T, Nakano A, Noji H (2009) Visualization of cargo concentration by COPII minimal machinery in a planar lipid membrane. EMBO J 28:3279–3289
14. Ide T, Yanagida T (1999) An artificial lipid bilayer formed on an agarose-coated glass for simultaneous electrical and optical measurement of single ion channels. Biochem Biophys Res Commun 265:595–599
15. Cherry RJ, Godfrey RE, Peters R (1982) Mobility of bacteriorhodopsin in lipid vesicles. Biochem Soc Trans 10:342–343

16. Peters R, Cherry RJ (1982) Lateral and rotational diffusion of bacteriorhodopsin in lipid bilayers: Experimental test of the Saffman-Delbruck equations. Proc Natl Acad Sci USA 79:4317–4321
17. Gambin Y, Lopez-Esparza R, Reffay M, Sierecki E, Gov NS et al (2006) Lateral mobility of proteins in liquid membranes revisited. Proc Natl Acad Sci USA 103:2098–2102
18. Ichikawa T, Aoki T, Takeuchi Y, Yanagida T, Ide T (2006) Immobilizing single lipid and channel molecules in artificial lipid bilayers with annexin A5. Langmuir 22:6302–6307
19. Stagg SM, Gurkan C, Fowler DM, LaPointe P, Foss TR et al (2006) Structure of the Sec13/31 COPII coat cage. Nature 439:234–238
20. Stagg SM, LaPointe P, Razvi A, Gurkan C, Potter CS et al (2008) Structural basis for cargo regulation of COPII coat assembly. Cell 134:474–484
21. Fath S, Mancias JD, Bi X, Goldberg J (2007) Structure and organization of coat proteins in the COPII cage. Cell 129:1325–1336
22. Matsuoka K, Morimitsu Y, Uchida K, Schekman R (1998) Coat assembly directs v-SNARE concentration into synthetic COPII vesicles. Mol Cell 2:703–708
23. Sato K, Nakano A (2004) Reconstitution of coat protein complex II (COPII) vesicle formation from cargo-reconstituted proteoliposomes reveals the potential role of GTP hydrolysis by Sar1p in protein sorting. J Biol Chem 279:1330–1335
24. Sato K, Nakano A (2007) Mechanisms of COPII vesicle formation and protein sorting. FEBS Lett 581:2076–2082
25. Lederkremer GZ, Cheng YF, Petre BM, Vogan E, Springer S et al (2001) Structure of the Sec23p/24p and Sec13p/31p complexes of COPII. Proc Natl Acad Sci USA 98:10704–10709
26. Kreis TE, Lowe M, Pepperkok R (1995) COPs regulating membrane traffic. Annu Rev Cell Dev Biol 11:677–706
27. Ballensiefen W, Ossipov D, Schmitt HD (1998) Recycling of the yeast v-SNARE Sec22p involves COPI-proteins and the ER transmembrane proteins Ufe1p and Sec20p. J Cell Sci 111 (Pt 11):1507–1520

Chapter 9
In Vivo Single-Molecule Microscopy Using the Zebrafish Model System

Marcel J. M. Schaaf and Thomas S. Schmidt

Abstract In recent years, several groups have succeeded in extending single-molecule microscopy technology to the level of a living vertebrate organism using the zebrafish embryo as a model system. In this chapter an overview will be presented of these studies and three lines of research will be discussed. First, work will be presented in which fluorescent proteins have been imaged at the single-molecule level in the epidermis of zebrafish embryos using total internal reflection fluorescence (TIRF) microscopy. Second, investigations will be presented in which individual quantum dots have been imaged in zebrafish embryos by selective plane illumination microscopy (SPIM). Third, studies will be discussed in which fluorescence correlation spectroscopy (FCS) has been applied to fluorescent proteins in zebrafish embryos. All three research lines show discrepancies between results obtained in zebrafish embryos and data obtained in cell cultures, illustrating the relevance of performing these studies in an in vivo model.

Keywords Cdc42 · Epidermis · Fibroblast growth factor (Fgf) · Fibroblast growth factor receptor (Fgfr) · Fluorescence correlation spectroscopy (FCS) · Green fluorescent protein (GFP) · HRAS · Quantum dot (QD) · Scanning FCS · Selective plane Illumination microscopy (SPIM) · Single-molecule microscopy · Single wavelength fluorescence cross-correlation spectroscopy (SW-FCCS) · Total internal reflection fluorescence (TIRF) microscopy · Two-focus FCS · Zebrafish

M.J.M. Schaaf (✉)
Molecular Cell Biology, Institute of Biology, Leiden University, Einsteinweg 55, 2333CC, Leiden, The Netherlands
e-mail: m.j.m.schaaf@biology.leidenuniv.nl

T.S. Schmidt
Physics of Life Processes, Institute of Physics, Leiden University, Niels Bohrweg 2, 2333CA, Leiden, The Netherlands
e-mail: schmidt@physics.leidenuniv.nl

Y. Sako and M. Ueda (eds.), *Cell Signaling Reactions: Single-Molecukar Kinetic Analysis*, DOI 10.1007/978-90-481-9864-1_9,

List of Abbreviations

CCD	charge-coupled device
CHO	Chinese hamster ovary
dpf	days post fertilization
eYFP	enhanced yellow fluorescent protein
FCS	fluorescence correlation spectroscopy
Fgf	fibroblast growth factor
GFP	green fluorescent protein
hpf	hours post fertilization
HSPG	heparan sulphate proteoglycan
IQGAP1	IQ motif containing GTPase activating protein 1
LCK	lymphocyte-specific protein tyrosine kinase
mRFP	monomeric red fluorescent protein
PTU	phenylthiourea
QD	quantum dot
SPIM	selective plane illumination microscopy
SW-FCCS	Single Wavelength Fluorescence Cross-Correlation Spectroscopy
TIRF	total internal reflection fluorescence
TMR	tetramethylrhodamine

9.1 Introduction

Studying the dynamics of individual molecules in living cells provides a wealth of knowledge about processes that take place. Application of this type of technology provides a detailed insight into the mobility pattern of specific subpopulations of molecules in the cell. Since alterations in the molecular mobility pattern and domain organization due to extracellular signals are accurately measured, a detailed view of all individual molecular events during signal transduction processes is provided. In addition, by studying the movements of molecules in a cellular environment, information can be gathered about the organization of the cell, for example the occurrence and nature of several types of domains in the plasma membrane.

Using fluorescent labeling techniques, molecules can be imaged and tracked using a laser-based fluorescence microscopy setup equipped with a high sensitivity charge-coupled device (CCD) camera [38, 40]. This technique has provided insight into the mobility of several types of proteins, at a time resolution of $\sim$5 ms and a positional accuracy of $\sim$40 nm in most studies. The labeling of proteins and lipids can be done using small fluorescent dye molecules which is generally applied for labeling of membrane proteins on an extracellular domain [36, 39]. For visualization of intracellular proteins labeling reactions are mostly performed in vitro, and the labeled protein is subsequently microinjected into cultured cells [6]. A non-invasive technique for labeling of intracellular proteins is provided by the application of autofluorescent proteins like green fluorescent protein (GFP) which can genetically be fused to an endogenous protein [8, 14, 15, 23]. This approach has the advantage that the protein of interest is always coupled to the fluorescent molecule in a one-to-one ratio. Only a few of the commercially available autofluorescent proteins are suitable for use in this technology, and their relatively high bleaching efficiency significantly limits the time period for which an individual protein can be traced [9].

Until recently, all single-molecule microscopy studies on living cells had been applied to cultured eukaryotic cells, often derived from immortalized cell lines. It is likely that the condition of these cells does not represent the situation of cells that constitute a specific tissue in a living multicellular organism. In addition, using cultured cells it is impossible to study the behavior of individual signaling molecules in relationship to processes like development and pathogenesis of various diseases. It is therefore desirable that the application of single-molecule microscopy techniques is extended to the level of a living organism. In many studies non-vertebrate model organisms like *Caenorhabditis elegans* or *Drosophila melanogaster* are unsuitable because of their evolutionary distance to vertebrates, whereas vertebrate model organisms like the mouse are more difficult to manipulate, mainly because of their relatively large size. A suitable vertebrate model system for in vivo single-molecule microscopy is the zebrafish embryo system.

The zebrafish has many advantages over other vertebrate animal model systems [11, 20, 21, 44]. It is small, easily maintained and breeds well under laboratory conditions. Each female can produce hundreds of eggs per day that are fertilized externally. Upon fertilization, the embryos develop rapidly and most organ systems have been formed a few days later (Fig. 9.1). The ex utero development makes the zebrafish embryos easily accessible for genetic manipulation by microinjection of DNA or mRNA in order to overexpress proteins in the developing embryo (Fig. 9.2a, b). Furthermore, an increasing number of transgenic zebrafish lines are

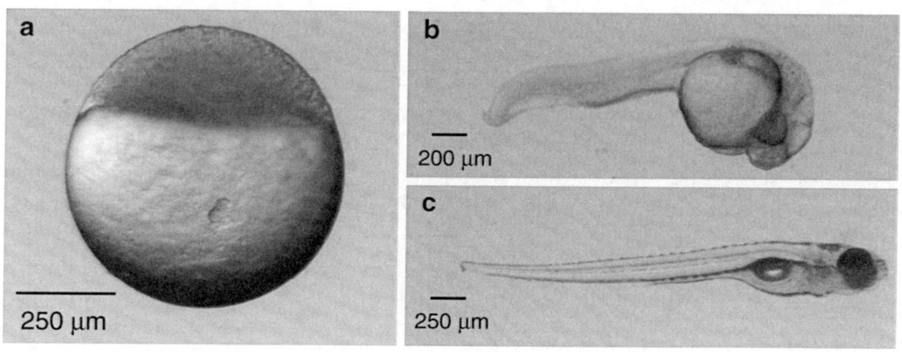

Fig. 9.1 Pictures of zebrafish developmental stages, taken with a stereo light microscope. (**a**) Sphere stage embryo, approximately 4 h post fertilization (hpf), consisting of several thousands of undifferentiated cells on top of the yolk sac. (**b**) Embryo at 1 day post fertilization (dpf), when most major organs are already recognizable. (**c**) Larva at 7 dpf, which is able to feed and swim.

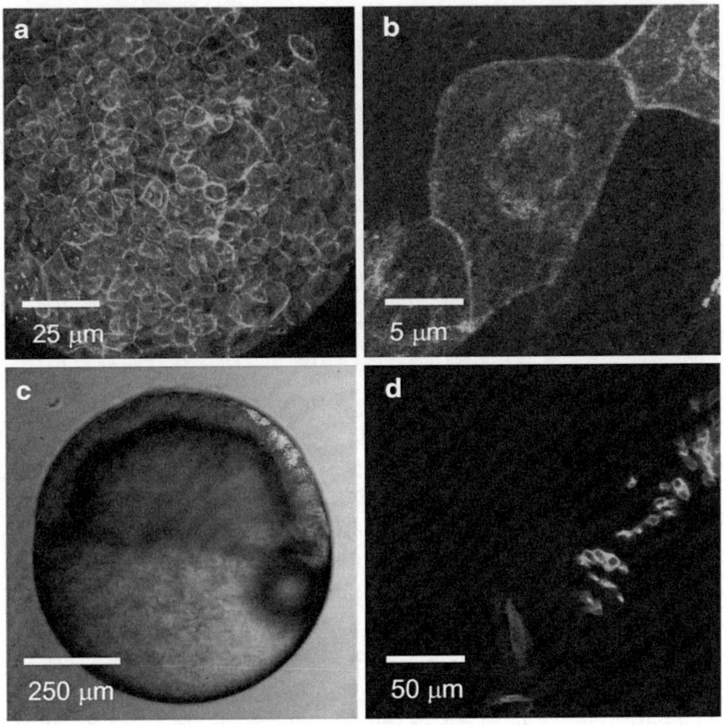

Fig. 9.2 Confocal microscopy pictures of eYFP-C10HRAS expression in manipulated zebrafish embryos. (**a**) *Top view* of sphere stage embryo, injected with eYFP-C10HRAS mRNA at the 1–2 cell stage. All cells of the embryo show expression of eYFP-C10HRAS, localized at the plasma membrane. (**b**) Detail of 24 hpf embryo, injected with eYFP-C10HRAS DNA at the 1–2 cell stage. The expression pattern is mosaic, as illustrated by the individual cells expressing eYFP-C10HRAS located in the trunk region of the embryo. (**c**) Embryo (6 hpf) transplanted with cells from donor embryos injected with eYFP-C10HRAS mRNA. The presence of individual eYFP-C10HRAS expressing cells is visible. (**d**) Individual eYFP-C10HRAS expressing cells in transplanted embryo as shown in (**c**).

available, which can overexpress proteins in all cells or just in a subset of cells, depending on the promoter that is used to drive the expression of the transgene. Zebrafish embryos are transparent, which allows for microscopic imaging at the cellular or subcellular level when performed in combination with fluorescent labeling of specific cells or proteins. This way, real-time imaging of GFP-labeled cells has been performed successfully and has provided detailed information of the movements of specific individual cells in these developing organisms, for example during early embryonic development [30], migration of primordial germ cells [2], development of the lateral line system [7], and in response to wounding [32]. Even dynamic processes at the subcellular level, like synapse formation, have been imaged using GFP-tagging of specific proteins in a living embryo [26].

Recently, several studies have applied single-molecule microscopy techniques in zebrafish embryos. First, total internal reflection fluorescence (TIRF) microscopy

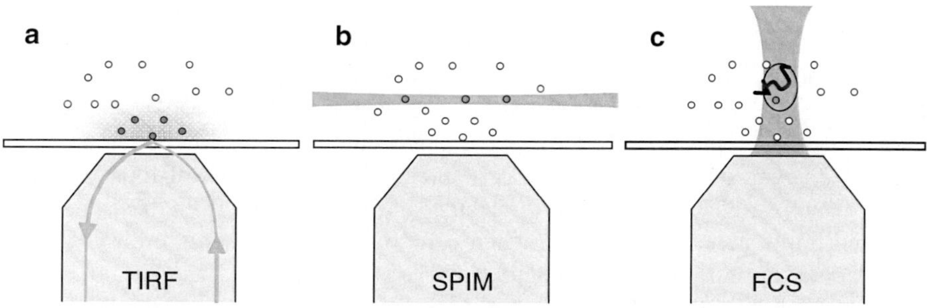

Fig. 9.3 Techniques used to detect single molecules in zebrafish embryos. (**a**) Total internal reflection fluorescence (TIRF) microscopy. In so-called objective-based TIRF microscopy, the laser beam is positioned at the periphery of a high numerical aperture objective, and leaves the objective at such an angle that, when it reaches the interface between the coverglass and the aqueous environment of the specimen, it will be reflected totally. An evanescent field is created that only excites molecules present in the region less than approximately 100 nm above the coverglass. (**b**) Selective plane illumination microscopy (SPIM). In SPIM, a sample is illuminated from the side in a well-defined volume around the focal plane of the detection optics. This way only molecules in the plane of interest are excited. (**c**) Fluorescence correlation spectroscopy (FCS). In FCS, the fluorescence intensity in a fixed confocal detection volume is measured. Molecules diffusing through this volume cause fluctuations in this intensity, enabling the analysis of the diffusion pattern of these molecules using the temporal autocorrelation.

has been used to detect and trace membrane proteins in zebrafish embryos (Fig. 9.3a). Second, by selective plane illumination microscopy (SPIM) single quantum dots were visualized in living embryos (Fig. 9.3b). Third, using fluorescence correlation spectroscopy (FCS) of fluorescent proteins in embryos, the mobility pattern of these proteins was determined in vivo (Fig. 9.3c). In this review, we will describe the different technologies used to detect single molecules in these studies, and discuss the applications and limitations of these methods.

9.2 Total Internal Reflection Fluorescence Microscopy of Fluorescent Proteins in Zebrafish Embryos

Recently, we have successfully detected and tracked single fluorescent proteins in living zebrafish embryos [37]. At the 1-cell stage, embryos were injected with mRNA encoding enhanced yellow fluorescent protein (eYFP) fused with the ten most C-terminal amino acids of the human HRAS protein which constitute its membrane anchor. By post-translational modification, the three most C-terminal amino acids are cleaved off this eYFP-C10HRAS protein and to the remaining part an S-farnesyl and two S-palmitoyl groups are attached. These lipid groups anchor the protein in the cytoplasmic leaflet of the plasma membrane lipid bilayer. The eYFP-C10HRAS protein is a well-characterized molecule in single-molecule microscopy applied in mammalian cell cultures [23] and shows a diffusion pattern similar to the human lymphocyte-specific protein tyrosine kinase (LCK) and KRAS membrane anchors [25] and the activated full length HRAS protein [23–25].

Subsequently, the resulting embryos, which express eYFP-C10HRAS in all cells during the first 2 days of development, were used for single-molecule microscopy on a wide-field epi-fluorescence microscopy setup at different developmental stages. However, due to the high background signal from the out-of-focus fluorescence, visualization of single eYFP molecules in the field of focus in these embryos appeared impossible. Furthermore, when cells from injected donor embryos were transplanted into non-injected acceptor embryos (Fig. 9.2c, d), detection of single eYFP molecules in the transplanted embryos was prevented by the high level of autofluorescence in the donor embryo. For example, the autofluorescence in embryos at 8 h post fertilization (hpf) was three to five times higher than the autofluorescence in cultured cells from a zebrafish cell line. The quantum yield of eYFP is already high (~0.6), such that a required three to five fold higher quantum yield necessary for single molecule detection zebrafish embryos cannot be achieved. Therefore, an alternative approach in this type of study must be employed that requires a different fluorescent protein that has excitation/emission spectra shifted towards longer wavelengths. At longer wavelengths the autofluorescence appears to be dramatically reduced (Fig. 9.4). However, a fluorophore that can be excited at these wavelengths, that is monomeric and has a sufficiently high quantum yield and low bleaching efficiency does not exist yet.

An alternative spectroscopic approach was used in order to diminish background fluorescence. We used total internal reflection fluorescence (TIRF) microscopy, a technique in which only molecules present in the region less than approximately 100 nm above the coverglass are excited. Hence molecules outside this plane are not illuminated and remain invisible. In our study objective-based TIRF was used, in which the laser beam is positioned at the periphery of a high numerical aperture objective. The excitation beam will leave the objective at such an angle that, when it reaches the interface between the coverglass and the aqueous environment of the specimen, it will be reflected totally. Thus, the light no longer propagates into the

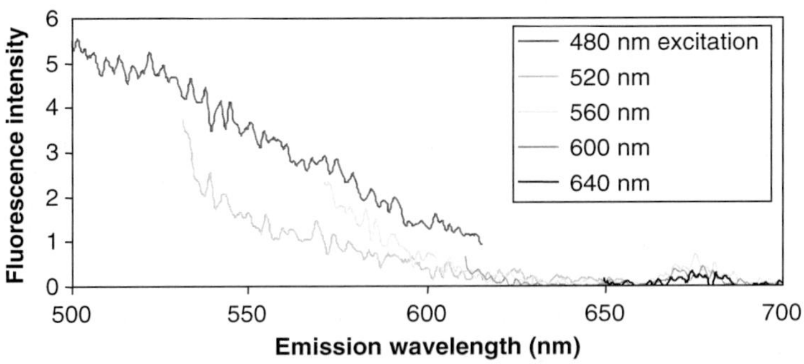

Fig. 9.4 Analysis of the autofluorescence spectrum in zebrafish embryonic cells. Zebrafish embryos were dissociated between 3.5 and 4 h after fertilization and lysed using a detergent buffer. The fluorescence spectrum of the lysate was measured using a fluorometer. Data show that the intensity of this signal significantly decreases with increasing excitation and emission wavelengths.

aqueous medium, but there is a small amount of the electromagnetic field that penetrates perpendicular to the interface. This field, also termed the evanescent field, has the same wavelength as the light that creates it, and is capable of exciting fluorophores. TIRF microscopy has previously been used to image single fluorescently labeled kinesin molecules [45] and ATP molecules [43] in vitro, and for imaging of individual membrane proteins in cell cultures [14, 36].

Detection of single eYFP molecules using TIRF microscopy on zebrafish embryos injected with eYFP-C10HRAS mRNA was optimal when applied to cells in the tail region of 2-day-old embryos (Fig. 9.5). In older embryos expression of the fluorescent protein is not detectable due to degradation of the injected mRNA. Generation of a transgenic zebrafish line would overcome this temporal limitation. At the age of 2 days, the two outer cell layers of the fish skin are homogenous layers of living cells forming the so-called superficial stratum (in contrast to the skin of terrestrial vertebrates in which the outer layer consists of dead keratinized cells). The tissue in the tail region is solid enough at this stage, allowing a thin layer of agarose to be placed on top of the embryo in order to press the zebrafish tail against the coverglass without damaging the tissue. This way, the outer membrane of the outer cell layer of the epidermis is positioned in the evanescent field, enabling visualization of single eYFP-C10HRAS molecules with a signal to noise ratio which is superior to wide-field imaging in cultured cells. The high signal-to-noise ratio resulted in a positional accuracy of single-molecule detection of ~23 nm.

By tracking the eYFP-C10HRAS molecules in the membrane of the outer epidermal cell layer, their diffusion pattern was analyzed. This revealed the presence of two fractions of molecules: a fast fraction containing ~75% of molecules and a slow fraction constituting ~25%. The fast fraction was freely diffusing through the membrane ($D \sim 0.5\ \mu m^2/s$), whereas the slow fraction showed confined diffusion ($D \sim 0.04\ \mu m^2/s$) inside a domain of ~120 nm. Data obtained in primary cell

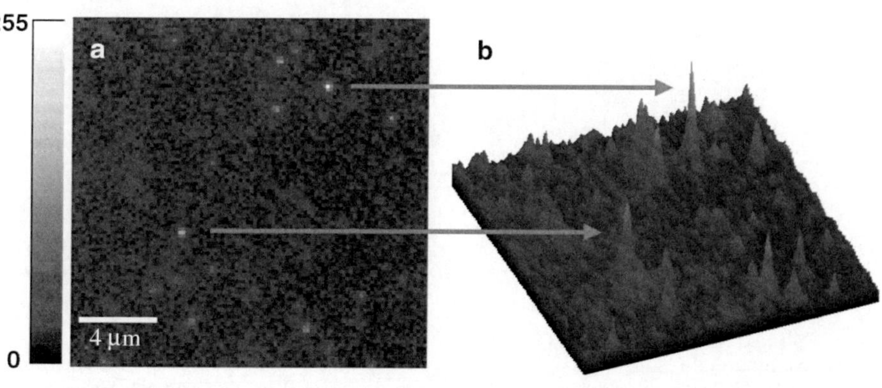

Fig. 9.5 Single-molecule microscopy of YFP-C10HRAS molecules in the membrane of epidermal cells in 2 dpf embryos. (**a**) Image of apical membrane of a YFP-C10HRAS-expressing captured by TIRF microscopy. (**b**) 3D representation of fluorescence intensities of image shown in (**a**). Two fluorescence intensity peaks that can be attributed to a single YFP molecule are indicated by *arrows*.

cultures derived by dissociation of 4 hpf embryos and in a zebrafish fibroblast cell line (ZF4) were similar to the data obtained in the 2-day-old embryos, but surprisingly the fast fraction showed confined diffusion in both cell culture systems (Fig. 9.6). This confinement is suggested to be a result of the organization of the cytoskeleton, which forms barriers limiting the diffusion of membrane proteins, and it has been shown to display large variations between cell types in mammalian cell culture studies [19]. Our studies further showed that membrane organization might also be dependent on the local context in an intact living organism.

9.3 Selective Plane Illumination Microscopy of Quantum Dots in Zebrafish Embryos

Detection and tracking of single particles in living zebrafish embryos was performed by Friedrich et al. [5] Single quantum dots (QDs) could be visualized in living zebrafish embryos. QDs are semiconductor nanocrystals that can be as small as 10–20 nm. QDs most commonly used in biological research consist of a cadmium selenide core coated with an additional zinc sulfide shell to improve the optical properties of the material. This core-shell material is further coated with a polymer shell that allows the materials to be conjugated to biological molecules and to retain their optical properties in the intracellular environment. Because the emission

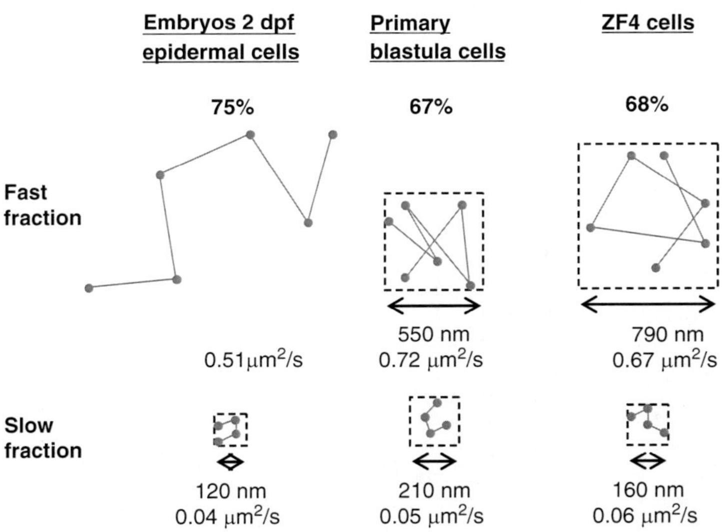

Fig. 9.6 Summary of results from eYFP-C10HRAS diffusion pattern analysis in zebrafish cells. Data were obtained in epidermal cells of 2 dpf embryos, in primary cell cultures derived from blastula stage embryos (~4 hpf), and in a fibroblast cell line (ZF4). The size of the fast fraction (%) is indicated, as well as the (initial) diffusion coefficients ($\mu m^2/s$) and domain size (nm) for both fractions. Data are very similar between the different systems, except for the confinement of the fast fraction, which is absent in the epidermal cells of the 2 dpf embryos.

spectrum is solely dependent on the size of the QD, a wide range of fluorophores has been generated with non-overlapping spectra. Their brightness and photostablity that outcompetes all other fluorescent dyes makes them ideal tools for single-molecule imaging [3]. Rieger et al. [33] explored the possibility of using QDs in zebrafish embryos, and found that injection of up to 10^8 QDs at the 2-cell stage did not affect the subsequent development of the embryos. The injected streptavidin-conjugated QDs were excluded from the cell nuclei, and remained within targeted cells since they do not cross gap junctions.

Friedrich et al. [5] performed the imaging of the QDs by selective plane illumination microscopy (SPIM). In SPIM, a sample is illuminated from the side in a well-defined volume around the focal plane of the detection optics [12]. This way only the plane of interest is excited, resulting in less background signal from other focal planes as compared to epifluorescence microscopy, especially when images are processed using deconvolution software [13]. As compared to confocal microscopy, SPIM results in less photobleaching and – damage and since a charge-coupled device (CCD) camera collects all the fluorescence from the sample at once, scanning is not required. Therefore image acquisition is very fast. Another important advantage of SPIM is that the excitation light is shaped and positioned independent of the detection optics in terms of numerical aperture and working distance. The latter is especially important in application of SPIM and related techniques in larger samples like fish embryos [13, 16].

Indeed, Friedrich et al. [5] were able to detect individual QDs in zebrafish embryos at the 32- and 64-cell stage ($\sim$2 hpf) at a tissue depth beyond 500 μm. QDs were tracked with a maximal time resolution of $\sim$10 ms. Although the QDs had a tendency to form aggregates which was reflected in a secondary intensity shoulder in the fluorescence intensity distribution, lower intensity signals could be attributed to individual QDs based on the blinking pattern that is typical for single QDs. The tracking of the QDs showed at the 32-cell stage a very high motility ($D \sim 0.08\ \mu m^2/s$), reflecting a directed flow of QDs in the embryos, whereas at the 64-cell stage the QDs showed a significantly lower motility in line with a confined diffusion model in which the QDs were freely diffusing ($D \sim 0.11\ \mu m^2/s$) but were confined within a domain of approximately 120 nm in size. This apparently sudden decrease in motility probably reflects the cessation of the cytoplasmic flux from the yolk into the embryonic cells and the closure of the cytoplasmic bridges between the cells, which is characteristic for the 32- to 64-cell stage transition [18].

9.4 Fluorescence Correlation Spectroscopy of Fluorescent Proteins in Zebrafish Embryos

Fluorescence Correlation Spectroscopy (FCS) has been applied to living cells for several years [1, 17]. Recently, several papers have been published describing the application of FCS in vivo using zebrafish embryos [35, 41, 42, 46]. In these studies, the concentration and diffusion coefficients of fluorescently labeled molecules were determined by FCS at specific locations in the embryos, requiring a setup that combines FCS with confocal laser scanning microscopy.

Shi et al. [41, 42] have performed FCS on 3-day-old embryos. First, they assessed the autofluorescence in these embryos, showing that the eye and the yolk sac are the regions with very high levels of autofluorescence compared to the rest of the body, on excitation at 488 nm [42]. The autofluorescence in the muscle fibers of the trunk region (which is the middle part of the embryo's body, posterior to the brain and anterior to the tail, which mainly contains epidermal, muscular, vascular and neuronal tissue) showed emission maxima at 570 and 630 nm. Pigmentation of the embryos, a significant source of autofluorescence, was suppressed by treatment with phenylthiourea (PTU). Alternatively, mutant strains can be used that show low pigmentation levels (albino, nacre, casper [10, 22]). Second, they analyzed the maximal depth in the tissue at which reliable FCS data can still be generated [42]. Using both single- and two photon excitation, FCS could be applied reliably up to a depth of approximately 50 μm. Beyond 80 μm no FCS data could be obtained at all. It should be noted that in this study eGFP was used, which is not a very efficient fluorophore for two-photon microscopy applications. When tetramethylrhodamine (TMR) was used instead, FCS data could be obtained up to a tissue depth of approximately 200 μm using two-photon excitation [42].

Subsequently, the authors determined the diffusion coefficient of eGFP expressed in motor neurons and in muscle fibers in 3-day-old embryos (45.3 ± 9.4 and 36.5 ± 4.1 $\mu m^2/s$, respectively), that were injected with the expression vector plasmid in a single cell of a 16-cell stage embryo resulting in a mosaic expression pattern 3 days later. In a similar way the diffusion coefficient was measured of the eGFP-tagged chemokine receptor CXCR4b in the plasma membranes of muscle fibers (0.60 ± 0.19 $\mu m^2/s$).

An extension of this work is the application of Single Wavelength Fluorescence Cross-Correlation Spectroscopy (SW-FCCS) in 3 days post fertilization (dpf) zebrafish embryos [41]. Classically, cross-correlation studies in FCS are done using two different fluorophores that are excited by the use of two different laser light wavelengths. In SW-FCCS, the two fluorophores are excited using only one laser line, simplifying alignment and reducing problems of spherical aberrations. However, this approach requires the use of fluorophores with significant overlap in their excitation spectrum, but emission spectra that are different enough to enable separation of the emitted light using appropriate filters. Shi et al. have used the combination of eGFP and monomeric red fluorescent protein (mRFP) as labels for SW-FCCS (excitation at 514 nm) in order to study the interaction between the small Rho-GTPase Cdc42 and the actin-binding scaffolding protein IQGAP1 (IQ motif containing GTPase activating protein 1). The concentration of the bound and free protein fraction was calculated based on data obtained from auto- and cross-correlation. The dissociation constant (k_d) for the interaction with IQGAP1 and a constitutively active and a dominant-negative Cdc42 mutant were determined, and the values determined were ~100 nM for the active mutant and ~1,500 nM for the dominant-negative mutant. This same experiment was performed in Chinese hamster ovary (CHO) cell cultures which differed significantly from data obtained in zebrafish embryos. This difference illustrated the relevance of performing these studies in a physiologically relevant system in comparison to cell culture studies.

In addition, the group of Wohland were able to measure blood flow in the dorsal aorta and cardinal vein in the trunk region of zebrafish embryos in 3-day-old embryos, taking advantage of the autofluorescence of the serum [42]. They measured the diastolic and systolic flow velocities at different positions in the blood vessels, using a 1.5 μm step size, showing that the flow velocity is maximal in the center of the vessel and close to zero at the vessel wall. By scanning the laser beam with a defined speed and direction the flow direction was determined [29]. This approach was recently extended to the third dimension, using a scan length of 0.5 μm in all dimensions [28].

Yu et al. [46] measured the concentration and diffusion coefficient of the eGFP-tagged fibroblast growth factor (Fgf)8 in the extracellular space of gastrulating embryos (sphere to germ ring stage (~4–6 hpf)), after injecting mRNA encoding this protein in a single cell of 32-cell stage embryos. This way they demonstrated that the occurrence of a concentration gradient of Fgf8 is a result of free diffusion of the protein away from the source and subsequent uptake and degradation by embryonic cells. They found a high diffusion coefficient for eGFP-Fgf8 (53 $\mu m^2/s$), comparable to that observed for eGFP secreted in the extracellular space (86 $\mu m^2/s$). These values indicate that this protein is diffusing freely in the extracellular space of the embryos. Furthermore, two-focus FCS, in which the cross-correlation between two confocal volumes that are laterally separated by a known fixed distance, was used to investigate the characteristics of the Fgf8 mobility. A small fraction of Fgf8 molecules moved slowly ($D \sim 4\ \mu m^2/s$), and this fraction appears to be associated with the extracellular matrix component heparan sulphate proteoglycan (HSPG), which was demonstrated by a decrease in the size of this fraction of molecules upon injection of the enzyme heparinase I into the extracellular space, which cleaves the sugar chains of HSPGs. The shape of the concentration gradient was analyzed by measuring the concentration of Fgf8 at different locations in the embryo. The gradient appeared to depend on endocytosis, since expression of a dominant-negative version of the GTPase Dynamin, which is required for endocytosis, resulted in a shallower gradient. In addition, overexpression of another GTPase required for endocytosis, RAB5c, induced a steeper gradient.

In another report, Ries et al. [35] analyzed the interaction between Fgf8 and its receptors, fibroblast growth factor receptor(Fgfr)1 and Fgfr3, in the membrane of cells that take part in gastrulation by using modular scanning FCS. These proteins were expressed in the embryos by injection of the mRNA encoding the receptor at the 1 cell stage and the ligand mRNA in a single cell of a 32 cell-stage embryo. In this study the authors applied different extension of the FCS technique in a modular fashion. First, in order to prevent artifacts due to cell and membrane movement scanning FCS was used in which the detection volume is repeatedly shifted along a line perpendicular to the plasma membrane [31, 34]. The position of the plasma membrane was determined based on the maximal fluorescence intensity along this line. This information was subsequently used to correct for cell and membrane movements. Second, two-focus FCS was applied [4]. By combining scanning and two-focus FCS the detection volume could be determined which enabled the determination of the absolute protein concentration. Third, using alternating dual-color excitation spectral cross-talk was prevented. By measuring auto- and cross-correlations of mRFP-labeled receptors and eGFP-labeled ligands the concentrations of receptor-ligand complexes and unoccupied receptors was calculated, which enabled the in vivo determination of the dissociation constants (~100 nM for Fgf8 binding to Fgfr1 and ~46 nM for Fgf8 binding to Fgfr4), indicating preferred binding of Fgf8 to Fgfr4. Surprisingly, these data are not in line with previously published data obtained in cell cultures, in which potencies for these receptor-ligand combinations were lower than 200 nM and a larger difference (20-fold) in potency between Fgfr1 and Fgfr4 was observed [27]. The authors explain this discrepancy by pointing out that molecules in the extracellular matrix or unlabeled endogenous ligand could interfere with ligand binding in their in vivo experiments, making their in vivo model more physiologically relevant.

9.5 Conclusions

In three different types of experiments fluorescent molecules have recently been observed at the single-molecule level in living zebrafish embryos and their mobility pattern has been analyzed. Given the versatility of the zebrafish model system, these techniques offer a wealth of opportunities for studying the behavior of single molecules during embryonic development and their role in the pathogenesis of various diseases. However, each of these techniques has its limitations. Using TIRF microscopy, single-molecule microscopy can be utilized to study eYFP-tagged membrane proteins expressed in the skin of embryos that are 2 days of age or older. SPIM has been used to visualize QDs deep inside the embryos, but it is yet unclear if this method can be applied to visualize autofluorescent proteins at the single-molecule level as well. Using FCS, the concentration and diffusion coefficient of eGFP and mRFP labeled proteins was analyzed up to a tissue depth of ~50 μm, but this method does not provide detailed information about the diffusion pattern, e.g. the confinement.

Interestingly, in three cases in which data from the in vivo experiments were compared to those obtained from cell culture studies, the results differed largely. The eYFP-C10HRAS protein appeared to diffuse freely in the membrane of epidermal cells in living embryos, whereas in cultured cells it showed confined diffusion [37]. In addition, FCS data showed a discrepancy between the k_ds measured in zebrafish embryos for the IQGAP1-Cdc42 [41] and the Fgfr1–Fgf8 and Fgfr4–Fgf8 [35] interactions and those measured in cell culture systems. These discrepancies illustrate the relevance of single-molecule microscopy in an in vivo system, indicating that future single-molecule microscopy studies should preferably be performed in living organisms rather than in individual cells.

References

1. Bacia K, Kim SA, Schwille P (2006) Fluorescence cross-correlation spectroscopy in living cells. Nat Methods 3:83–89
2. Blaser H, Eisenbeiss S, Neumann M, Reichman-Fried M, Thisse B, Thisse C, Raz E (2005) Transition from non-motile behaviour to directed migration during early PGC development in zebrafish. J Cell Sci 118:4027–4038
3. Dahan M, Levi S, Luccardini C, Rostaing P, Riveau B, Triller A (2003) Diffusion dynamics of glycine receptors revealed by single-quantum dot tracking. Science 302:442–445
4. Dertinger T, Pacheco V, von der Hocht I, Hartmann R, Gregor I, Enderlein J (2007) Two-focus fluorescence correlation spectroscopy: a new tool for accurate and absolute diffusion measurements. Chemphyschem 8:433–443
5. Friedrich M, Nozadze R, Gan Q, Zelman-Femiak M, Ermolayev V, Wagner TU, Harms GS (2009) Detection of single quantum dots in model organisms with sheet illumination microscopy. Biochem Biophys Res Commun 390:722–727
6. Grunwald D, Martin RM, Buschmann V, Bazett-Jones DP, Leonhardt H, Kubitscheck U, Cardoso MC (2008) Probing intranuclear environments at the single-molecule level. Biophys J 94:2847–2858
7. Haas P, Gilmour D (2006) Chemokine signaling mediates self-organizing tissue migration in the zebrafish lateral line. Dev Cell 10:673–680
8. Harms GS, Cognet L, Lommerse PH, Blab GA, Kahr H, Gamsjager R, Spaink HP, Soldatov NM, Romanin C, Schmidt T (2001) Single-molecule imaging of l-type Ca(2+) channels in live cells. Biophys J 81:2639–2646
9. Harms GS, Cognet L, Lommerse PH, Blab GA, Schmidt T (2001) Autofluorescent proteins in single-molecule research: applications to live cell imaging microscopy. Biophys J 80:2396–2408
10. Henion PD, Raible DW, Beattie CE, Stoesser KL, Weston JA, Eisen JS (1996) Screen for mutations affecting development of Zebrafish neural crest. Dev Genet 18:11–17
11. Hsu CH, Wen ZH, Lin CS, Chakraborty C (2007) The zebrafish model: use in studying cellular mechanisms for a spectrum of clinical disease entities. Curr Neurovasc Res 4:111–120
12. Huisken J, Stainier DY (2009) Selective plane illumination microscopy techniques in developmental biology. Development 136:1963–1975
13. Huisken J, Swoger J, Del Bene F, Wittbrodt J, Stelzer EH (2004) Optical sectioning deep inside live embryos by selective plane illumination microscopy. Science 305:1007–1009
14. Iino R, Koyama I, Kusumi A (2001) Single molecule imaging of green fluorescent proteins in living cells: E-cadherin forms oligomers on the free cell surface. Biophys J 80:2667–2677
15. Ike H, Kosugi A, Kato A, Iino R, Hirano H, Fujiwara T, Ritchie K, Kusumi A (2003) Mechanism of Lck recruitment to the T-cell receptor cluster as studied by single-molecule-fluorescence video imaging. Chemphyschem 4:620–626
16. Keller PJ, Schmidt AD, Wittbrodt J, Stelzer EH (2008) Reconstruction of zebrafish early embryonic development by scanned light sheet microscopy. Science 322:1065–1069
17. Kim SA, Heinze KG, Schwille P (2007) Fluorescence correlation spectroscopy in living cells. Nat Methods 4:963–973
18. Kimmel CB, Ballard WW, Kimmel SR, Ullmann B, Schilling TF (1995) Stages of embryonic development of the zebrafish. Dev Dyn 203:253–310
19. Kusumi A, Ike H, Nakada C, Murase K, Fujiwara T (2005) Single-molecule tracking of membrane molecules: plasma membrane compartmentalization and dynamic assembly of raft-philic signaling molecules. Semin Immunol 17:3–21
20. Levraud JP, Colucci-Guyon E, Redd MJ, Lutfalla G, Herbomel P (2008) In vivo analysis of zebrafish innate immunity. Methods Mol Biol 415:337–363
21. Lieschke GJ, Currie PD (2007) Animal models of human disease: zebrafish swim into view. Nat Rev Genet 8:353–367

22. Lister JA, Robertson CP, Lepage T, Johnson SL, Raible DW (1999) nacre encodes a zebrafish microphthalmia-related protein that regulates neural-crest-derived pigment cell fate. Development 126:3757–3767
23. Lommerse PH, Blab GA, Cognet L, Harms GS, Snaar-Jagalska BE, Spaink HP, Schmidt T (2004) Single-molecule imaging of the H-ras membrane-anchor reveals domains in the cytoplasmic leaflet of the cell membrane. Biophys J 86:609–616
24. Lommerse PH, Snaar-Jagalska BE, Spaink HP, Schmidt T (2005) Single-molecule diffusion measurements of H-Ras at the plasma membrane of live cells reveal microdomain localization upon activation. J Cell Sci 118:1799–1809
25. Lommerse PH, Vastenhoud K, Pirinen NJ, Magee AI, Spaink HP, Schmidt T (2006) Single-molecule diffusion reveals similar mobility for the Lck, H-ras, and K-ras membrane anchors. Biophys J 91:1090–1097
26. Niell CM, Meyer MP, Smith SJ (2004) In vivo imaging of synapse formation on a growing dendritic arbor. Nat Neurosci 7:254–260
27. Ornitz DM, Xu J, Colvin JS, McEwen DG, MacArthur CA, Coulier F, Gao G, Goldfarb M (1996) Receptor specificity of the fibroblast growth factor family. J Biol Chem 271: 15292–15297
28. Pan X, Shi X, Korzh V, Yu H, Wohland T (2009) Line scan fluorescence correlation spectroscopy for three-dimensional microfluidic flow velocity measurements. J Biomed Opt 14:024049
29. Pan X, Yu H, Shi X, Korzh V, Wohland T (2007) Characterization of flow direction in microchannels and zebrafish blood vessels by scanning fluorescence correlation spectroscopy. J Biomed Opt 12:014034
30. Pauls S, Geldmacher-Voss B, Campos-Ortega JA (2001) A zebrafish histone variant H2A.F/Z and a transgenic H2A.F/Z:GFP fusion protein for in vivo studies of embryonic development. Dev Genes Evol 211:603–610
31. Petrasek Z, Schwille P (2008) Precise measurement of diffusion coefficients using scanning fluorescence correlation spectroscopy. Biophys J 94:1437–1448
32. Renshaw SA, Loynes CA, Trushell DM, Elworthy S, Ingham PW, Whyte MK (2006) A transgenic zebrafish model of neutrophilic inflammation. Blood 108:3976–3978
33. Rieger S, Kulkarni RP, Darcy D, Fraser SE, Koster RW (2005) Quantum dots are powerful multipurpose vital labeling agents in zebrafish embryos. Dev Dyn 234:670–681
34. Ries J, Schwille P (2006) Studying slow membrane dynamics with continuous wave scanning fluorescence correlation spectroscopy. Biophys J 91:1915–1924
35. Ries J, Yu SR, Burkhardt M, Brand M, Schwille P (2009) Modular scanning FCS quantifies receptor-ligand interactions in living multicellular organisms. Nat Methods 6:643–645
36. Sako Y, Minoghchi S, Yanagida T (2000) Single-molecule imaging of EGFR signalling on the surface of living cells. Nat Cell Biol 2:168–172
37. Schaaf MJ, Koopmans WJ, Meckel T, van Noort J, Snaar-Jagalska BE, Schmidt TS, Spaink HP (2009) Single-molecule microscopy reveals membrane microdomain organization of cells in a living vertebrate. Biophys J 97:1206–1214
38. Schmidt T, Schutz GJ, Baumgartner W, Gruber HJ, Schindler H (1996) Imaging of single molecule diffusion. Proc Natl Acad Sci USA 93:2926–2929
39. Schutz GJ, Kada G, Pastushenko VP, Schindler H (2000) Properties of lipid microdomains in a muscle cell membrane visualized by single molecule microscopy. EMBO J 19:892–901
40. Schutz GJ, Schindler H, Schmidt T (1997) Single-molecule microscopy on model membranes reveals anomalous diffusion. Biophys J 73:1073–1080
41. Shi X, Foo YH, Sudhaharan T, Chong SW, Korzh V, Ahmed S, Wohland T (2009a) Determination of dissociation constants in living zebrafish embryos with single wavelength fluorescence cross-correlation spectroscopy. Biophys J 97:678–686
42. Shi X, Teo LS, Pan X, Chong SW, Kraut R, Korzh V, Wohland T (2009b) Probing events with single molecule sensitivity in zebrafish and Drosophila embryos by fluorescence correlation spectroscopy. Dev Dyn 238:3156–3167

43. Tokunaga M, Kitamura K, Saito K, Iwane AH, Yanagida T (1997) Single molecule imaging of fluorophores and enzymatic reactions achieved by objective-type total internal reflection fluorescence microscopy. Biochem Biophys Res Commun 235:47–53
44. Trede NS, Langenau DM, Traver D, Look AT, Zon LI (2004) The use of zebrafish to understand immunity. Immunity 20:367–379
45. Vale RD, Funatsu T, Pierce DW, Romberg L, Harada Y, Yanagida T (1996) Direct observation of single kinesin molecules moving along microtubules. Nature 380:451–453
46. Yu SR, Burkhardt M, Nowak M, Ries J, Petrasek Z, Scholpp S, Schwille P, Brand M (2009) Fgf8 morphogen gradient forms by a source-sink mechanism with freely diffusing molecules. Nature 461:533–536

Chapter 10
Analysis of Large-Amplitude Conformational Transition Dynamics in Proteins at the Single-Molecule Level

Haw Yang

Abstract By monitoring the processes of individual, immobilized molecules in real time, it is possible to capture transient and stochastic events that cannot be detected using conventional ensemble-averaged methods. Such rare events on the molecular level are believed to have significant consequences in biological functions. The single-molecule approach therefore offers promising new routes to uncovering the physical and chemical transformations underlying cellular responses. Dynamics and distribution are two unique pieces of information provided by time-dependent single-molecule spectroscopy. However, to extract these pieces of information from the noisy single-molecule time series in an unbiased way is very challenging because single-molecule signals are *qualitatively* different from ensemble-averaged experiments. With an overarching goal of formulating a predictive understanding of protein molecular machines, this chapter outlines a framework that affords a quantitative and objective analysis of single-molecule signals, with an emphasis on Förster-type energy transfer. Both computer simulations and experimental results are used to illustrate the ideas and practical protocols.

Keywords Maximum Likelihood Estimate (MLE) · Correlation function · Fisher information · Maximum Entropy · Cramér-Rao bound · Local unfolding · Induced fit · Dynamically induced fit · Maximum-information algorithm · Orientation factor, (κ^2) · Dynamical depolarization · Förster-Type Resonance Energy Transfer (FRET) · Ergodic · Conformation distribution · Molecular machine · Emergence

H. Yang (✉)
Department of Chemistry, Princeton University, Princeton NJ 08544, USA
e-mail: hawyang@princeton.edu

Y. Sako and M. Ueda (eds.), *Cell Signaling Reactions: Single-Molecular Kinetic Analysis*, 199
DOI 10.1007/978-90-481-9864-1_10,

10.1 Introduction

Proteins are molecular machines, carrying out a diverse array of tasks in a cell. These tasks include catalysis, ligand binding, mechanical work, as well protein-protein interactions for signaling and other biochemical and biological processes [1]. In sharp contrast to man-made machines for which low friction is an important design criterion, nanoscale protein machines are immersed in the hustle and bustle of surrounding water molecules and are able to operate in a highly dissipative environment. Thermal fluctuations become significant on this length scale, stochastically driving the different types of molecular motions in a protein. These molecular motions spread over many decades of timescales and substantially overlap with the timescales of protein functions (cf. Fig. 10.1).

In this context, it would seem that proteins have evolved to harness thermal fluctuations, rather than frustrated by them, to carry out chemical transformations and mechanical work. Yet, it is not apparent how the function of a protein emerges from the comprising amino acids, even though the physical and chemical properties of amino acids are well known. Such phenomena of "molecular emergence," where novel functions appear in a molecular system but cannot yet be predicted from knowledge of the chemical moieties and their three-dimensional arrangements is one characteristic of complex systems.

Under the premise that there exist general principles underlying molecular emergence, one is interested in uncovering the operation and design principles of protein machines. To frame the problem in a tractable way, several basic questions have been formulated to guide the experimental design [12, 16]. They are: (a) How many conformational states can a protein sample on the functionally important timescale?

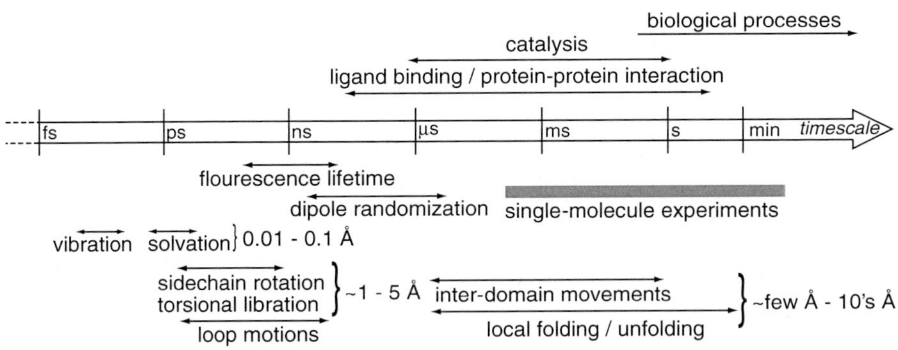

Fig. 10.1 Time and length scales of molecular motions in proteins compared to timescales for biological events [36]. Examples of "biological processes" include cell division and ageing, to be distinguished from biochemical and biophysical events. The types of molecular motion in proteins and corresponding length scales are summarized below the time axis. Tasks that a protein may perform are indicated above the time axis. Also included are the processes that are relevant to single-molecule measurements of Förster-type resonance energy transfer (FRET): fluorescence lifetime, dipole randomization, and single-molecule measurement time [36].

(b) What are the inter-conversion rates between states? (c) How do ligand binding or interactions with other proteins modulate the motions? (d) What functional roles do these motions play? (e) What are the structural basis of flexibility and its underlying molecular mechanics? It is important to note that only when the range of motions and the kinetics of conformational transitions (questions a and b) are simultaneously determined can the dynamics of the system be properly defined [14]. Indeed, the entangled timescales for stochastic fluctuations and protein function (cf. Fig. 10.1) strongly argue that dynamics need to be explicitly articulated in order to fully develop a predictive understanding of the intricate interplay between structure and function in proteins. Toward this goal, optical single-molecule spectroscopy [25, 26] with Förster-type resonance energy transfer (FRET) [15] should in principle provide a unique measure of both the timescale and the magnitude of structural fluctuations and thus is the method of choice for this particular problem. For instance, guided by the aforementioned basic questions, Hanson et al. used *E. coli* adenylate kinase as model and discovered that ligand binding accelerates the rates of protein dynamics, generalizing Koshland's induced-fit picture [23] to "dynamically induced fit" [16]. A recent experiment by Flynn et al. suggested that one of the mechanisms regulating large-amplitude conformational changes in protein tyrosine phosphatase B of *M. tuberculosis* is the folding-unfolding transition of a local helix segment [12]. These are but two examples demonstrating how quantitative high-resolution single-molecule FRET experiments [33, 34] can provide novel insights into the functional roles of protein conformational dynamics and enable unexpected discoveries. While the problem of reaching a predictive understanding of protein machines is far from being solved, quantitative measurements should help accelerating the research pace. In this regard, a fundamental understanding of single-molecule measurements and signals is critical [3].

In this chapter, we discuss some basics of single-molecule measurements as well as practical procedures to overcome certain difficulties that an experimentalist may encounter in the evaluation of fluorescence single-molecule data. Much of the ideas, materials, and wordings have been published previously in the forms of original research or review articles though with contents more focused on specific topics [35, 19, 36, 37]. Here, we attempt to present the key ideas in those works in a concise and cohesive way. We emphasize approaches that do not require an experimentalist to assume a distribution model or a kinetic scheme in data reduction. Readers who are interested in technical details are encouraged to consult the original articles, which will be cited where appropriate in the remainders of this chapter.

10.2 Bulk Versus Single-Molecule Measurements

Single-molecule measurements are fundamentally and qualitatively different from ensemble-averaged bulk measurements in that the former is time averaging whereas the latter is number averaging. It should be recognized that in both cases it necessarily takes finite time duration to make any meaningful measurement. Consider two

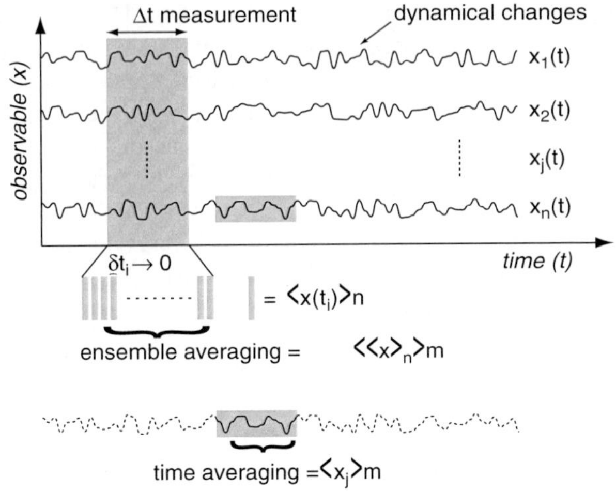

Fig. 10.2 Number averaging in ensemble-averaged experiments versus time averaging in single-molecule experiments. The main difference is that bulk experiments, $\overline{\langle x \rangle_n}$, include number averaging followed by time average, in this specific order, whereas single-molecule measurement only involves time averaging, $\overline{x_j}$ [36].

sub-domains of a protein undergoing large-amplitude conformational transitions. The parameter of interest in this example is the distance x between the two sub-domains. The magnitude of x fluctuates stochastically as a function of time because of thermal agitation. Let $x_j(t)$ denote the time-dependent distance with $j = 1, \ldots, \mathrm{n}$ indicating the j-th protein in an ensemble of n molecules. Suppose it takes Δt seconds to make an ensemble-averaged measurement of the bulk sample from which a mean distance $\langle x \rangle_{\text{ensemble}}$ can be obtained (cf. Fig. 10.2)

The ensemble averaging $\langle \cdots \rangle_{\text{ensemble}}$ actually contains both number and time averaging operations. Conceptually, this can be seen by breaking Δt into m sequential "instantaneous snapshots" of the entire ensemble where each snapshot takes $\delta t = \Delta t / m$ seconds to complete. Here, δt is taken to be much shorter than the timescale of any relevant molecular motions. Therefore, each molecule in the ensemble can be viewed as "frozen" during the snapshot time. The appropriate average for the i-th snapshot is therefore,

$$\langle x(t_i) \rangle_n = \frac{1}{n} \sum_{j=1}^{n} x_j(t_i).$$

An ensemble-averaged measurement is therefore the time average over Δt of the number averaging shown above:

$$\langle x \rangle_{\text{ensemble}} = \lim_{m \to \infty} \frac{1}{m\delta t} \sum_{i=1}^{m} \langle x(t_i)\delta t \rangle_n = \frac{1}{\Delta t} \int_0^{\Delta t} \langle x(t) \rangle_n \, dt = \overline{\langle x(t) \rangle_n},$$

where the overbar denotes time averaging. The order of averaging in ensemble-averaged experiment is crucial – first number averaging followed by time averaging. We will come back to this point later when discussing bulk versus single-molecule FRET measurements.

On the other hand, as opposed to ensemble-averaged experiments, there is only time-averaging operation in single-molecule experiments. Following the notation in Fig. 10.2, this understanding can be expressed as,

$$\langle x_j \rangle_{\text{single-molecule}} = \frac{1}{m\delta t} \sum_{i=1}^{m} x_j(t_i)\delta t = \frac{1}{\Delta t} \int_0^{\Delta t} x_j(t)\, dt = \bar{x}_j,$$

where $\langle x_j \rangle_{\text{single-molecule}}$ reads as the time-averaged single-molecule measurement of the jth molecule over the data acquisition time Δt. Since we will be dealing with a single molecule, we will drop the j subscript in the remaining discussion for simplicity. The data that one would acquire in such an experiment is of the form, $\{\bar{x}(t_1), \bar{x}(t_2), \ldots, \bar{x}(t_m)\}$, where it is understood that for $\bar{x}(t_i)$ a time average over a period of Δt has been performed at around time t_i. The dynamics are contained in the time series sequence of $\{\bar{x}(t_1), \bar{x}(t_2), \ldots, \bar{x}(t_m)\}$. A major task of single-molecule analysis is to extract physical parameters from such experimentally measured time series.

10.3 Time Correlation Functions

The dynamics of a system contained in the time series of x are usually characterized by the time correlation function of x:

$$C_{xx}(s) = \langle x(t)x(t+s) \rangle_{\text{ensemble}} - \langle x \rangle^2_{\text{ensemble}}. \tag{10.1}$$

The time correlation can be recovered from a prolonged, time-averaged study of a single system if the system is ergodic and when the duration of observation, T, approaches infinity:

$$C_{xx}(s) = \lim_{T\to\infty} \frac{1}{T} \int_0^T x(t)x(t+\tau)dt - \left[\lim_{T\to\infty} \frac{1}{T} \int_0^T x(t)dt \right]^2.$$

In practice, the correlation function is calculated from an experimentally recorded time series, $\{\bar{x}(t_1), \bar{x}(t_2), \ldots, \bar{x}(t_m)\}$.

There are two common ways of calculating the empirical correlation function. The moving-average approach is appropriate for time series that are potentially aperiodic and is given by:

$$\bar{C}_{xx}(s) = \frac{1}{m - q_s} \sum_{i=1}^{m-q_s} \bar{x}_i \bar{x}_{i+q_s} - \bar{x}^2, \tag{10.2}$$

where $\bar{x}_i$ is a shorthand expression for $\bar{x}(t_i)$, $q_s = s/\Delta t$, and $\bar{x} = \sum_i \bar{x}_i/m$. If the time series can be assumed to exhibit a periodicity of m, i.e., $\bar{x}_i = \bar{x}_{i+m}$, then the correlation function can be calculated using discrete Fourier transform (DFT) and its inverse operation (iDFT), $\bar{C}_{xx}(s) = iDFT\{\tilde{x}_i^* \tilde{x}_j\}_s$, where $\tilde{x}_j \equiv DFT\{x_i\}_j$ denotes the i-th Fourier transform element, and "*" indicates complex conjugate. The Fourier transform method is equivalent to calculating:

$$\bar{C}_{xx}^{FT}(s) = \frac{1}{m} \sum_{i=1}^{m} \bar{x}_i \bar{x}_{i+q_s} - \bar{x}^2. \tag{10.3}$$

Both Eqs. 10.2 and 10.3 can be considered as a finite-sample estimation of Eq. 10.1, and contain uncertainties due to insufficient sampling of the time correlation function. For optical single-molecule experiments where the detection noise follows Poisson statistics, the uncertainties related to the autocorrelation function can be derived analytically [17], given below. For moving-average correlation functions, the time-dependent variance for the correlation function of a Poisson variable is given by,

$$\mathrm{var}\{\overline{C}_{xx}(s)\} = h_1(\bar{x}, m) + h_2(\bar{x}, m, q_s), \tag{10.4}$$

where

$$h_1(\bar{x}, m) = -\left(\frac{4\bar{x}^3}{m} + \frac{2\bar{x}^2}{m^2} - \frac{2\bar{x}}{m^3}\right),$$

$$h_2(\bar{x}, m, q_s) = \begin{cases} \frac{\bar{x}^2+6\bar{x}^3}{m-q_s} - \frac{2m\bar{x}^3}{(m-q_s)^2}, & 1 \leqslant q_s < (m/2); \\ \frac{\bar{x}^2+2\bar{x}^3}{m-q_s}, & (m/2) \leqslant q_s < m. \end{cases}$$

For Fourier-transform calculated correlation functions, the time-dependent variance is,

$$\mathrm{var}\left\{\bar{C}_{xx}^{FT}(s)\right\} = \frac{\bar{x}^2}{m} + \frac{2\bar{x}^2}{m^2} - \frac{\bar{x}}{m^3}. \tag{10.5}$$

The analytical expressions in Eqs. 10.4 and 10.5 allow one to quantitatively access the uncertainties in an experimentally measured time series even for exceptionally noisy or short data. This is particularly useful in fluorescence single-molecule experiments where shot-noise is pervasive and irreversible photo-bleaching of the probes limits the length of each experiment. Furthermore, the two common ways of calculating the auto-correlation, moving-average and Fourier transform, exhibit different uncertainty characteristics. For periodic time series, the Fourier transform method is preferred because it gives smaller uncertainties that are uniform through all time lags. An important corollary of these results is that one can easily derive a

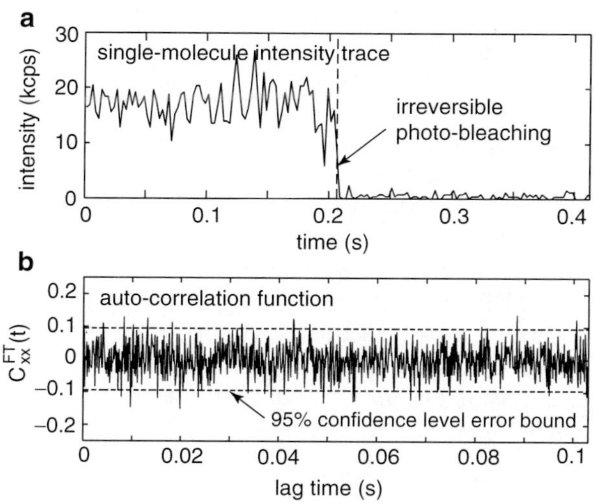

Fig. 10.3 Auto-correlation analysis of an experimental single-molecule fluorescence trajectory. (**a**) A representative single-molecule fluorescence trajectory. Irreversible photo-bleaching is indicated by a *vertical dashed line*. The trajectory is binned at 2.5 ms. (**b**) Fourier-transform auto-correlation analysis of the single-molecule trajectory. The statistical error bounds at 95% confidence level were calculated using $1.96 \times \sqrt{\mathrm{var}\{\tilde{C}_{xx}^{FT}(s)\}}$, Eq. 10.5. Fourier-transform auto-correlation analysis was performed with a bin size of 100 μs [17].

statistically robust method to test the existence of correlations in a time series [17]. Applications include (i) testing whether a single-molecule time series signal contains significant correlation, and (ii) testing whether the donor and acceptor channels from a single-molecule FRET experiment are anti-correlated.

In application (i), one may examine the orientational freedom of a chromophore attached to a protein on the experimental timescale (~ms). As an example, Fig. 10.3a shows an experimental time trace of a single dye attached to an immobilized adenylate kinase molecule. This trajectory illustrates the challenges underlying the quantitative analysis of auto-correlation functions from single-molecule fluorescence data: they are short and inherently noisy. In that experiment, a Pockel's cell was used to rotate the orientation of the polarized excitation source by 90° at a frequency of 1 kHz. If the dye is freely rotating with respect to the immobilized protein on timescales faster than the Pockel's cell alternation frequency, the auto-correlation function should be a Gaussian with zero mean. A Fourier-transform auto-correlation of this experimental trajectory is presented in Fig. 10.3b. Error bars were calculated according to the model-independent Eq. 10.5 and had been plotted to 95% confidence. The lack of correlation in this trajectory indicates that there are no transient interactions between the dye and the protein on the millisecond timescale and that these types of interactions do not contribute to the millisecond dynamics that are measured in FRET experiments.

In application (ii), it is common to reject invalid time traces and not include them in further data analysis for single-molecule FRET experiments. In experiments investigating immobilized molecules, one typically selects single-molecule traces that exhibit anti-correlated donor and acceptor emission pattern based on visual inspection. However, selection (and rejection) of single-molecule traces based on the subjective visual inspection alone can be ambiguous. This problem is illustrated by simulated two-channel detection experiment where both the donor and the acceptor can be quenched non-specifically via a quencher in the vicinity of the macromolecule under investigation. The quenching is time dependent because of the slow conformational fluctuations of the macromolecule. It is possible that two singly labeled macromolecules co-localize within the same detection spot. To an experimental observer, the FRET intensity trace from this configuration cannot be distinguished from a true donor-acceptor doubly labeled molecule without further analysis. This difficulty is illustrated in Fig. 10.4a where the simulated donor (dark gray line) and acceptor (light gray line) traces appear as if they indeed come from FRET; the inset further shows how they can appear to be anti-correlated. One might think that cross-correlation analysis will help to resolve this problem; yet, without a

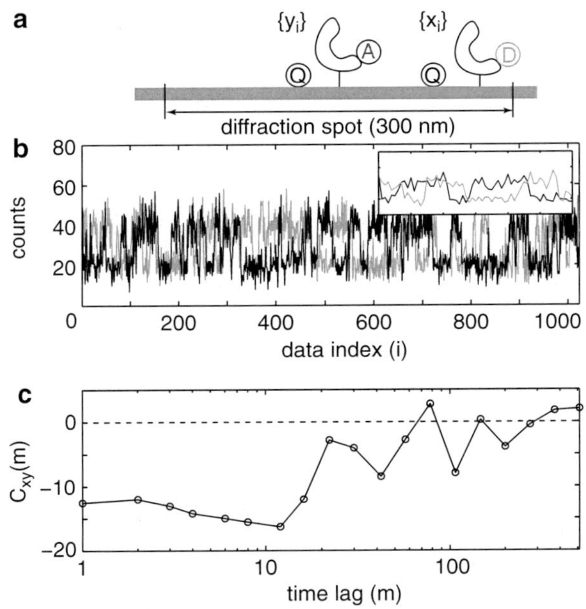

Fig. 10.4 Identification of anti-correlated FRET traces based on visual inspection alone can be misleading. (**a**) A scenario that can lead to seemingly anti-correlated FRET traces which, in fact, should be uncorrelated. The circled letters are: A, acceptor; D, donor; and Q, quencher. fxig and fyig represent time series data from the two distinct molecules. (**b**) Simulated intensity time traces for the donor (gray) and the acceptor (black) appear as if they are anti-correlated. The *inset* displays a zoom-in for the traces between the 230–290 index range. (**c**) A cross-correlation analysis of the donor and acceptor traces, and presented after log averaging – a practice not encouraged for visualizing data without proper error bounds (please see text for definition of log averaging). Without a quantitative assessment, the cross-correlation curve may lead one to believe that the two traces in (b) are anti-correlated [18].

quantitative assessment, visual examination combined with inappropriate data presentation can further exacerbate the problem. The cross-correlation function is calculated by,

$$\bar{C}_{xy}(s) \simeq \frac{1}{m} \sum_{i=1}^{m} (\bar{x}_i - \bar{x})(\bar{y}_{i+q_s} - \bar{y}),$$

where the definition of $\bar{y}_i$ is similar to that of $\bar{x}_i$ but come from another channel. Figure 10.4c displays a cross-correlation curve for the traces shown in Fig. 10.4b with log averaging. In log averaging, the correlation time points are regrouped on a log-10 time scale. One then sets up a linear grid on the log-scale and averages over the time points in each grid. This way, the number of time points in each grid increases exponentially at longer lag time and will give better statistics at long-lag time for over-sampled data [28, 29]. Figure 10.4c appears anti-correlated even though, by construction, the donor and acceptor signals should be uncorrelated. Therefore, Fig. 10.4b, c clearly demonstrate the difficulties of evaluating single-molecule time traces based on visual assessment alone.

It turns out that the analytical expression for the time-dependent variance of the cross-correlation is simpler than that of auto-correlation. It is given by [18],

$$\operatorname{var}\{\bar{C}_{yx}\} \simeq \frac{m_{eff} - 1}{m_{eff}^2} \operatorname{var}\{\bar{x}\} \operatorname{var}\{\bar{y}\}, \tag{10.6}$$

where $m_{eff} = m/m_\tau$ with $m_\tau = \max\{m_\tau^{(x)}, m_\tau^{(y)}\}$ being the larger of the two decay-constants, $m_\tau^{(x)}$ and $m_\tau^{(y)}$, in the auto-correlation of the x and y series, $\bar{C}_{xx}(m)$ and $\bar{C}_{yy}(m)$, respectively. Applying Eq. 10.6 to the simulated data shown in Fig. 10.4, we see that the correctly calculated error bounds allow us to reject this data set as anti-correlated (see Fig. 10.5).

Up to this point, we have discussed how to detect dynamical fluctuations in the experimentally measured signals, the recorded photon intensities. These raw experimental data can be further analyzed to extract time-dependent distance changes using the energy transfer theory developed by Förster. We start with a brief review of the theory and point out the important differences when dealing with single-molecule data.

10.4 Single-Molecule Förster-Type Resonance Energy Transfer

Förster's formulation [13] for the energy transfer rate (k_T) from an energy donor at the excited state to the acceptor is [6],

$$k_T = C k_D \kappa^2 R^{-6}$$

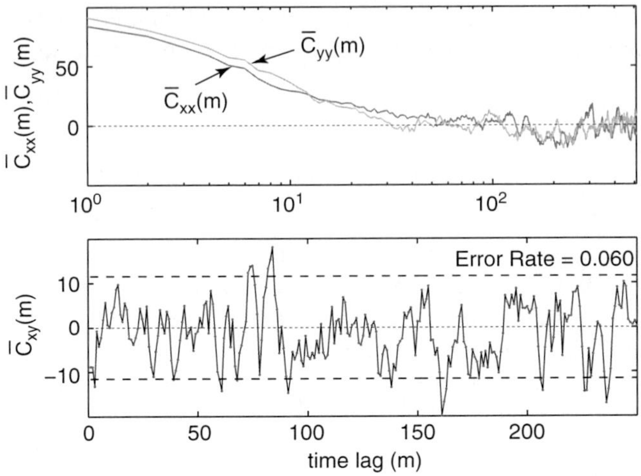

Fig. 10.5 (**a**) The auto-correlation functions of the donor ($\bar{C}_{xx}$, *dark gray line*) and the acceptor ($\bar{C}_{yy}$, *light gray line*) traces for the traces shown in Fig. 10.4b, showing that $\{x_i\}$ and $\{y_i\}$ are not mutually independent within their respective set. The dependence within each data set has to be taken into account. (**b**) The cross-correlation, $\bar{C}_{xy}$ (*solid line*), for the data shown in Fig. 10.4b and 95% confidence region calculated using Eq. 10.6 (*dashed lines*). The error rate is calculated the number of the $\bar{C}_{xy}$ data points that exceed the confidence region divided by the total number of $\bar{C}_{xy}$ data points. For an infinitely long trajectory, one would expect an error rate of 0.05 for 95% error bounds. This indicates that Eq. 10.6 is an excellent approximation for calculating cross-correlation error bounds.

with

$$C = \frac{9(\ln 10)\Phi_D}{128\pi^5 N' n^4} J(\tilde{\upsilon})$$

where N' is Avogadro's number per cubic centimeter, n the refractive index of the medium, R the donor-acceptor distance, and Φ_D and $\tau_D = k_D^{-1}$ are the donor emission quantum yield and lifetime in the absence of the acceptor, respectively. $J(\tilde{\upsilon})$ is the spectral overlap integral of donor emission and acceptor absorption, defined by

$$J(\tilde{\upsilon}) = \int_0^\infty \varepsilon_A(\tilde{\upsilon}) f_D(\tilde{\upsilon}) \tilde{\upsilon}^4 d\tilde{\upsilon} = \int_0^\infty \varepsilon_A(\lambda) f_D'(\lambda) \lambda^4 d\lambda$$

where $\varepsilon_A(\tilde{\upsilon}) = \varepsilon_A(\lambda)$ is the molar absorbance of the acceptor at frequency $\tilde{\upsilon} = 1/\lambda$ and $f_D(\tilde{\upsilon})$ (or $f_D'(\lambda)$) is the normalized fluorescence spectrum of the donor. The orientation factor, κ^2, is defined by,

$$\kappa^2 \equiv \left[\hat{r}_D \cdot \hat{r}_A - 3\left(\hat{r}_D \cdot \hat{R}\right)\left(\hat{r}_A \cdot \hat{R}\right)\right]^2,$$

where $\hat{r}_D$ and $\hat{r}_A$ are the unit vectors for the transition dipoles of the donor and acceptor, respectively, and $\hat{R}$ is the unit vector for the vector defined by the

$\overrightarrow{DA}$ vector. The numerical value for $\langle\kappa^2\rangle$ is 2/3 when averaged over all $\hat{r}_D$ and $\hat{r}_A$ orientations. The experimental observable is the efficiency of energy transfer, defined by,

$$E \equiv \frac{k_T}{k_T + k_D} = \frac{\kappa^2}{\frac{2}{3}x^6 + \kappa^2}, \tag{10.7}$$

where $x \equiv R/R_0$ is the normalized FRET distance and R_0 is the donor-acceptor separation (Förster radius) when $k_T = k_D$ under the assumption that $\langle\kappa^2\rangle = 2/3$.

The key difference between ensemble-averaged and single-molecule measurements of $\langle E\rangle$ is how the averaging $\langle\cdots\rangle$ is carried out. Following Eq. 10.7, one sees that the FRET efficiency contains contributions from both the distance x and the orientation factor κ^2. In other words, ensemble-averaged FRET efficiency is a convoluted average of both distance and orientation *distributions* of the donor and acceptor chromophores:

$$E = \left\langle \frac{\kappa^2}{\frac{2}{3}x^6 + \kappa^2} \right\rangle \sim \frac{\langle\kappa^2\rangle_n}{\frac{2}{3}\langle x^6\rangle_n + \langle\kappa^2\rangle_n}, \tag{10.8}$$

The two contributions are difficult to separate, especially in cases where the conformational distribution in x is broad and inter-converts very slowly. However, the last term in Eq. 10.8 becomes a good approximation when both the donor and acceptor chromophores are able to dynamically sample all the allowed orientations during the respective excited-state lifetime ("dynamical depolarization"). That is, every molecule in the ensemble has exactly the same κ^2 distribution.

A notion related to this understanding, primarily due to Dale and Eisinger [6, 7], and later also with Blumberg [8], is that $\langle\kappa^2\rangle = 2/3$ is a good approximation only when the dynamical depolarization conditions are met for the chromophore transition dipoles. Otherwise, the orientation factor may carry significant uncertainties because the tethered chromophores can no longer sample the entire 4π solid angle within the excited-state lifetime of the chromophores. These views have dominated the design and interpretation of ensemble-averaged FRET experiments, and often extended to single-molecule measurements without recognizing the fundamental differences between these two types of experiments. The issue of the orientation factor in single-molecule measurements is discussed below.

10.5 The Orientation Factor, κ^2

The orientation factor is of particular interest in single-molecule FRET because, if clarified, will allow one to directly relate FRET signals to distances. In a recent work, it has been shown that dynamical depolarization is not a necessary requirement for quantitative FRET distance measurements due to the fundamental differences in single-molecule measurement versus ensemble-averaged experiments [36].

Following Eq. 10.8, by definition, the single-molecule measurement guarantees that the averaging operation samples from the same κ^2 distribution because they are from the same molecule, regardless of the chromophores' excited-state lifetime. Moreover, since the relevant randomization timescale is no longer limited by the excited-state lifetime, the notion of dynamical depolarization as conventionally defined is no longer a necessary requirement. The remaining issues are the extent to which κ^2 can be replaced by $\langle\kappa^2\rangle$ and the numerical value for $\langle\kappa^2\rangle$

For the first issue, one recognizes that the dynamics measured by single-molecule experiments are coarse-grained in time because a meaningful measurement requires averaging over many detected photons which, in turn, requires an integration time in data acquisition on the millisecond timescale for typical single organic chromophores (see also information bounds in single-molecule measurements to be discussed later). It follows that κ^2 can be replaced by $\langle\kappa^2\rangle$ as long as the donor and acceptor transition dipoles can sample sufficient different orientations. In other words, the relevant timescale when considering the orientation factor in single-molecule experiments is the data acquisition time, as opposed to the excited-state lifetime in ensemble-averaged experiments. In general, if one were to consider each detected photon as a data point, the numerical value of κ^2 converges very rapidly to its mean value, $\langle\kappa^2\rangle$. For example, as shown in Fig. 10.6, it takes only 15 photons for κ^2 to converge to $\langle\kappa^2\rangle = 2/3$ within 10% relative error [33].

The numerical value of $\langle\kappa^2\rangle$ generally depends both on the solid angles over which the donor and acceptor transition dipoles can sample and on the average relative orientation of the two chromophores. A detailed analysis has revealed that the $\langle\kappa^2\rangle \sim 2/3$ is a good approximation to about 10% relative error when two conditions are met [34]: (i) The solid angle over which the chromophores can sample is greater than 1.83 π. This corresponds to a maximum zenithal angle of $\theta_{\max}>85^0$ over which the transition dipoles can sample. (ii) The donor and acceptor average transition dipoles are not parallel to the $\vec{R}$, the vector connecting the donor dipole to the acceptor

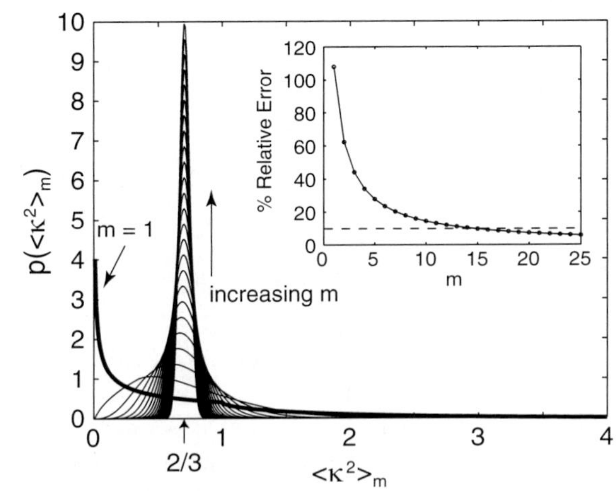

Fig. 10.6 Converging rate of $\langle\kappa^2\rangle_m$ averaging over m photons [33].

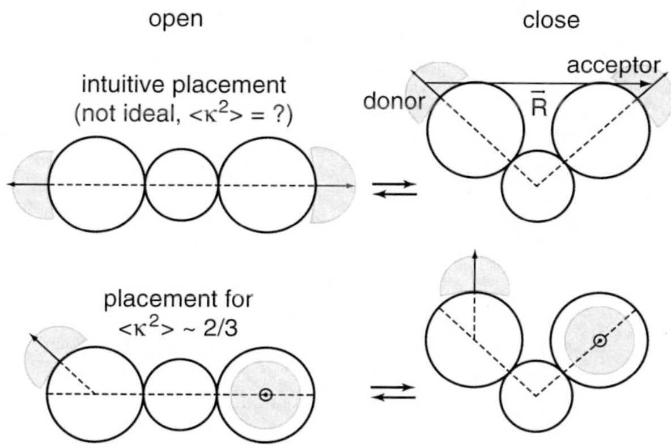

Fig. 10.7 Cartoon contrasting different chromophore placement designs for accurately measuring single-molecule FRET distances during protein conformational changes. The *shaded areas* indicate the solid angle over which the tethered chromophore can sample. The ⊙ icon in the cartoon indicates vectors pointing out of the paper plane [36].

dipole. In the context of measuring time-dependent large-amplitude conformational transitions in proteins, condition (i) can usually be fulfilled by carefully choosing the labeling sites, making sure that the sites are solvent exposed and belong to the more robust α-helix or β-sheet structures. Condition (ii), on the other hand, provides guidelines for placing the chromophores, illustrated by Fig. 10.7.

At this point, we have established that it is possible to unambiguously relate single-molecule FRET signals to distances. We next discuss an approach derived from information theory to extract the most amount of information from single-molecule photon detection time series.

10.6 Information Bounds and Photon-by-Photon Analysis

In principle, the time-dependent energy transfer efficiency in Eq. 10.7 or 10.8 can be calculated from FRET experiments using the intensities from the donor and acceptor channels. Very qualitatively, it is given by,

$$E(t) \sim \frac{I_A(t)}{I_D(t) + I_A(t)},$$

where $I_D(t)$ and $I_A(t)$ are the detected photon intensities at time t for the donor and acceptor cannels, respectively. Note that this equation is shown here only to illustrate the basic idea of how efficiency is related to experimental observable and should not be used "as is" because it does not consider background, cross talk, as well as differing detection efficiency in the donor and acceptor detectors; a more rigorous expression is given by Eq. 10.10. For "intensity" to be meaningful, it is

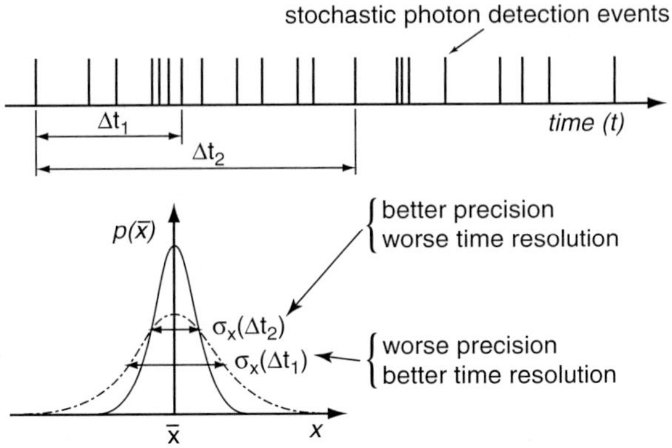

Fig. 10.8 Schematic illustration for the information bounds and the tradeoff between time resolution and measurement uncertainties in single-molecule measurements.

necessary to average over a finite time period. Since single-molecule experiments rely on time averaging, one can make a more precise measurement by averaging more (longer binning time), but one loses time resolution that way. On the other hand, one can improve the time resolution by averaging less (shorter binning time), but the measurement will contain significant amount of noise; to an extreme, the measurement may become meaningless (cf. Fig. 10.8). The challenge is thus to strike a balance between time resolution and measurement uncertainties.

This problem can be addressed using ideas from applied statistics and information theory. The idea is that one would analyze the single-molecule time trajectory in a way that each distance measurement from a "bin" of photons will give the same uncertainty [33]. In other words, the time series is "binned" adaptively with the bin size determined by the amount of information contained in each time bin. Each bin, Δt_j, satisfies the equation for Fisher information, $J(\bar{x}_j)$ [4],

$$J(\bar{x}_j) = \frac{36\bar{x}_j^{10}}{(1+\bar{x}_j^6)^3}\left[I_D^\beta \Delta t_j \frac{(1-\beta_D^{-1})^2}{(\bar{x}_j^6+\beta_D^{-1})} + I_A^\beta \Delta t_j \frac{(1-\beta_A^{-1})^2}{(1+\bar{x}_j^6\beta_A^{-1})}\right] = \frac{1}{\mathrm{var}\{\bar{x}_j\}}, \tag{10.9}$$

where $\bar{x}_j \equiv \bar{R}_j/R_0$ is the normalized distance estimated using a maximum-likelihood estimator (MLE), with $\bar{R}_j$ being the donor-acceptor distance and R_0 the Förster radius as defined earlier where the energy transfer efficiency is 50%. The MLE expression for $\bar{x}_j$ is,

$$\mathrm{MLE}\{\mathrm{x_j}\} = \frac{\beta_\mathrm{A}}{\beta_\mathrm{D}} \times \frac{\mathrm{I_D^\beta n_A - I_A^\beta n_D \beta_D}}{\mathrm{I_A^\beta n_D - I_D^\beta n_A \beta_A}}, \tag{10.10}$$

where I_D^β (I_A^β) is the donor (acceptor) intensity in the absence of the acceptor (donor) with β_D (β_A) being the signal-to-background ratio for the donor (acceptor) channel and n_D (n_A) the number of donor (acceptor) photons collected within the time bin.

One anticipates that the more information for $\bar{x}$ contained in a time bin, the higher precision (low uncertainty) the $\bar{x}$ measurement should be. In information theory, this intuitive reasoning is quantified by the Cramér-Rao bound [5, 27], $\mathrm{var}\{\hat{x}_j\} \geqslant J(x_j)^{-1}$, which states that the variance of a statistical estimator is bound by the inverse of Fisher information where the equality "=" is true when the estimator is unbiased such as MLE [30].

Depending on the nature of the question that the experiment is designed to address, an experimentalist may decide on the precision ($\mathrm{var}\,\{\bar{x}_j\}$) with which the measurement is to be accomplished. Eq. 10.9 also provides a quantitative relationship between measurement uncertainties, $\mathrm{var}\,\{\bar{x}_j\}$, and time resolution for realistic experimental conditions, Δt_j, for the jth time bin. From an information theory viewpoint, this method yields the optimal achievable time resolution. A photon-by-photon analysis algorithm that achieves this information bound (maximum-information algorithm) is illustrated in Fig. 10.9.

With the capability to measure the time-dependent changes in intra-molecular distances quantitatively, we are now at a stage to start address the scientific questions posed in the Introduction section. As a final example, we next discuss how molecular conformational distributions can be extracted from single-molecule traces without assuming any models.

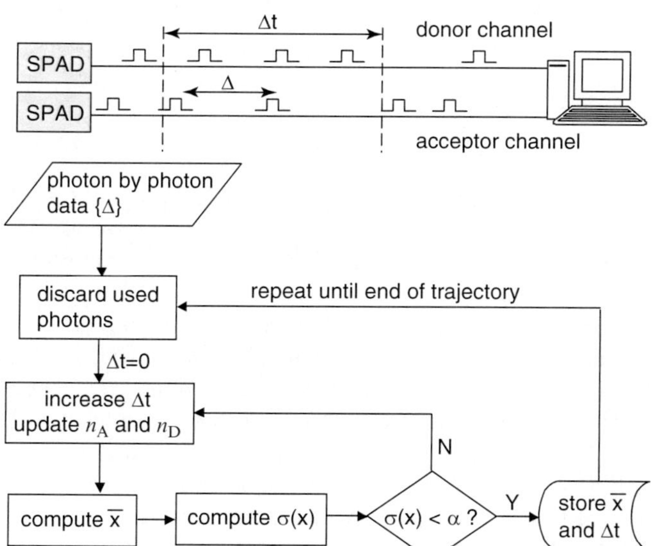

Fig. 10.9 Flowchart of the maximum information algorithm. SPAD stands for single-photon avalanche photodiode. The uncertainty is $\sigma(x) = \sqrt{\mathrm{var}\,\{x\}}$ and is compared with a pre-defined tolerance, α. For example, $\alpha = 0.1$ corresponding to a distance uncertainty of 0.5 Å for $R_0 = 50$ Å [34].

10.7 Extracting Conformational Distributions

One of the unique information provided by single-molecule experiments is the distribution of molecular properties. In the present context, the distance distribution will allow one to measure the apparent intra-molecular potential of mean force (Helmholtz free energy) for large-amplitude conformational transitions of a protein. The apparent potential of mean force $H(\bar{x})$ is related to the distance distribution $p(\bar{x})$ by $H(\bar{x}) = -k_B T \ln p(\bar{x})$ with k_B being Boltzmann's constant and T temperature in Kelvin.

There are however two immediate complications if one were to use simple histogramatic methods to construct distributions from experimentally measured $\bar{x}$ (cf. Fig. 10.10). First, the distribution will depend on the bin time. Second, the distribution is greatly broadened by photon counting noise. These two related issues will prevent one from making a scientifically sound statement if they are not addressed properly.

The fact that the photon-counting noise can be quantified in FRET distances (cf. the Fisher information in Eq. 10.9) permits one to effectively remove the broadening by deconvolution. The Maximum Entropy Method (MEM) developed by Jaynes [20–22] allows us to quantitatively recover the underlying distribution within the experimental error without assuming any model. The "model-free" approach is important because it will obviate the need to make judicial assumptions and enable an experimentalist to explore the molecular system as presented by the data. This is achieved by finding the noise-removed distribution $p(\bar{x})$ that minimizes the merit function [32],

$$M\left[p(\bar{x}), \Lambda\right] = \chi^2 + \Lambda \int_{-\infty}^{\infty} p(\bar{x}) \ln p(\bar{x})\, d\bar{x},$$

where χ^2 is the well-known chi-squared measure, Λ is the Lagrange undetermined multiplier to be optimized during deconvolution. Since the deconvolved distribution, $p(\bar{x})$, is derived from experimental measurements, it is important that they carry proper uncertainties.

To determine the uncertainties [31], one can use the non-parametric (model-free) Bootstrap method [9–11] to resample the single-molecule trajectories. Assuming that the collected single-molecule trajectories are from an independently and identically distributed population, the bootstrap method samples the collected trajectories with replacement and forms a re-sampled population from which many properties (such as variance) of the statistical estimator – in the present case the density estimator and the subsequent MEM noise removal – can be effectively evaluated. Here, each re-sampled single-molecule trajectory is then subjected to the same MEM deconvolution procedure to give a re-sampled distribution, $p_i(\bar{x})$. These bootstrapped distributions are then used to estimate the uncertainties in $p(\bar{x})$ at a given distance $\bar{x}$ by $\mathrm{var}\left\{p(\bar{x})\right\} = \frac{1}{n}\sum_{i=1}^{n}\left(p_i(\bar{x}) - \langle p_i(\bar{x})\rangle\right)^2$.

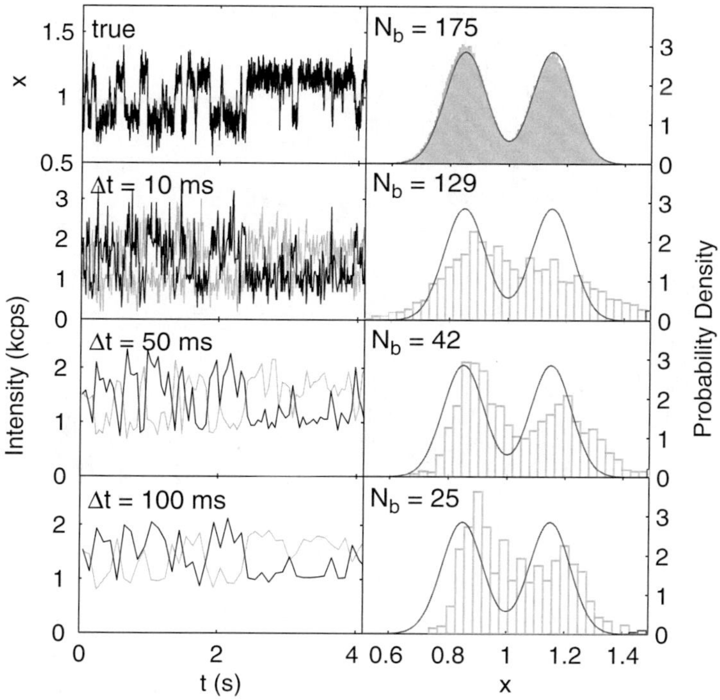

Fig. 10.10 Issues involved in recovering distance distribution from single-molecule FRET measurements. The computer-generated trajectory simulates a Langevin dynamics between two wells, for which the "true" position trajectory and corresponding distribution are displayed on the *left* and *right panels* in the *top row*, respectively. Panels 2–3 in the *left column* show how the apparent FRET intensity traces, also simulated using realistic experimental conditions, are changed by different binning time (donor: *gray*; acceptor: *black*). Panels 2–3 in the *right column* show how different binning time (Δt) along the FRET trajectory, and the number of bins (N_b) used in constructing the distribution (*vertical bars with gray outlines*), can dramatically alter the resulting distribution. The solid line in black represents the true distribution used to generate the Langevin dynamics [33]. The second panel, when compared to the first, noise-less panel on the right column, shows that the distribution is broadened by photon counting noise. The increased bin time Δt averages away not only the counting noise but also the intra-well and inter-well transitions dynamics. The result is narrowed apparent distributions at longer bin times, a phenomenon that shares the same physical idea as the Kubo-Anderson motional narrowing in NMR spectroscopy [2, 24]. At longer bin times, furthermore, the number of bins N_b allowed for constructing the distribution also decreases. This is because longer bin time produces smaller number of data points along the trajectory, which in turn can only afford reduced number of grid points (N_b) for building statistically meaningful histograms [31].

10.8 Example

We illustrate the ideas discussed in this chapter with an example. The system is the protein tyrosine phosphatase B from *M. tuberculosis*, PtpB. All the known structures of the PtpB enzyme are in closed conformation in which the active site is

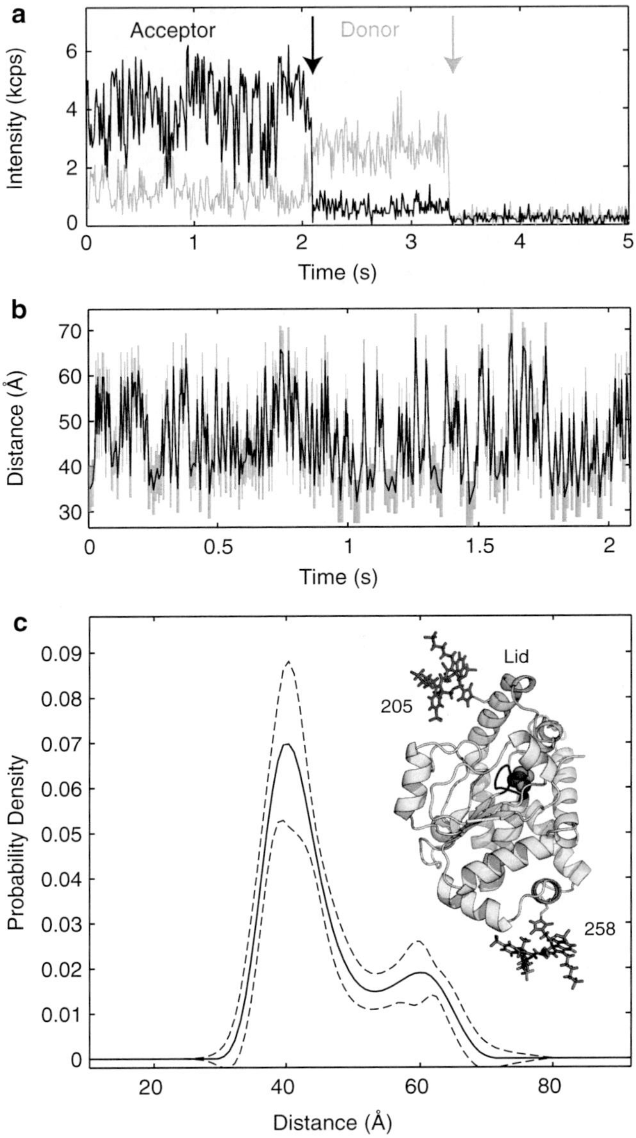

Fig. 10.11 An example of single-molecule FRET data from *M. Tuberculosis* protein tyrosine phosphatase B, PtpB, with structure shown in the insert of panel c. (**a**) Time-dependent fluorescence emission intensity for a single PtpB molecule at a 10 ms bin size. Acceptor emission is denoted by the black line and donor emission by gray. *Arrows* indicate the time at which each dye bleached. (**b**) Visualization of the lid dynamically switching between the closed and the open conformations. Distance trajectory created from the emission intensity using the maximum-information method, where the boxes represent the uncertainty of the time of each measurement along the x-axis and uncertainty in the position of each measurement along the y-axis. (**c**) Probability density functions (3-ms time resolution) from more than 150 individual single-molecule fluorescence trajectories using the maximum entropy deconvolution method. *Dashed lines* represent the error bounds for the distribution, showing that the bimodal distributions are indeed statistically significant [12].

buried inside a lid motif. Yet, biochemically, this enzyme is supposed to act on relatively large protein substrates. Therefore, it has been hypothesized that the lid of the PtpB enzyme can spontaneously open in room-temperature solution. Single-molecule experiments were carried out to test this hypothesis, where a pair of dye molecules was attached to the PtpB molecule for FRET measurements (see insert in Fig. 10.11c).

Figure 10.11a displays a typical single-molecule FRET trajectory from a single PtpB protein that undergoes large-amplitude conformational transitions. The anti-correlated donor and acceptor intensity changes indicate stochastic distance fluctuations on the millisecond timescale. To convert the intensity trace into distance time trace, the maximum information method has been used. The result is displayed on Fig. 10.11b. Based on this analysis, one can state with high confidence that indeed the protein lid is able to open spontaneously. To answer the question of the number of conformational states PtpB can sample, the MEM deconvolution method has been used to construct a noise-removed distance distribution. The result is displayed in Fig. 10.11c. The dashed lines represent one standard deviation error bounds. Taken together, Fig. 10.11 provides quantitative experimental evidence that PtpB can sample both the closed and the open states in room-temperature solutions [12]. A surprising discovery made in the work by Flynn et al. is that the two helices that form the lid (marked by dark gray in the insert of Fig. 10.11c) move at different rates. It turned out that local helix folding-unfolding transitions could regulate the kinetics of the conformational dynamics. These findings have been made possible by the quantitative high-resolution approach described in this chapter.

10.9 Concluding Remarks

Starting with an articulation of the fundamental differences between ensemble-averaged experiments and single-molecule measurements – the former involves number and then time averaging, in this specific order, whereas the latter only time averaging – this chapter has outlined ideas and practical protocols that will allow one to make accurate single-molecule measurements. With the theoretical foundation established, it is now possible to unambiguously relate single-molecule FRET signals to distances and distance distributions. The capability to quantitatively investigate time-dependent distance changes within individual molecules without having to assume any conformational distribution or kinetic models will certainly help when resolving competing models for complicated chemical and biological processes. Importantly, it will allow tackling problems that are currently beyond the scope of hypothesis testing, since a significant amount of knowledge has to be accumulated before an experimentally testable hypothesis can be formed. Understanding how protein nano-machines utilize local thermal fluctuations to accomplish work, and eventually fabricating man-made equivalents de novo, is one such problem for which further new discoveries are surely to come.

Acknowledgments The work presented here would not have been possible without the contributions from the students and post-doctoral associates with whom the author has had the good fortune to work. Princeton University and the U.S. National Institutes of Health are gratefully acknowledged for their financial support.

References

1. Alberts B (1998) The cell as a collection of protein machines: preparing the next generation of molecular biologists. Cell 92(3):291–294
2. Anderson PW (1954) A mathematical model for the narrowing of spectral lines by exchange or motion. J Phys Soc Jpn 9(3):316–339
3. Barkai E, Brown FLH, Orrit M et al (eds) (2008) Theory and evaluation of single-molecule signals. World Scientific, Singapore
4. Cover TM, Thomas JA (1991) Elements of information theory. Wiley, New York
5. Cramér H (1946) Mathematical methods of statistics. Princeton University Press, Princeton, NJ
6. Dale RE, Eisinger J (1975) In: Chen RF, Edelhoch H (eds) Polarized excitation energy transfer. Biochemical fluorescence: concepts, vol 1. Marcel Dekker, New York, pp 115–284
7. Dale RE, Eisinger J (1976) Intramolecular energy-transfer and molecular-conformation. P Natl Acad Sci USA 73(2):271–273
8. Dale RE, Eisinger J, Blumberg WE (1979) Orientational freedom of molecular probes – orientation factor in intra-molecular energy transfer. Biophys J 26(2):161–193
9. DiCiccio TJ, Efron B (1996) Bootstrap confidence intervals. Stat Sci 11(3):189–212
10. Efron B (1979) 1977 Rietz lecture – bootstrap methods – another look at the Jackknife. Ann Stat 7(1):1–26
11. Efron B, Gong G (1983) A leisurely look at the bootstrap, the Jackknife, and cross-validation. Am Stat 37(1):36–48
12. Flynn EM, Hanson JA, Alber T et al (2010) Dynamic active-site protection by the M. tuberculosis protein tyrosine phosphatase PtpB Lid domain. J Am Chem Soc 132:47–72
13. Förster T (1948) Zwischenmolekulare Energiewanderung und Fluoreszenz. Ann Phys-Berlin 2(1–2):55–75
14. Goldstein H (1980) Classical mechanics. Addison-Wesley, Reading, MA
15. Ha T, Enderle T, Ogletree DF et al (1996) Probing the interaction between two single molecules: Fluorescence resonance energy transfer between a single donor and a single acceptor. Proc Natl Acad Sci USA 93(13):6264–6268
16. Hanson JA, Duderstadt K, Watkins LP et al (2007) Illuminating the mechanistic roles of enzyme conformational dynamics. Proc Natl Acad Sci USA 104(46):18055–18060
17. Hanson JA, Yang H (2008a) A general statistical test for correlations in a finite-length time series. J Chem Phys 128:214101
18. Hanson JA, Yang H (2008b) Quantitative evaluation of cross-correlation between two finite-length time series with applications to single-molecule FRET. J Phys Chem B 112:13962–13970
19. Hanson JA, Tan Y-W, Yang H (2009) Conformation studies of protien dynamics using single-molecule FRET. In: *Single Particle Tracking and Single Molecule Energy Transfer: Application in the Bio and Nano Sciences* edited by Christoph Bräuchle, Don Lamb and Jens Michaelis. Wiley-VC H
20. Jaynes ET (1957a) Information theory and statistical mechanics. Phys Rev 106(4):620–630
21. Jaynes ET (1957b) Information theory and statistical mechanics. 2. Phys Rev 108(2):171–190
22. Jaynes ET (1982) On the rationale of maximum-entropy methods. Proc IEEE 70(9):939–952

23. Koshland DE (1958) Application of a theory of enzyme specificity to protein synthesis. Proc Natl Acad Sci USA 44(2):98–104
24. Kubo R (1954) Note on the stochastic theory of resonance absorption. J Phys Soc Jpn 9(6):935–944
25. Moerner WE, Kador L (1989) Optical-detection and spectroscopy of single molecules in a solid. Phys Rev Lett 62(21):2535–2538
26. Orrit M, Bernard J (1990) Single pentacene molecules detected by fluorescence excitation in a para-terphenyl crystal. Phys Rev Lett 65(21):2716–2719
27. Rao CR (1949) Sufficient statistics and minimum variance estimates. Proc Cambridge Philos Soc 45(2):213–218
28. Saffarian S, Elson EL (2003) Statistical analysis of fluorescence correlation spectroscopy: the standard deviation and bias. Biophys J 84(3):2030–2042
29. Schatzel K, Drewel M, Stimac S (1988) Photon-correlation measurements at large lag times: improving statistical accuracy. J Mod Opt 35(4):711–718
30. Schervish MJ (1995) Theory of statistics. Springer, New York
31. Silverman BW (1986) Density estimation for statistics and data analysis. Chapman & Hall, New York
32. Skilling J, Bryan RK (1984) Maximum-entropy image-reconstruction – general algorithm. Mon Not R Astron Soc 211(1):111–124
33. Watkins LP, Chang H, Yang H (2006) Quantitative single-molecule conformational distributions: a case study with poly-l-proline. J Phys Chem A 110(15):5191–5203
34. Watkins LP, Yang H (2004) Information bounds and optimal analysis of dynamic single molecule measurements. Biophys J 86(6):4015–4029
35. Yang H (2008) Model-Free Statistical Reduction of Single-Molecule Time Series. In: *Theory and Evaluation of Single-Molecule Signals* edited by E. Barkai, F. Brown, M. Orrit, and H. Yang, World Scientific publishing
36. Yang H (2009) The orientation factor in single-molecule förster-type resonance energy transfer with examples for conformational transitions in proteins. Israel J Chem 49: 313–322
37. Yang H (2010) Change-Point Localization and Wavelet Spectral Analysis of Single-Molecule Time Series. In: *Single Molecule Biophysics: Theories and Experiments*, a special volume in *Advances in Chemical Physics*, edited by Tamiki Komatsuzaki, Haw Yang, and Robert Silbey, with targeted publication year 2010

Chapter 11
Extracting the Underlying Unique Reaction Scheme from a Single-Molecule Time Series

Chun Biu Li and Tamiki Komatsuzaki

Abstract Single molecule spectroscopy provides us with a new means to look deeply into the question of how an individual molecule behaves when performing biological functions in a thermally fluctuating environment. However, what information one can extract from the observed data is still an open question. We overview our new method which extracts the underlying reaction scheme, a state-space network (SSN), from the time series data of an experimental measurement. We demand that a time series analysis should provide not only an interpretation of the dynamical behavior but also provide new insights into biological functions buried in ensemble-based measurements. Our method is based on the combination of information theory and Wavelet multiresolution decomposition analysis. The resultant reaction scheme does not rely on an a priori ansatz like local equilibrium and detailed balance. It is mathematically assured as unique, minimally complex and stochastic, but best predictive. We demonstrate the potential of this method by applying it to the analysis of an anomalous conformation in Flavin oxidoreductase dependent on the timescale of observation. We also discuss future perspectives concerning its use as a new means for the exploration of single molecule biophysics.

C.B. Li
Molecule & Life Nonlinear Sciences Laboratory, Research Institute for Electronic Science, Hokkaido University, Kita 20, Nishi 10, Kita-ku, Sapporo 001-0020, Japan
e-mail: cbli@es.hokudai.ac.jp

T. Komatsuzaki (✉)
Molecule & Life Nonlinear Sciences Laboratory, Research Institute for Electronic Science, Hokkaido University, Kita 20, Nishi 10, Kita-ku, Sapporo 001-0020, Japan
and
Core Reseach for Evolution Science and Technology(CREST),
Japan Science and Technology Agency(IST), Kawaguchi,
Saitama 332–0012, Japan
e-mail: tamiki@es.hokudai.ac.jp

Y. Sako and M. Ueda (eds.), *Cell Signaling Reactions: Single-Molecular Kinetic Analysis*, 221
DOI 10.1007/978-90-481-9864-1_11, © Springer Science+Business Media B.V. 2011

Keywords Complex networks · Dynamical heterogeneity · Free energy landscape · Information theory · Memory effects · Hierarchical organization · Kinetic schemes · Time series analysis · State-space network (SSN) · Computational mechanics

11.1 Introduction

Biological systems such as cells are complex – in response to an input stimulus on the membrane of a cell, signals are transmitted into the downstream part of a reaction network in the cytoplasmic space, resulting in robust functions in the cell in a thermally fluctuating and congested environment. Such functions are composed of a 'sequence' of structural changes involving chemical reactions triggered by the stimulus across hierarchies of time and space scales. In general, there exist two distinct approaches to scrutinizing the underlying mechanisms: one is an anatomical bottom-up approach in which one investigates the complex system from the microscopic molecular basis, and the other is a constructive top-down approach in which one develops mathematical models/frameworks in order to grasp features of the complexity at the system level. Both approaches have their merits and demerits. For example, molecular dynamics simulation, classified as the former approach, can obtain detailed dynamic information by decomposing the system into components at a certain level of approximation, but the time range of the computation is far from the timescale of interest. Single molecule experiments can enlarge the timescale of observation although the accessible information about the system is rather limited by the projection onto a single observable. The latter top-down approach has no limit of time and space scales and can refer to experimental phenomena to some extent in the modeling, but it does not exclude the possibility of ending up with unrealistic models far from the actual realm in biology.

Drastic revolutions in natural science have often been triggered by a new observation. The idea of energy quantization by Planck, for example, was stimulated by the discovery of black body radiation by Kirchhoff. Single molecule experiments such as optical single-molecule spectroscopy have provided a new scope in biology with unique insights into not only just the distribution of molecular properties but also their dynamic behavior at the single molecule level, which cannot be accessed by any ensemble averaged measurement. If people could access the detailed information of molecular dynamics in terms of computer simulation in the day of Boltzmann and Gibbs, people might have devoted more of their time to investigating the ergodicity hypothesis in the stream of the actual multivariate data, and might not have come up with the idea of constructing statistical mechanics. Accordingly, the observation of single molecule events can provide us with a new device for unveiling the origin of the mechanisms of functions of biological systems but also might prevent us from taking some other possible paths or branches in the evolution of science. In this sense, we should be responsible not just only for observing new phenomena buried in the ensemble experiments but also for establishing a new analytical and theoretical platform to extract unique insights to shed light on new concepts or theories relying on the actual observation from single molecule experiments.

In this chapter, we overview our recent studies aimed at bridging the two approaches of bottom-up and top-down approaches, that is, the extraction of the underlying unique reaction scheme, the state-space network (SSN), to capture the complexity of kinetics observed in single molecule time series. Information from the time series provides us with a "piece" of the underlying multidimensional dynamics of single molecules that is projected onto an observable such as donor–acceptor distance. It is of essential importance for the study of complex systems to develop a platform for analysis which can extract the underlying foundations/mechanisms from the observed data stream by "*letting the system speak for itself* " without a priori assumptions. The analysis should not only interpret the observed kinetics to reproduce the experimental data but more importantly unveil the origin of the complexity in a noisy environment.

11.2 Complex Network

Kinetic schemes may be regarded as Markovian networks composed of nodes (metastable states) and links (transitions). Irrespective of Markovianity in transitions, the global feature of the dynamics in general can be highly complex; a survival probability distribution within a subset of the network can be non-single-exponential. This may remind us that in classical mechanics the Liouville equation is linear (i.e., a Markovian process) in the probability densities but nonlinear equations in hydrodynamics can be derived from this equation. The network properties of biological systems can provide us with a new perspective for dissecting their hierarchical organization in multidimensional state space [1, 12, 13, 26, 34].

The dynamical evolution of a complex network of biomolecules can be regarded as itinerant motions traversing from one state (node) to another on a multi-scale complex network in the conformation space or, more generally, in the state space. Here we illustrate a conformational space network (CSN) of a small polypeptide on the multidimensional energy landscape [2–4, 11, 13, 24, 34, 41, 44]. By means of computer simulations, Caflisch and his coworkers derived the CSN for beta3s (a 20-residue anti-parallel β-sheet peptide) [13, 34]. The CSN is composed of nodes (the set of snapshots recorded along the trajectory grouped according to the secondary structure) and the links (transitions) between them. The CSN showed that the underlying multidimensional energy landscapes are much more complex than what one could deduce from a funnel-like landscape. It was also revealed that the projection onto a single variable such as the number of native contacts, masks the complexity so that the profile apparently looks simple on the free energy landscape along the variable, as the funnel landscape does. However, the CSN showed that the denatured state actually consists of not only entropically-favored conformations as the funnel landscape provides but also enthalpically-favored ones arising from being trapped in deep superbasins (see Figs. 11.1 and 11.2). It was also found that the CSN of beta3s exhibits scale-free characteristics [1, 5] (power-law behavior of the degree (k) distribution, i.e., $P(k) \sim k^{-\nu}$ with $\nu > 0$), similar to other real world networks such as the World Wide Web (WWW), protein interaction networks, and metabolic networks [12]. It was argued that the scale-free properties of the CSN network

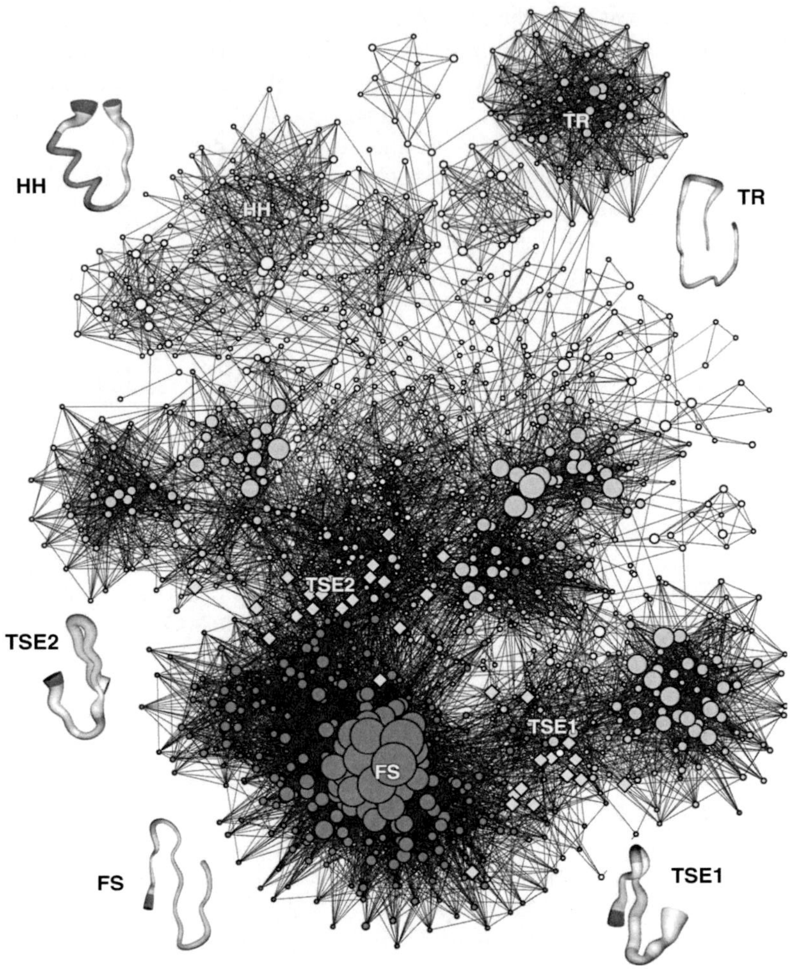

Fig. 11.1 A CSN of a small polypeptide, β3s. Nodes represent conformations and links represent transitions between them at the melting temperature of 330 K. The size of the (circle) nodes represents the statistical weight. Representative conformations are shown by a pipe along which the radius reflects the size of conformation fluctuation within the node. The diamonds are folding transition state conformations. HH, TR, TSE and FS are the helical, trap, transition state ensemble and folded states, respectively. Figure reprinted with permission from F. Rao and A. Caflisch *J. Mol. Biol.* (2004) **342**, 299-306. Copyright 2004 by Elsevier.

originate from the hierarchical organization of the native basin in the conformation space [34].

The most important message of this example is the possibility of missing the actual nature by postulating the underlying scenario for the observed kinetics, in this case, the energy landscape for the folding. The actual conformation space is much more complex than the funnel landscape of proteins even while the projection of the

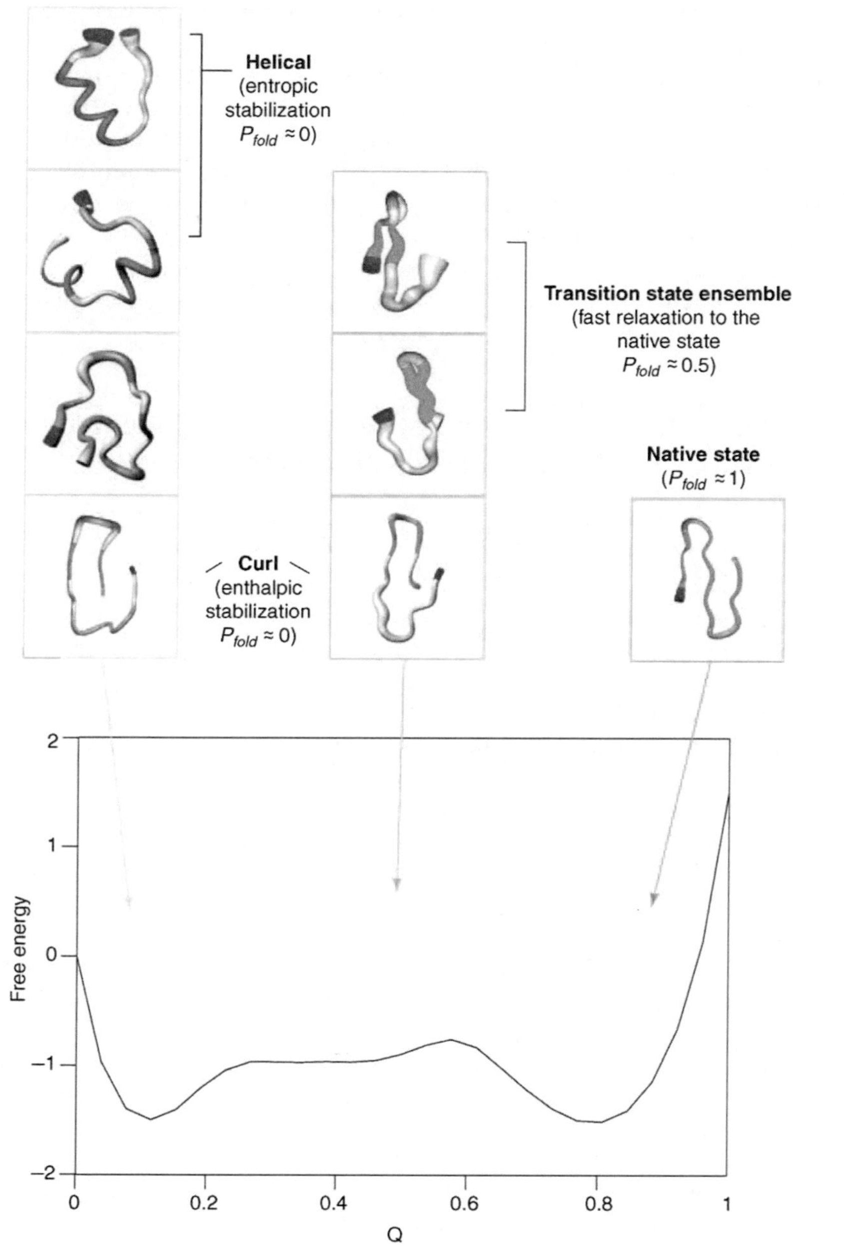

Fig. 11.2 A "free energy landscape" projected onto the fraction of the number of native contacts at 330 K where the stability of the native and the non-native states are comparable. The "free energy landscape" is apparently smooth in contrast to the network representation. Note that for $Q < 0.8$ the projection masks the complexity of the non-native states so that structurally-different conformations are grouped together in the transition state and denatured state ensemble. Figure reprinted with permission from A. Caflisch *Current Opinion in Struct. Bio.* (2006) **16**, 71-78. Copyright 2006 by Elsevier.

network onto the number of the native contacts does not seemingly contradict the funnel picture.

It should be noted, however, that most studies on complex networks have focused on the so-called binary networks where only the topological features of the links (transitions) among nodes (states) are taken into account. This corresponds to the case that the resident probabilities on the nodes and the probabilities of transitions among the links are regarded as equally weighted. Moreover, no transition directions are assigned to the links. In reality, transitions from one metastable state to another can be, in general, non-Markovian and directed. Further, the existing discussions of conformation networks are limited to computer simulations.

11.3 Time-Dependent Nature of Conformation Fluctuation, Energy Landscape and Reaction Network

11.3.1 Time-Dependence of Conformation Fluctuation

Recent developments in single-molecule spectroscopy have revealed new features of the dynamic behavior at the single molecule level, which are inaccessible by ensemble-averaged measurements [6, 20, 28, 30, 32, 35, 36, 42, 46, 48]. For example, a single-molecule electron transfer experiment [48] revealed the complexity of protein fluctuations of the NADH:flavin oxidoreductase (Fre) complex. It was found that the distance between flavin adenine dinucleotide (FAD) and a nearby tyrosine (Tyr) in a single Fre molecule fluctuates on a broad range of timescales (10^{-3} s – 1 s). As shown in Fig. 11.3, a strange, non-Brownian kinetics was observed in the interdye distance fluctuation on a wide range of the timescale, but it turns into a normal diffusion on longer timescales >10 s. The potential of mean force averaged over the whole time series was found to simply fall into a harmonic potential. (See Fig. 11.3b where the harmonic potential was used to make the theoretical plot for Brownian kinetics.) The authors conjectured that the morphological feature of the underlying energy landscape that the system actually "feels" at the single molecule level depends on the timescale, and exhibits a frustrated transient landscape with multiple timescales of inter-conversions among the basins for the timescale of non-Brownian kinetics.

To understand such anomalous behavior for conformational fluctuations, several analytical models have been proposed in terms of the generalized Langevin equation with fractional Gaussian noise [31], and the simplified discrete [43] and continuous [9] chain models. All these attempts are classified as top-down approaches where all the features are model-dependent. As for the bottom-up approach, an all-atoms simulation [29] was performed to extrapolate the physical origin of the anomalous FAD-Tyr distance fluctuation ($>10^{-3}$ s) from the simulation timescales in nano-seconds. These different theoretical models clearly

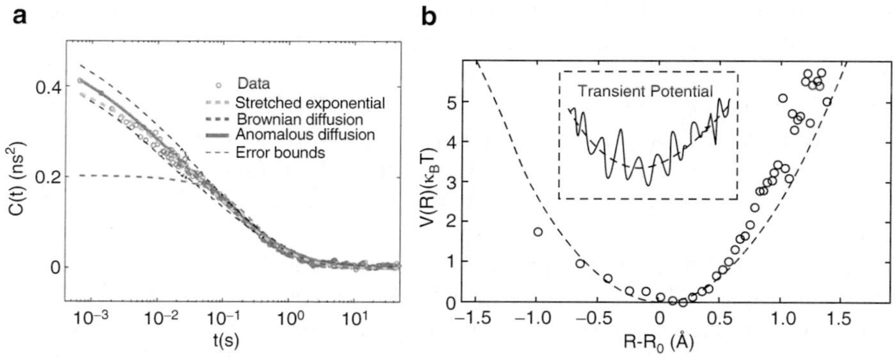

Fig. 11.3 (**a**) The autocorrelation of fluorescence lifetime fluctuation of the Fre/FAD complex showing the dynamical correlation of the conformation fluctuation between Fre and FAD. (**b**) The potential of mean force averaged over the whole time series with an assumption of relationship between the fluorescence lifetime γ^{-1} and the interdye distance R: $\gamma^{-1} \approx \exp(-\lambda R)$ where $\lambda = 1.4$ Å. The *inset* illustrates a transient potential at shorter timescale. Figures reprinted with permission from H. Yang, G. Luo, P. Karnchanaphanurach, T.-M. Louie, I. Rech, S. Cova, L. Xun, & X. S. Xie, *Science* **302**, 262-266 (2003).

demonstrate the difficulty in establishing a minimal, unique physical model for revealing the origin of complexity in the kinetics of biomolecules.

11.3.2 Revisiting the Concept of Free Energy Landscape – Its Time Dependency

All the complexity of kinetics and dynamics are, in principle, governed by the underlying energy landscape. A kinetic analysis should enable us to capture the energy landscape behind the observation. It is worth revisiting the concept of free energy landscape in order to consider what type of energy landscape a single molecule actually feels.

The most commonly used definition of "free energy landscape" $F(\mathbf{Q})$ is given as a function of m-dimensional progress variables $\mathbf{Q}$ by

$$Z(\mathbf{Q}) = \int\int\cdots\int d\mathbf{q}d\mathbf{p}\delta^m(\mathbf{Q}(\mathbf{q}) - \mathbf{Q})\exp(-\frac{E(\mathbf{p},\mathbf{q})}{k_BT}), \tag{11.1}$$

$$F(\mathbf{Q}) = -k_BT\log Z(\mathbf{Q}), \tag{11.2}$$

where $E(\mathbf{p}, \mathbf{q})$ denotes the total energy of the system as a function of its momenta $\mathbf{p}$ and coordinates $\mathbf{q}$ coupled with the surrounding heat bath of temperature T. k_B, δ^m, $\mathbf{Q}(\mathbf{q})$ and $Z(\mathbf{Q})$ are Boltzmann constant, a multidimensional Dirac's delta function defined by $\delta^m(\mathbf{z}) = \delta(z_1)\delta(z_2)\cdots\delta(z_m)$, some progress variables on which free energy landscape is depicted (usually a certain function of $\mathbf{q}$), and the partition function with respect to **Q**, respectively. The physical interpretation is that all the degrees of freedom except a set of "quasi-constant" **Q** are distributed according to the Boltzmann distribution and the characteristic timescale with respect to the **Q** motions is much longer than those of the other unseen degrees of freedom so that the system can move about "ergodically" in the complementary space at each "quasi-constant" **Q**. Furthermore, such a timescale separation between **Q** and the other unseen degrees of freedom should remain irrespective of the region in the conformation space. Baba and Komatsuzaki [2, 3], followed Evans and Wales [10], have clearly shown that the topography of the energy landscape depends on the timescale of observation. Note that such time dependent nature of the underlying energy landscape have also explored in a similar context of single molecule detections [14]. We briefly explain this time dependency of free energy landscape. Suppose there are two metastable states. They will be unified as one when the timescale is longer than the typical timescale of the inter-conversion for which the system can visit both. This implies a decrease of the number of metastable states as the timescale of observation increases and makes the landscape smoother with a lower dimension. Here the decrease of the dimension results from the fact that some degrees of freedom used for describing the energy landscape at the shorter timescale can be "thermalized" within the longer timescale (Fig. 11.4).

It should be noted that there exists a clear distinction between a potential of mean force and a free energy landscape. The free energy landscape, in principle, requires detailed balance between stable states which are considered to be locally equilibrated in the **Q**-space. Suppose the existence of an Arrhenius relation of the reaction rate, i.e.,

$$k_{i\to j} = A\exp(-\Delta F^{\ddagger}_{i\to j}/k_{\mathrm{B}}T), \tag{11.3}$$

where A, $\Delta F^{\ddagger}_{i\to j}$, and T, denote the pre-factor constant, the free energy barrier height from the ith local equilibrium state (LES) to the jth LES, and absolute temperature,

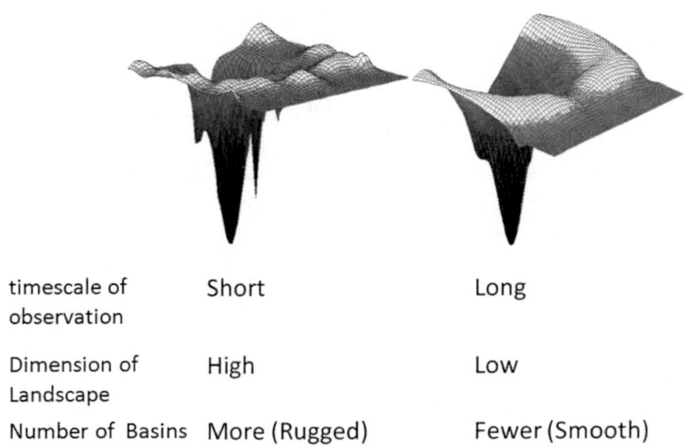

Fig. 11.4 A schematic picture of an energy landscape as a function of timescale of observation. 'Energy landscape' means either a free energy landscape or a potential of mean force. Although these landscapes are depicted in two dimensions, in general, the dimension is not necessarily two.

Fig. 11.5 A free energy landscape.

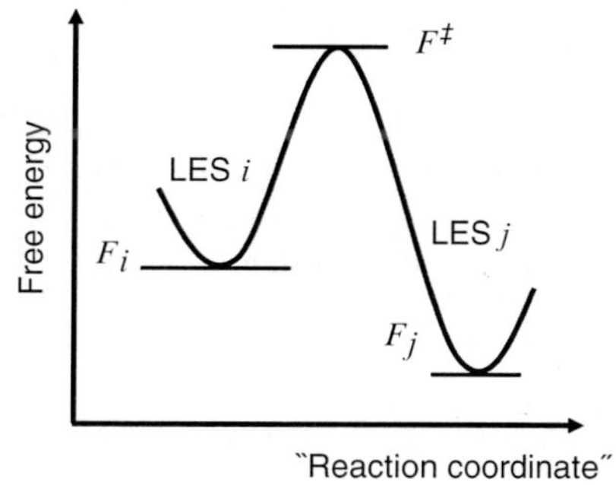

respectively. (Also see a recent review [3, 22] which overviews the concept of local equilibrium covering from the historical background and the applications to single molecule biophysics.) Then one can evaluate the (relative) free energy of the barrier $F^{\ddagger}$ linking the free energy minima F_i and F_j of the ith LES and the jth LES (see Fig. 11.5):

$$F_i = -k_{\mathrm{B}}T \ln P_i, \tag{11.4}$$

and

$$F^{\ddagger} = F_i - k_{\mathrm{B}}T \ln\left(\frac{1}{A}k_{i\to j}\right) = F_j - k_{\mathrm{B}}T \ln\left(\frac{1}{A}k_{j\to i}\right), \tag{11.5}$$

where $k_{i\to j}$, P_i, F_i, and $F^{\ddagger}$ denote the rate constant from the ith LES to the jth LES, the resident probability of the ith LES, the relative free energy of the ith LES, and the relative free energy of the barrier linking the ith and jth LES, respectively. Note here that the second equality in Eq. 11.5 is based on the assumption of the existence of free energy barrier $F^{\ddagger}$ acting on the reaction bottleneck of the forward and backward reactions between these two LES. The condition to make this assumption validated is called the (local) detailed balance given by

$$k_{i\to j}P_i \simeq k_{j\to i}P_j \tag{11.6}$$

(One can find Eq. 11.6 by substituting Eq. 11.4 into Eq. 11.5.) In the other words, unless the (local) detailed balance is satisfied, the second equality in Eq. 11.5 does not hold. This implies that one can neither identify the relative free energy of the barrier $F^{\ddagger}$ nor construct the landscape of *free energy* in the **Q**-space unless the detailed balance holds (The accuracy of the free energy barrier depends on, for example, how the pre-factor constant A depends on viscosity exerted by the environment [15, 21, 23, 36, 39]).

Again, the essential difference from "free energy landscape" defined by Eqs. 11.1 and 11.2 is the requirement of two conditions, the local equilibrium and the local detailed balance in the space used to describe the landscape (e.g., **Q**). The term "local" is also an important concept for the exploration of single molecule dynamics because people have not paid attention to the difference between "local" and "global" when investigating the ensemble behavior of the system. Note that Eqs. 11.1 and 11.2 has no-assumption on the dynamics of **Q** except the existence of the timescale separation with respect to the complementary subspace. The "free energy landscape" by Eqs. 11.1 and 11.2 is more appropriately referred to as 'the potential of mean force landscape.'

11.4 Extraction of the Unique Reaction Scheme

Recently, we developed a novel method to extract a unique, minimal but best predictive reaction scheme, the state space network (SSN), from a single molecule time series spanning several decades of timescales [26, 27]. (The mathematical definition of the term "minimal" and "best predictive" are presented in Section A4.) It was shown that the topology and topography of the network naturally depend on the timescale. The states are defined not only in terms of the *present value* of the observable at each time but also the *past information* in the time series. The states are connected with each other in such a way that each transition takes place as a Markovian process. If there exists a certain memory in the process, the memory effects are automatically incorporated into the content of the states or their network topology (the information of the components in states is equivalent to that of network connectivity [7, 37]). Our method can resolve degeneracy – different physical states having the same value for a measured observable – as much as possible with the limited information available from a scalar time series.

Let us begin to illustrate what we can extract from a single molecule time series by the single-molecule electron transfer (ET) time series of Fre/FAD complex [48] described in Section 11.3.1. Figure 11.6 shows, again, the autocorrelation function of the lifetime fluctuation $C(t) = \langle \delta\gamma^{-1}(t)\delta\gamma^{-1}(0) \rangle$ obtained from the experiment, but now with the values analytically derived using our multiscale state space network (SSN) (indicated by the small circles in the figure). Because the SSN is designed so as to be Markovian for the state-to-state transition, analytical expressions can be formulated for any quantities such as autocorrelation functions of any order, e.g., $\langle \delta\gamma^{-1}(t_1)\delta\gamma^{-1}(t_2)\delta\gamma^{-1}(0) \rangle$. This is because of the Markovian nature. One can see that our SSN approach is capable of reproducing the hierarchical diffusion kinetics although the normal Brownian model fails to capture this complexity of kinetics.

Figure 11.7 shows a visualization of the extracted SSNs at three different timescales, constructed *directly* from the observed time series in the single molecule spectroscopy. The description of the procedure to construct the SSNs will be described in the later sections and in Appendix. Here we rather focus on what one can learn from the topographical features of the constructed SSNs. In the figure, the abscissa corresponds to the average electron donor–acceptor distance $R_I - R_0$ of the state I and the ordinate a quantity to measure how the pattern of transitions (i.e., the destination of the transitions and their transition probabilities) from the

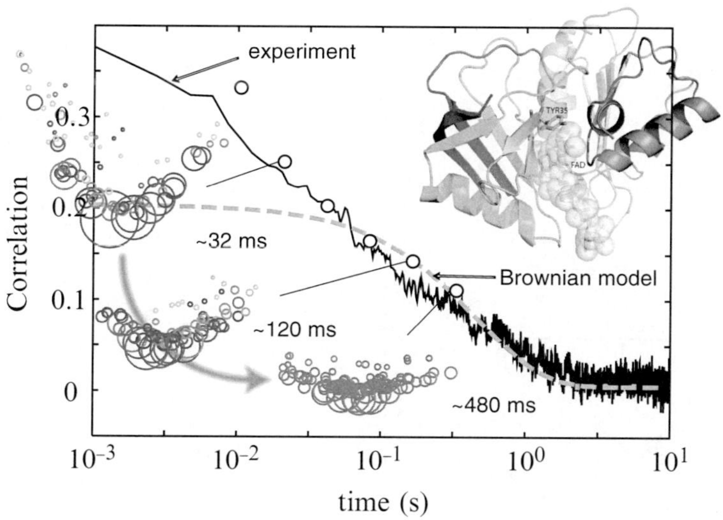

Fig. 11.6 The autocorrelation function of fluorescence lifetime fluctuation $\langle \delta\gamma^{-1}(t)\delta\gamma^{-1}(0) \rangle$ evaluated using the SSNs. The solid and gray dashed lines, respectively, denote the numerical result from the photon-by-photon-measurement and a normal Brownian diffusion model represented by an overdamped Langevin equation on a harmonic potential well whose curvature is explicitly determined from the experimental observation using a histogram of the observed interdye distance R along the time series [48]. In the *inset*, the corresponding SSNs are also depicted at the three distinct timescales.

state I to another is close to the averaged pattern of transitions taken over the entire set of states. (See the mathematical definition in the caption of Fig. 11.7.) Each circle represents a state reconstructed from the observed time series, whose area is proportional to the resident probability. That is, the larger the circle, the more often the system (re)visits in the given time trace. The (gray) color codings represent the topographical features of links/transitions between states. The color coding in Fig. 11.7a–c represents the degree (i.e., the number of links/transitions) of each state normalized by the maximum value of the degree in the SSN: when a state connects to almost all states, the normalized degree is close to unity. The color coding in Fig. 11.7d–f represents the so-called normalized transition entropy, which measures the uniformity of the transition probabilities: The closer the quantity to unity, the more uniform the transition probabilities. (See Section A5 for the mathematical description.) The deviation from unity indicates that the transitions are directional (= the existence of preferred transitions). The arrow indicates the (global) variance of the pattern of transitions, that is, the diversity of the transition patterns in the SSN.

What can we learn from these visualizations? One can see that, as a function of timescale, the topography of the multiscale SSN changes. Namely, as the time scale approaches the Brownian diffusion regime $\sim$480 ms, the SSN becomes simpler, i.e., the pattern of the transitions from each state becomes more or less uniform, as indicated by the smaller (global) variance of the pattern of transitions. In addition,

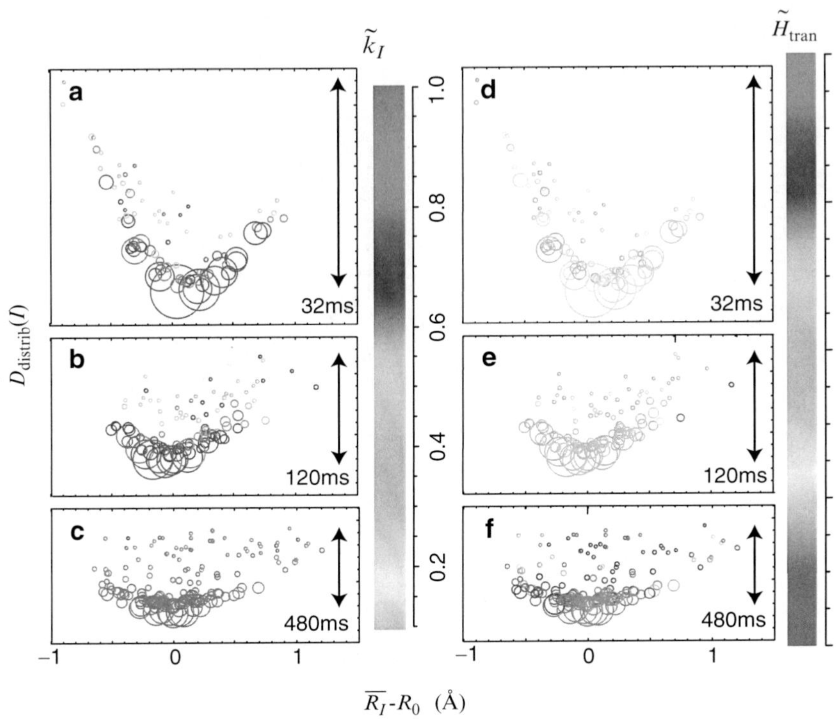

Fig. 11.7 SSNs obtained at three different timescales of (**a, d**) 32, (**b, e**) 120, and (**c, f**) 480 ms and quantification of topographical features of the SSNs. The abscissa and ordinate denote, respectively, the average electron donor–acceptor distance $R_I - R_0$ and a quantity associated with the state I in each SSN (denoted here by $D_{\text{distrib}}(I)$) measuring the closeness of the pattern of transitions (i.e., the destination of the transitions and their transition probabilities) to the average pattern of transitions taken over all states. $D_{\text{distrib}}(I)$ is defined by $D_{\text{distrib}}(I) = \sum_J P(S_J) d_H(I,J)$ where $P(S_J)$ denotes the resident probability of the state J, S_J, and the summation is taken over all the states in the SSN. $d_H(I,J)$ is the Hellinger distance [25] between two transition probabilities from S_I and from S_J to all the other states defined by $[\sum_K(\sqrt{P(S_K|S_I)} - \sqrt{P(S_K|S_J)})]^{1/2}$, where, e.g., $P(S_K|S_I)$ is the transition probability from S_I to S_K. The variance of $D_{\text{distrib}}(I)$ over the set of states in the network (see the black arrows in the figure) measures how diverse the transition probabilities of the states are. $R_I - R_0$ is evaluated by using the lifetime γ^{-1} assigned for each state in the SSNs. The lifetime $\gamma^{-1}(t)$ of the excited state of the acceptor molecule is composed of the contributions from the fluorescence decay rate in the absence of quencher(s) γ_0 and the electron transfer (ET) rate between the two dye molecules k_{ET}: $\gamma^{-1}(t) = [\gamma_0 + k_{ET}(t)]^{-1} \approx k_{ET}{}^{-1}(t)$. The averaged R for the state I, $\overline{R_I}$, is evaluated by $\overline{R_I} \approx R_0 - \beta^{-1} \log \gamma_I$ under the assumption of $k_{ET}(t) \approx k^0_{ET} \exp[-\beta R(t)]$ with $\beta = 1.4$ Å^{-1} for proteins [33] where $R_0 = \log k^0_{ET}/\beta$. In rows (**a**)–(**c**), the (gray) color coding corresponds to the degree (number of links/transitions) of each state normalized by the maximum value of the degree, $\tilde{k}_I$. In rows (**d**)–(**f**), the coding corresponds to a quantity called the normalized transition entropy, $\tilde{H}_{tran}$ (See Section A5), which measures the uniformity of transition probabilities of links from a state: this quantity is unity when all the transition probabilities are the same. In all panels, the area of the circle is proportional to the resident probability of the state (states with resident probability $< 10^{-4}$ are not displayed for clarity) and the arrow indicates the (global) variance of $D_{\text{distrib}}(I)$.

as seen in Fig. 11.7a–c, the existence of the nonuniform-colored circles at the timescale of the non-Brownian regime ~ 32–120 ms ceases at the time scale of the Brownian diffusion regime, implying that the system can make direct transitions from any state to all the other states within the timescale. As inferred from Fig. 11.7d–f, the transitions seem to be more nonuniform and directional at the timescale of non-Brownian regime compared with the transitions in the Brownian regime. Note however that, even in the Brownian regime, the directionality of transitions is not uniform even though all states have direct transitions with each other in the SSN: the more transitions become uniform and non-directional the closer the states are to the global minimum at $R_I \simeq R_0$.

Figure 11.8 plots the relationship between the stability of states and the number of links/transitions in the SSNs at the three different timescales. The stability is evaluated by log P_I where P_I is the resident probability of state I: the larger the value, the more stable the state (Recall that $-k_B T \log P_I$ may correspond to the free energy of the state I if both the detailed balance and the local equilibrium hold). Roughly, the figure shows that as the number of the links increases, the state becomes more stable. It is worth noting that, at the range of log $P_I > -10$, a set of states exhibits almost the same stability but with a different number of links at the timescales of non-Brownian regime (see the states indicated by the two ellipses in the figure) while most of all states monotonically increases as a function of the degree at the timescale of Brownian regime. The states in the top ellipse have more links in the SSNs compared with those in the bottom ellipse, while their stabilities are almost the same. This may indicate that the former states are stabilized entropically because of the higher chance to move around the SSN, while the latter are stabilized enthalpically (c.f. the work by Rao and Caflisch using computer simulation [34] in Section 11.2). Such information buried in the observed single-molecule time series can be retrieved naturally by using our theory.

It is also straightforward to analyze the degree distribution of the SSNs at different timescales. We found that the degree distributions of the SSNs for the

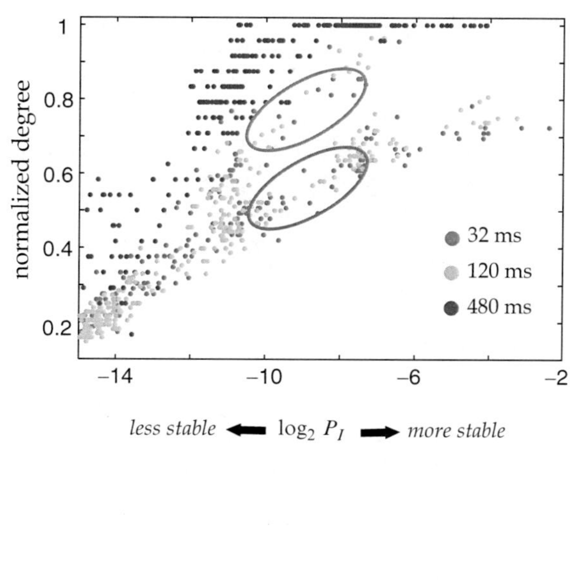

Fig. 11.8 The stability of states logP_I (P_I: resident probability of state I) and the number of the transitions/links from the state normalized by the maximum value of the degree in the SSNs obtained at the three different timescales.

protein conformation of Flavin oxidoreductase is almost the same at $\sim$480 ms with that of the SSN constructed from a time series made from an overdamped Langevin model, while those at the timescales of the non-Brownian regime deviate from the corresponding overdamped Langevin model [27].

Our method of time series analysis based on information theory is not only capable of analytically reproducing physical quantities of hierarchical kinetics but also of capturing the underlying mechanism in terms of morphologies of SSNs that are dependent on timescale.

11.4.1 The Construction of the State-Space Network

In this subsection, we present the main procedure to construct the SSN from an experimental time series based on the combination of Wavelet multiscale decomposition [8] and computational mechanics (CM) developed by Crutchfield et al. using information theory [7, 37]. An illustrative example for the SSN construction, detailed properties and various measures to quantify the structural and dynamical properties of the constructed SSN are presented in the Sections A4 and A5.

11.4.1.1 Evaluating the Transition Probabilities

Without loss of generality, we present the construction of SSN from a one-dimensional time series. The generalization to multi-dimensional time series obtained from multi-channel measurements is straightforward.

Given a time series of length N of a certain physical observable obtained from an experiment $\mathbf{x} = \{x(t_1), x(t_2), ..., x(t_N)\}$ (Fig. 11.9a), such as the donor–acceptor distance, fluorescence intensity, enzymatic turnover rate, and so forth, the construction of the state-space network starts from discretizing the continuous observable $\mathbf{x}$ into a symbolic sequence $\mathbf{s} = \{s(t_1), s(t_2), ..., s(t_N)\}$ (Fig. 11.9b), in which $s(t_i)$ denotes the symbolized observable at time t_i. Symbolization is a crucial step in the construction of the state-space network because it allows us to obtain good statistics when evaluating transition probabilities by sampling along the time series (The choice of symbolization will be discussed more in detail in Section A2).

After symbolization, the next step in the construction of the state-space network is to evaluate the transition probabilities from different subsequences, called *past* subsequences, to the future symbols. Figure 11.9b illustrates an example for a particular subsequence $s_3s_2s_2$ of length three (indicated by the open dots) that can make transition to all the three symbols (s_1, s_2, s_3) along the time series. By tracing through the whole time series, one can evaluate $P(s_i|s_3s_2s_2)$, the transition probability from the past subsequence $s_3s_2s_2$ to the future symbol s_i, as shown in Fig. 11.9c. The transition probabilities of all other past subsequences $s_js_ks_l$ with length three to the next symbol s_i can be evaluated similarly from the time series (see Fig. 11.9d).

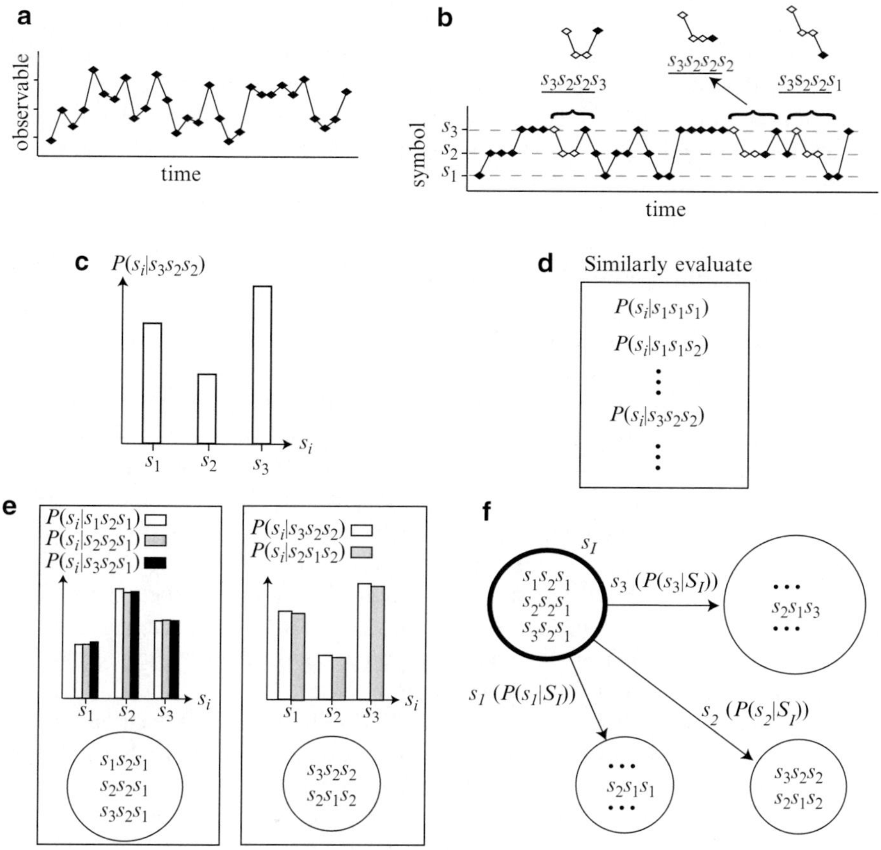

Fig. 11.9 Procedures to construct the state-space network from time series data (see text for detail). Given the time series of a continuous observable (**a**), the time series is discretized to a symbolic sequence. (**b**) An example of discretization with three symbols s_1, s_2, s_3 from which the transition probabilities can be evaluated. (**c**) The transition probabilities of a particular past subsequence (e.g. $s_3s_2s_2$ indicated by the open dots in (**b**) to the future symbol s_i, $P(s_i|s_3s_2s_2)$, can be determined by tracing through the whole symbolic sequence. (**d**) The transition probabilities of all possible past subsequences to the future symbols are evaluated similarly. (**e**) A state in the SSN is defined as a collection of past subsequences with the same (or approximately the same) transition probabilities to the next symbols. Two states containing three (*left panel*) and two (*right panel*) past subsequences are shown as examples. (**f**) The transition between the state S_I (*thick circle*) to another state (*thin circle*) by producing a particular symbol is shown by an arrow pointing from S_I to the target state containing the corresponding past subsequence after a one-step shift in time. Note that the transition from the state S_I with a symbol s_2 ends up with a state composed of not only $s_2s_1s_2$ which results from the one-step shift in time from the past sequences in the state S_I but also $s_3s_2s_2$. This means that a transition or a link to this state will be added from another state containing $s_*s_3s_2$ ($*$ means a "blank card" such as 1, 2, or 3) with producing the symbol s_2 through the process of constructing the state transitions (For the sake of simplicity arrows (links) departing from the state S_I are only depicted). The symbol produced in the transition (s_i) and the weight of the transition ($P(s_i|S_I)$) are indicated with the labels next to the arrows.

We note here that a new parameter, the length of the past subsequences denoted by L_{past}, was introduced and was set equal to three (as an example) at this stage of the construction. The suitable value of L_{past} depends on the nature of the underlying dynamics of the time series and can be determined by choosing a L_{past} such that the structural properties of the SSN do not change as L_{past} increases. We will come back to the discussion for the choice of L_{past} in Section A3 and its implication later in this section.

11.4.1.2 Determining the States of the SSN

Now we are ready to define the states of the SSN from the list of transition probabilities. In the SSN, a state is defined as a collection of past subsequences with the same (or approximately the same) transition probabilities to the next symbols. This is illustrated in Fig. 11.9e in which the three past subsequences $s_1s_2s_1$, $s_2s_2s_1$ and $s_3s_2s_1$ with almost equal transition probabilities to the next symbols are grouped to form a state (left panel). The right panel of Fig. 11.9e shows another state that is formed by grouping two past subsequences $s_3s_2s_2$ and $s_2s_1s_2$ with almost equal transition probabilities. This grouping procedure is performed for all past subsequences resulting in a list of states, and each of them has distinct probabilities of transition to the future symbols. Accordingly, the constructed states are termed "causal states" in computational mechanics as they represent, in a certain sense, the "causes" in the time series that result in distinct futures.

The (resident) probability of a state in the SSN is defined to be the summation of the occurrence probabilities of all the past subsequences contained in that state. For example, the probability of the state in the right panel of Fig. 11.9e is simply given by $P(s_3s_2s_2)+P(s_2s_1s_2)$. On the other hand one can see that, instead of postulating the number of states a priori, the number of states is an outcome of the construction which is determined by the diversity of the transition patterns in the time series. For example, the number of states is one if all past subsequences have the same transition probabilities to the future symbols, whereas the number of states is large if there are many distinct transition probabilities of the past subsequences. This feature is in big contrast to most fitting and modeling schemes in which the number of states are pre-assumed.

11.4.1.3 State Transitions in the SSN

With the states defined, the next stage of the SSN construction is to link the states together to form a network which is able to describe time evolution (i.e., dynamics). In Fig. 11.9f, we demonstrate how state transitions are determined. As an example we consider a state called S_I (denoted by the thick circle) containing the past subsequences $\{s_1s_2s_1, s_2s_2s_1, s_3s_2s_1\}$. In particular, starting from the state S_I, one can make a transition to other states by producing one of the three symbols $\{s_1, s_2, s_3\}$. For example, when the symbol s_1 is produced in the transition from the state S_I which corresponds to one of the three transitions $s_1s_2s_1 \rightarrow s_1$, $s_2s_2s_1 \rightarrow s_1$ and $s_3s_2s_1 \rightarrow s_1$, one ends up with a new past subsequence of $s_2s_1s_1$ with $L_{past} = 3$ due to a one-step shift in time. Therefore, the transition from the state S_I by producing the symbol s_1 is indicated by an arrow pointing from S_I to the state that contains the past subsequence $s_2s_1s_1$ as shown in Fig. 11.9f. Moreover, the strength (or weight) of the transition is characterized by the transition probability $P(s_1|S_I)$, which is equal to $P(s_1|s_1s_2s_1) = P(s_1|s_2s_2s_1) = P(s_1|s_3s_2s_1)$ by the definition of causal state. The two characteristics of a transition, namely the produced symbol and the weight of the transition, are indicated as labels in the form "$s_1(P(s_1|S_I))$" next to the arrow. The two other transitions from S_I corresponding to production of the symbols s_2 or s_3 can be determined similarly as shown in Fig. 11.9f. Finally, the above procedure for determining the transitions of a given state are carried out for all causal states and results in the SSN as a directed weighted network. It was mathematically proved that there exists an equivalence between the topological nature of the SSN (i.e., how each state is connected with each other, resulting in one symbol for each transition from one state to another) and the components (a set of time segments) belonging to each state. That is, once one knows the components, one can *uniquely* identify all the connections among the states in the network. Likewise, once one knows the connections with their (one-symbol) transition outputs, one can uniquely identify what time segments should belong to each state. As we described above, in the CM procedure, memory in the process of **s**, if it exists, is manifest in the length of the optimal past sequences L_{past} used in defining the states **S**. This equivalence relation implies that memory in the process is expressed either by the components of the states or by the topological nature of the SSN. In the next subsection, we describe several serious problems of the original formulation of CM in application to single-molecule time series of a wide class of biological systems.

11.4.2 Wavelet Multi-timescale Decomposition

The SSN derived by CM is regarded as an optimal statistical "equation of motion" inferred from the time series. The SSN can, in principle, incorporate memory effects naturally into the topological nature of the network and capture the underlying mechanism of the complexity in kinetics. Furthermore it can analytically predict any physical quantities in kinetics thanks to the Markovian nature of the network. It can naturally be expected that the potential of CM makes it a very attractive tool in single molecule experiments. However, there are several practical difficulties in applying the current formulation of CM to single molecule time series. These difficulties and limitations are summarized as follows:

1. The original formulation of CM is based on time series of stationary processes or, in other words, time series with nonstationary behavior

(continued)

which is not significant or which changes only slowly during the (finite) whole length of the time series from which the SSN (originally called ε-machine) is constructed [7, 37]. However, this is not the case of most biological systems in which hierarchies of non-stationarity can exist that spoil the convergence of the SSN. Therefore, a hierarchical decomposition of the time series into a set of stationary (and non-stationary) processes with different time scales is necessary to justify the application of CM across different time scales of the system.

2. Another major difficulty in the original formulation of CM arises when the length of the past sequences L_{past} increases. In this case, the number of possible past sequences $\mathbf{s}_{\text{past}}$ grows exponentially with the size of L_{past} and the statistical accuracy in sampling of $\mathbf{s}_{\text{past}}$ becomes rapidly worse due to the finite length of the time series. As a consequence, with the original procedure it may be too hard to properly capture long-time memory effects if they exist.
3. The other common difficulty inherent in such a task of extracting 'states' from a scalar time series is that since the number of physical observables measured from experiments is limited, *degeneracy* (different physical states having the same value in the measured observable) can occur, which is known to give rise to apparent "memory."

In order to resolve these difficulties, we proposed a Wavelet based CM (WbCM), that is, the application of CM to a set of multiscale time series decomposed in terms of a discrete Wavelet decomposition.

Discrete Wavelet decomposition produces a family of hierarchically organized decompositions from a scalar time series. Figure 11.10 shows an example of a discrete Wavelet decomposition which transforms a scalar time series, denoted by $\tau = (\tau_1, \ldots, \tau_N)$ where the subscript $(1, \ldots, N)$ denotes the index of time step along the time series:

$$\tau = \underbrace{\mathbf{A}^{(n)} + \mathbf{D}^{(n)}}_{\mathbf{A}^{(n-1)}} + \mathbf{D}^{(n-1)} + \cdots + \mathbf{D}^{(1)}, \tag{11.7}$$

where $\mathbf{A}^{(j)} = (A_1{}^{(j)}, \ldots, A_N{}^{(j)})$ and $\mathbf{D}^{(j)} = (D_1{}^{(j)}, \ldots, D_N{}^{(j)})$, are given in terms of Haar Wavelet basis as

$$A_i^{(j)} = \left(\sum_{k=i}^{i+2^j-1} \tau_k\right)/2^j, D_i^{(j)} = \left(\sum_{k=i}^{i+2^{j-1}-1} \tau_k - \sum_{k=i+2^{j-1}}^{i+2^j-1} \tau_k\right)/2^j, \quad j \geq 1. \tag{11.8}$$

One can see that $A_i^{(j)}$ and $D_i^{(j)}$ are simply the mean and the mean fluctuation over a bin of 2^j time steps, respectively. Note that $A_i^{(j)}$ and $A_{i'}^{(j)}$ (or $D_i^{(j)}$ and $D_{i'}^{(j)}$) with $|i - i'| \leq 2^j$ are unphysically correlated since some common data points are

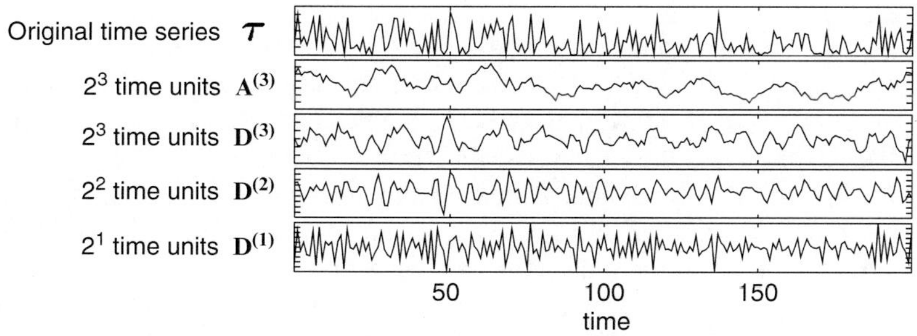

Fig. 11.10 A discrete Wavelet decomposition into three different timescales with $n = 3$ using Haar Wavelet basis (c.f. Eq. 11.7).

used in evaluating the two $A^{(j)}$'s (or $D^{(j)}$'s). Therefore, only $N/2^j$ points, e.g., $(A_1^{(j)}, A_{1+2^j}^{(j)}, \ldots, A_{1+2^j n}^{(j)}, \ldots)$, in the j–level should be taken into account to avoid apparent correlations. The larger the level j (i.e., the longer the timescale), the fewer the number in the sampling set. This is called the 'down-sampling problem,' leading to poor statistics in constructing the SSN, especially for processes of long timescale. This problem can be resolved by treating a set $\{(A_i^{(j)}, A_{i+2^j}^{(j)}, \ldots, A_{i+2^j n}^{(j)}, \ldots), i = 1, \ldots, 2^j\}$ (and similarly for $D^{(j)}$'s) as an ensemble of 2^j time series (each with $N/2^j$ data points).

The $\mathbf{A}^{(j)}$ and $\mathbf{D}^{(j)}$ are called the j-level 'approximation' and 'detail', respectively. The j-approximation $\mathbf{A}^{(j)}$ approximates τ with a time resolution of 2^j time steps by discarding fluctuations (details) with time scale smaller than 2^j time steps. The j-detail $\mathbf{D}^{(j)}$, on the other hand, captures the fluctuations of τ over the time scale of 2^j time steps. Equation 11.7 therefore implies that the time series can be reconstructed by adding back all fluctuations with time scales smaller than or equal to those of the approximation. Moreover, approximations of different time scales are related by $\mathbf{A}^{(j)} = \mathbf{A}^{(j+1)} + \mathbf{D}^{(j+1)}$ with $j \geq 0$.

1. The stationarity of the approximation and the details can be evaluated by their autocorrelations. The autocorrelation of $\mathbf{D}^{(j)}$ decays rapidly in a time scale of 2^j time steps so that the $\mathbf{D}^{(j)}$'s are approximately stationary within the time scale of 2^j. In contrast, the autocorrelation of $\mathbf{A}^{(j)}$ is approximately constant with similar behavior as that of τ for a time scale longer than 2^j. $\mathbf{A}^{(j)}$ captures nonstationary behavior of τ in a time scale longer than 2^j. That is, Eq. 11.7 is regarded as decomposing the original time series into a hierarchy of "stationary" processes (the details) at different time scales and their nonstationary counterpart (the approximation).
2. The WbCM allows us to properly quantify the characteristic length of memory by decomposing the original time series into a set of time series at different time scales. This avoids poor statistical accuracy in sampling of

(continued)

$\mathbf{s}_{\text{past}}$, in which the number of possible past sequences $\mathbf{s}_{\text{past}}$ grows exponentially with the size of L_{past}, especially when quantifying long term memory.

3. The WbCM can to some extent resolve the *degeneracy problem* inherent in observation since the original scalar time series τ is decomposed into a *vector* time series with approximation and details as components. In addition, more importantly, in defining "states" from scalar time series CM takes into account not only just the value itself at each instantaneous time step, but also the time sequence (i.e., history) near the instantaneous time step. The combination of CM with the Wavelet multiscale decomposition is thus expected to avoid the degeneracy problem more than just either the standard CM or Wavelet multiscale decomposition alone.

Here, we point out some remarks on Wavelet decomposition in comparison with Fourier transformation. Fourier transformation also decomposes time series into a set of time series with different frequency/time scales, but the decomposed time series cannot avoid apparent correlation along the time trace because of the global nature of the Fourier basis. Local Fourier transformation may be the next candidate to overcome this apparent correlation in the resultant time series. However, this requires to determine a priori a single local window size in time, which cannot avoid subjective choices. In contrast, Wavelet decomposition naturally provides us with a hierarchical decomposition of timescale. With the Haar mother Wavelet, $A_i^{(j)}$ and $D_i^{(j)}$ are simply interpreted as the mean and the mean fluctuation over a bin of 2^j time steps, respectively. The other choices of the mother Wavelet are also possible, although the interpretations of $A_i^{(j)}$ and $D_i^{(j)}$ may become obscure. Among them, the so-called empirical mode decomposition [16] does not require any form of mother Wavelet a priori and rather determines the form in an adaptive fashion for each time series of interest.

11.4.3 Combining Different SSNs Constructed by Hierarchical Time Series Components in Wavelet Multiscale Decomposition

The algorithm used in the original CM described in Section 11.4.1 is applied to the *ensemble* of $\mathbf{A}^{(n)}$ (as mentioned above) and that of $\mathbf{D}^{(j)}$ ($j \leq n$) to construct the *sub*-SSNs, denoted by $\epsilon^{A^{(n)}}$ and $\epsilon^{D^{(j)}}$, under the assumption that the decomposed time series can be regarded as approximately stationary. Since the approximation (the binned average) and its sub-SSN $\epsilon^{A^{(n)}}$ average out the information contained in each bin, it suppresses the noise but, on the other hand, suffers from information loss inside the bins. Therefore, the combination of these SSNs is highly desirable, and a combination can be constructed by adding back the SSNs of the fluctuations inside the bin $\epsilon^{D^{(j)}}$ to $\epsilon^{A^{(n)}}$; the incorporation of $\epsilon^{D^{(n)}}$ and $\epsilon^{D^{(n-1)}}$ into $\epsilon^{A^{(n)}}$ gives a network that describes kinetics with the time scale of 2^n by taking into account the fluctuations down to the bin size of 2^{n-1}. One can evaluate which sub-SSNs are correlated mutually, and so should be unified together into one SSN in underpinning the kinetics at the desired timescale, by their cross correlation defined by

$$C_{\mathbf{A}^{(n)},\mathbf{D}^{(j)}}(i) = \langle (A_{i_0}^{(n)} - \langle \mathbf{A}^{(n)} \rangle)(D_{i_0+i}^{(j)} - \langle \mathbf{D}^{(j)} \rangle) \rangle, \tag{11.9}$$

where $\langle \cdot \rangle$ denotes the time average. As one can expect, the cross correlations between the sub-SSNs tend to be more significant as the timescales of the sub-SSNs becomes closer (i.e., $|C_{\mathbf{A}^{(n)},\mathbf{D}^{(n)}}| > |C_{\mathbf{A}^{(n)},\mathbf{D}^{(n-1)}}| > |C_{\mathbf{A}^{(n)},\mathbf{D}^{(n-2)}}| > \cdots$).

The states of the sub-SSNs $\epsilon^{A^{(n)}}$ and $\epsilon^{D^{(j)}}$ are hereinafter denoted by $\{S_i^{A^{(n)}}, i = 1, \ldots, N^{A^{(n)}}\}$ and $\{S_i^{D^{(j)}}, i = 1, \ldots, N^{D^{(j)}}\}$, respectively. Suppose that we have the state sequences with the transition time steps of 2^n for $\mathbf{A}^{(n)}$ and $\mathbf{D}^{(n)}$ as shown in Fig. 11.11a.

The procedure for combining $\epsilon^{A^{(n)}}$ and $\epsilon^{D^{(n)}}$, both with the same time steps 2^n, is carried out as follows: The sequence of states from $\mathbf{A}^{(n)}$ visited at each 2^n-time step can be constructed as shown in Fig. 11.11a, and similarly for $\mathbf{D}^{(n)}$. The possible candidates of the states in the combined SSN $\epsilon^{A^{(n)},D^{(n)}}$ are given by the product set $S_{ij} \equiv \{S_i^{A^{(n)}}, S_j^{D^{(n)}}\}$. The probability of the combined state S_{ij} denoted by $P_{A^{(n)},D^{(n)}}(S_{ij})$ can then be computed from the two state sequences as

$$P_{A^{(n)},D^{(n)}}(S_{ij}) = N(S_{ij})/N, \tag{11.10}$$

by tracing through the state sequences in Fig. 11.11a. In Eq. 11.10, $N(S_{ij})$ is the number of simultaneous occurrences of the states $S_i^{A^{(n)}}$ and $S_j^{D^{(n)}}$, and N is the number of data points in the time series. One can expect that $P_{A^{(n)},D^{(n)}}(S_{ij}) \neq P_{A^{(n)}}(S_i^{A^{(n)}})P_{D^{(n)}}(S_j^{D^{(n)}})$ in general due to the fact that the two time series $\mathbf{A}^{(n)}$ and $\mathbf{D}^{(n)}$ are statistically correlated. On the other hand, the probability of state transition from S_{ij} to $S_{i'j'}$ can also be obtained by

$$P_{A^{(n)},D^{(n)}}(S_{i'j'}|S_{ij}) = N(S_{i'j'}, S_{ij})/N(S_{ij}), \tag{11.11}$$

where $N(S_{i'j'}, S_{ij})$ is the number of visits to the new state $S_{i'j'}$ at 2^n time steps after visiting the state S_{ij}. In general, a transition from one state to another in the combined SSN $\epsilon^{A^{(n)},D^{(n)}}$ takes 2^n time steps, which is the same as in $\epsilon^{A^{(n)}}$. Therefore, the combined SSN $\epsilon^{A^{(n)},D^{(n)}}$ corresponds to a "splitting" of the states $S_i^{A^{(n)}}$ to S_{ij} by incorporating the fluctuation inside the bin of 2^n.

Similarly, the other combined sub-SSNs ($\epsilon^{D^{(n-1)}}$ and $\epsilon^{D^{(n-2)}}$ and so forth) can be incorporated into $\epsilon^{A^{(n)},D^{(n)}}$ one by one. This procedure depends on the fineness of the fluctuations of the hierarchical dynamics in which one may be interested. By referring to the state sequences of the combined SSN $\epsilon^{A^{(n)},D^{(n)}}$ and the SSN $\epsilon^{D^{(n-1)}}$ to be incorporated, as shown in Fig. 11.11b, one can further evaluate the state

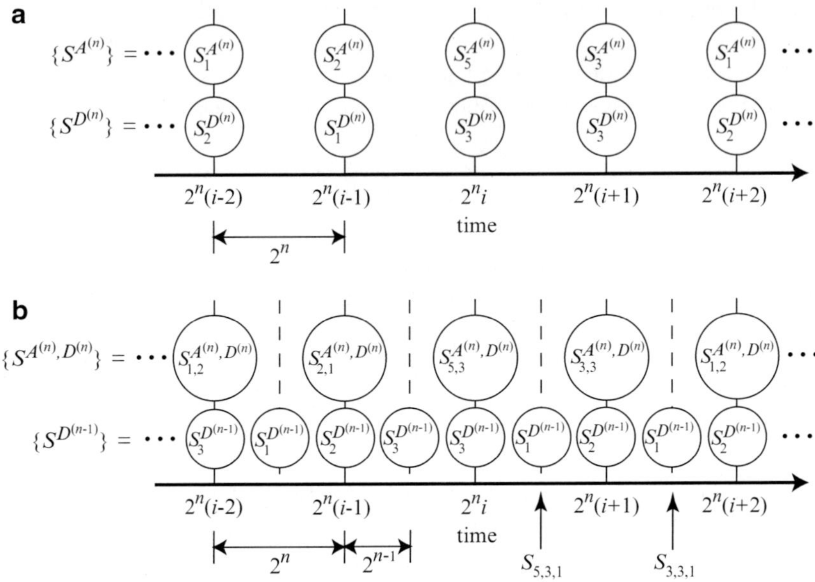

Fig. 11.11 (**a**) An example of the state sequences of $\epsilon^{A^{(n)}}$ and $\epsilon^{D^{(n)}}$ from which the combined SSN $\epsilon^{A^{(n)},D^{(n)}}$ is constructed. The transition time steps of both $\epsilon^{A^{(n)}}$ and $\epsilon^{D^{(n)}}$ are equal to 2^n. (**b**) An example of the state sequence of the $\epsilon^{A^{(n)},D^{(n)}}$ and $\epsilon^{D^{(n-1)}}$ from which the (next) combined SSN $\epsilon^{A^{(n)},D^{(n)},D^{(n-1)}}$ is constructed. Note that the transition step of $\epsilon^{D^{(n-1)}}$ is 2^{n-1}. At the half step of 2^{n-1}, the state of the combined $\epsilon^{A^{(n)},D^{(n)}}$ is assigned to be the same as that in the last half step, e.g., the states of the combined $\epsilon^{A^{(n)},D^{(n)}}$ at the time $2^n(i+1/2)$ and $2^n(i+3/2)$ are $S_{i=5,j=3}$ and $S_{i=3,j=3}$, so that the state of the combined $\epsilon^{A^{(n)},D^{(n)},D^{(n-1)}}$ at the time $2^n(i+1/2)$ and $2^n(i+3/2)$ are $S_{i=5,j=3,k=1}$ and $S_{i=3,j=3,k=1}$ (shown by the arrows in **b**), respectively.

probability $P_{A^{(n)},D^{(n)},D^{(n-1)}}(S_{ijk})$ and the transition probabilities $P_{A^{(n)},D^{(n)},D^{(n-1)}}(S_{i'j'k'}|S_{ijk})$ as in the case of Eqs. 11.10 and 11.11 with $S_{ijk} \equiv \{S_i^{A^{(n)}}, S_j^{D^{(n)}}, S_k^{D^{(n-1)}}\}$.

For a single molecule time series with multiple timescales the nonstationarity of $\mathbf{A}^{(n)}$ tends to be more pronounced as n increases. The number of partitions in the symbolization and the length of the past sub-sequence L_{past} are chosen such that the statistical complexity Eq. 11.27 of the sub-SSN does not change significantly.

11.4.4 An Analytical Expression of Autocorrelation Derived from SSN

As we explained in Fig. 11.6, our SSN can naturally reproduce a hierarchical diffusion process changing from subdiffusion to normal Brownian diffusion. Likewise, we can derive analytical expressions for kinetics because the constructed SSN is Markovian and all the memory effects are "imprinted" in either the composite elements of the states or the topological nature of the network (these two forms of information have been proven to be equivalent [7, 37]).

Here, as an example, we derive an expression for an autocorrelation function of an observable s in terms of SSN. First note that the correlation function $C(t)$ can be expressed by

$$C(t) = <\delta s(t)\delta s(0)> = <s(t)s(0)> - <s>^2$$
$$= \sum_{s_0,s_t} s_t P(s_t, s_0)s_0 - \left(\sum_s sP(s)\right)^2 \tag{11.12}$$

where $\delta s(t) = s(t) - \langle s \rangle$ and s_0 and s_t are the value of the observable at the current time and at t steps later, respectively. $\langle . \rangle$ means the average taken over time. $\sum_{s_0,s_t}$ means the summation over all possible pairs of s_0 and s_t. $P(s_t, s_0)$ denotes the joint probability of s_0 and s_t: the probability of finding s_0 at the current time *and* s_t at time t later. For a given SSN with timescale $t = 2^n$, the joint probability can be expressed as

$$P(s_{2^n}, s_0) = \sum_{I,J} P(s_{2^n}, s_0, S_J, S_I), \tag{11.13}$$

where $P(s_{2^n}, s_0, S_J, S_I)$ is the joint probability of visiting the state S_I at the current time with the value s_0 and visiting the state S_J at $t = 2^n$ time steps later with the value s_{2^n}. Therefore for $t = 2^n$ the first term on the right hand side of Eq. 11.12 becomes $\sum_{I,J}\sum_{s_0,s_{2^n}} s_{2^n}P(s_{2^n}, s_0, S_J, S_I)s_0$.

Note here that, in terms of the chain rule of joint probability, the joint probability can be decomposed as

$$P(s_{2^n}, s_0, S_J, S_I) = P(s_{2^n}|s_0, S_J, S_I)P(s_0|S_J, S_I)P_{2^n}(S_J, S_I), \tag{11.14}$$

where $P_{2^n}(S_J, S_I) = P_{2^n}(S_J|S_I)P(S_I)$ is the joint probability of visiting S_I followed by S_J after 2^n steps. The current value s_0 does not depend on the future state S_J due to causality. Thus, we have $P(s_0 \mid S_J, S_I) = P(s_0 \mid S_I)$. If s_{2^n} depends solely on the state S_J where the system resides at that time, such that $P(s_{2^n}|s_0, S_J, S_I) \approx P(s_{2^n}|S_J)$, the first term of Eq. 11.12 can then be estimated as

$$\sum_{s_0,s_{2^n}} s_{2^n}P(s_{2^n}, s_0)s_0 \approx \sum_{I,J}\sum_{s_0,s_{2^n}} s_{2^n}P(s_{2^n}|S_J)P(s_0|S_I)P_{2^n}(S_J, S_I)s_0$$
$$= \sum_{I,J} \overline{s_J} P_{2n}(S_J, S_I)\overline{s_I}, \tag{11.15}$$

where

$$\overline{s_J} = \sum_{s_{2^n}} s_{2^n} P(s_{2^n}|S_J), \tag{11.16}$$

$$\overline{s_I} = \sum_{s_0} s_0 P(s_0|S_I). \tag{11.17}$$

The second term on the right hand side of Eq. 11.12 can also be evaluated as follows:

$$\left(\sum_s sP(s)\right)^2 = \left(\sum_I \sum_s sP(s|S_I)P(S_I)\right)^2 = \sum_I \sum_J \overline{s_I}\;\overline{s_J} P(S_I)P(S_J). \tag{11.18}$$

By combining Eqs. 11.15 and 11.18, one can obtain

$$C(t = 2^n) = \sum_{I,J} \overline{s_J}\;\overline{s_I}(P_{2^n}(S_J, S_I) - P(S_J)P(S_I)). \tag{11.19}$$

The implication of Eq. 11.19 is that the autocorrelation function with respect to the (symbolized) observable s is represented solely in terms of the nature of the states and their transitions in the SSN and can be solved analytically thanks to the Markovian nature of the SSN. One can also generalize the above procedure to evaluate the multi-time correlation functions. For example, one can estimate the three-time correlation function $\langle\, s(2t)s(t)s(0)\,\rangle$ as $\sum_{I,J,K} \overline{s_K}\;\overline{s_J}\;\overline{s_I} P(S_K, S_J, S_I)$, where $P(S_K, S_J, S_I)$ is the joint probability of visiting the states S_I, S_J, and S_K at the current time, t steps later, and $2t$ steps later.

In the above derivation we assumed $P(s_{2^n}|s_0, S_J, S_I) \approx P(s_{2^n}|S_J)$. The physical implication of this assumption is as follows: For simplicity, let us rewrite S_I (the current state), S_J (the future state), s_0 (the symbol to be produced at S_I) and s_{2^n} (the symbol to be produced at S_J) as S, S', s and s', respectively. We can write

$$\begin{aligned} &P(s'|s, S, S') \\ &= \frac{P(s, s', S, S')}{P(s, S, S')} \\ &= \frac{P(s|s', S, S')P(s'|S, S')P(S'|S)P(S)}{P(S'|s, S)P(s|S)P(S)} \quad \text{chain rule used} \\ &= \frac{P(s|S)P(s'|S')P(S'|S)P(S)}{P(S'|s, S)P(s|S)P(S)} \quad \text{causality used} \\ &= P(s'|S')\frac{P(S'|S)}{P(S'|s, S)}. \end{aligned} \tag{11.20}$$

The third equality arises from causality: the probability of finding s at the current time depends neither on the future symbol s' nor the future state S', yielding $P(s\,|\,s',S,S')=P(s\,|\,S)$; the probability of finding s' depends not on the "past" state S but on the state S' at that time because of the Markovian nature of the state-to-state transitions resulting in $P(s'\,|\,S,S')=P(s'\,|\,S')$. Thus we see that $P(s'\,|\,s,S,S')=P(s'\,|\,S')$ holds if and only if $P(S'\,|\,S)=P(S'\,|\,s,S)$. Recall that $P(S'\,|\,S)$ is the total transition probability for the transition from S to S' in a certain time interval, starting from the state S at the time origin. $P(S'\,|\,s,S)$ is the transition probability for the transition from S to S' after the same time interval, departing from the state S and passing through certain pathways ($S\rightarrow S^{\dagger}\rightarrow\cdots\rightarrow S'$) which are restricted so that the departure from the state S to the next state $S^{\dagger}$ is only through a link which produces the symbol s, while the former transition is allowed to take all possible routes from S to S'. $P(S'\,|\,s,S)$ is in general smaller than $P(S'\,|\,S)$. As the time interval of the transition from the first state S to the final state S' increases, it is expected that the difference between the two probabilities decreases. As for the sufficient condition, if the time interval is much longer than the characteristic time scale of the relaxation to the steady state (if it exists), these two probabilities converge to the same value. For the SSNs in Fig. 11.7 constructed from the single electron transfer experiment of the Fre/FAD complex, $P(S'\,|\,S)$ approximates $P(S'\,|\,s,S)$ fairly well since most states S and the next state S' are only connected by a single symbol. At much longer timescales, e.g., beyond 480 ms in which the conformation dynamics is well approximated by the normal Brownian diffusion (see Fig. 11.6), $P(S'\,|\,S)$ can deviate from $P(S'\,|\,s,S)$ because connections between two different states can be mediated by producing several different symbols.

In the case of Fig. 11.6, we performed a slightly more complicated computation in terms of the SSN that we extracted from the single molecule time series [26]. We constructed the multiscale SSNs using a delay-time time series, that is, a series of time differences (delay times) between the (chronological) time of an excitation light pulse in the pulse train and that of the photon emitted from the system in response just after the excitation pulse. The autocorrelation in Fig. 11.3 discussed in the original paper of the experiment [48] is not of the delay-time time series τ itself they monitored but of the lifetime γ^{-1}. We therefore assigned the lifetime for each state I, γ_I^{-1}, in the multiscale SSNs because the delay time distribution at each state was found to be well approximated by an exponential function $\sim e^{-\tau/\gamma_I^{-1}}$ and replaced all s's in all above equations by γ^{-1} (see details in Ref. [26]).

11.5 Outlook and Future Perspectives

We summarize the nature and potential of our novel time series analysis method based on an information theoretic framework.

Basic Property of Our State-Space Network Our method is not based on any a priori assumption regarding, e.g., the number of states, the local equilibrium, and the detailed balance. Non-Markovian nature or memory effects in the multiscale process, if it exists in the time trace of an observable, is naturally incorporated into either the components of the states or the topographical nature of the state network. The computational mechanics used in our framework provides the firm mathematical foundation that the extracted SSN is a minimal but best predictive machinery (if it converges) (see Section A4).

States in Network The 'states' in the SSN are generally a set of subsequences of the values of the (symbolized) observable resulting in the same transition pattern (see Section 11.4.1 and Appendix). If and only if the process (i.e., the time evolution) along the measured observable is Markovian, each state is defined solely by the value of the observable.

One might imagine that any state in a reaction scheme should associate with a *single* conformation of the system as one might deduce from the network representation in the conformation space, i.e., conformation space network (CSN). In our framework, this is only the case when transitions among the conformations are Markovian. When the transitions in the conformation space are non-Markovian, our method assigns each state as a certain "time sequence" of conformational changes, whose sequence length originates from the characteristic timescale of memory that the conformational transitions acquire. Note that, whenever one wishes to define states in terms of single conformations, one should work on the evaluation of an appropriate memory kernel for reproducing the complex dynamics, which however does not help us get any insights about the underlying mechanism.

The striking consequence of our theory is that one can analytically derive any physical quantities without postulating a particular memory kernel by using the multiscale SSNs, because the resultant SSNs have the Markovian property in the state representation, i.e., transitions among the states are Markovian.

In order to shed light on the relationship between the states in the multiscale SSNs and the conformation of the system, it is necessary to construct a series of SSNs in terms of a set of different single-molecule time series data from systematic mutations of amino-residues. It is expected that one may identify which amino residues perform to yield non-Markovian kinetics by monitoring the morphological changes of the network.

Heterogeneity of Memory in Network Our method can naturally capture the heterogeneity of memory in the process dependent on each state. Our method can systematically quantify and list up in the underlying complex network which states are more responsible for non-Markovianity in the time course of the observable (We remind that even when the state-to-state

(continued)

transitions are Markovian, the time course of the observable can be non-Markovian). Figure 11.12 illustrates the situation of coexistence of states with short and long time memory in the time course of the observable. Suppose that the time series is binary (A or B) (e.g., on-off time series in patch clamp measurements), and, in the procedure to evaluate the appropriate past length L_{past}, one "state" AA at $L_{\mathrm{past}} = 2$ splits and diversifies into four states with different transition probabilities as L_{past} increases to 5, while the other state BA does not. This implies that, by looking into the splitting pattern as an increase of the length L_{past}, one can identify which states yield non-Markovian nature in the time course of the observable. Our preliminary studies on patch clamp time series of the gating kinetics of a mechanosensitive ion channel show the existence of heterogeneity of memory in the process in the network.

Complex Network and Energy Landscape Related with Section 11.3.2, the question to be addressed is what type of properties of complex networks or multidimensional energy landscapes might be acquired through the evolution with mutations. It is almost impossible, by using computer simulations, to assess the underlying energy landscape or network of complex systems in biology such as signal transductions from the extracellular to intracellular space. Our method is expected to enable us to address such questions combined with single molecule measurements. The establishing of a mutual relation between our network representation and the underlying multidimensional energy landscape, both dependent on the timescale, is also highly desired. The so-called max-flow and mini-cut theorem in network theory may help us to bridge the representations even when the detailed balance does not hold for each transition. We can also scrutinize the specificity or the role of a state regarded as a "hub-station" in a network, which has been thought of as a necessary condition for the global minimum of the energy landscapes of proteins.

Single Molecule Systems Biology How complex systems such as cell adapt to the change of the environment, or a certain stimulus at the membrane, to initiate consecutive 'signal transduction' to the cytoplasmic space is one of the most intriguing subjects in systems biology. How the underlying reaction network, or the multidimensional energy landscape, may adaptively change before and after the adaptation is of crucial importance for understanding the mechanism of adaptability of systems. More importantly, biological systems can robustly perform their functions even with an (apparently small) amount of chemical energy comparable to the thermal energy, $k_{\mathrm{B}}T$. The efficiency of the energy transfer of molecular machinery across different spatiotemporal scales in a thermally fluctuating dissipative environment is considered to be much higher than any artificial manufacturing machinery [47]. Other important properties of biological systems are plasticity and emergency, although neither firm definitions nor the means to quantify

(continued)

these measures have been established. We believe that our multiscale SSNs have large potentials that enable us to address these problems; our hierarchical SSNs can quantify the efficiency of information flow across different time and space scales by scrutinizing how and along which pathways information transduction takes place across hierarchies in the SSNs at different scales.

Finally, it should be noted that there exists a series of problems at different stages of single molecule biophysics before the application of our analysis; that is, all measured single molecule time series are contaminated by noise. Experimental noise is composed of external and internal noise. The former, for example, includes read-out noise from the Charge Coupled Devices (CCD) camera and shot noise of the image intensifier. The latter originates from fluorophore fluctuations with diverse mechanisms; photophysics such as blinking and bleaching, environmental changes around the fluorophore (hydrophobic or hydrophilic) in the process of measurement, polarization of dye molecules, different quantum yields of acceptor and donor molecules. Complementary to our studies, some studies have been devoted to extracting the desired time series of a physical quantity, such as an interdye distance from the observed time series contaminated by stochastic noise [45]. The integration of these complementary studies is highly desired to establish a solid framework of analysis for single molecule biophysics. Theory more meets experiments, and vice versa, in the forthcoming near future.

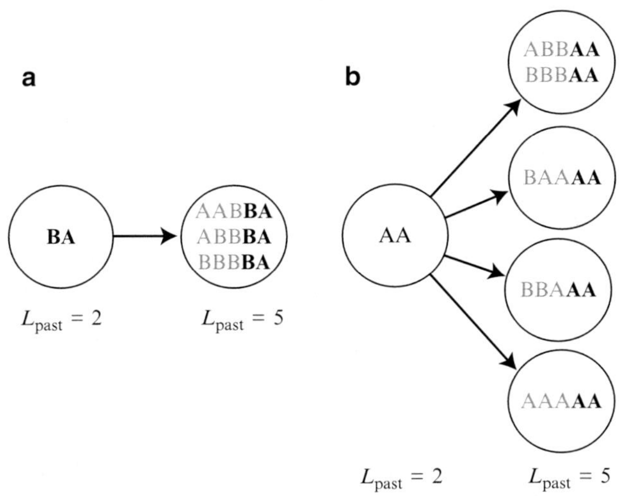

Fig. 11.12 A possible coexistence of distinct states associated with short- and long-term memory in the same SSN. Here the time series is assumed to be binary, A or B. The circle denotes a "state" evaluated with $L_{past} = 2$ and 5, and the two or five sequences inside each circle correspond to those which have the same, unique transition probabilities. (**a**) A "state" at $L_{past} = 2$ whose transition probabilities do not change even when the past length increases, $L_{past} = 5$. (**b**) A "state" at $L_{past} = 2$ whose transition probabilities change when the past length increases, splitting into four states with different transition probabilities.

11.6 Appendix

A1 Basics Concepts in Information Theory

A1.1 Shannon Entropy: The Measure of Missing Information

Imagine that one observes the probability, $P(x)$, of an observable x to occur (e.g., $x = 1, 2, \ldots, 6$ can be the six faces of a dice) from measurements. Given this probability function $P(x)$, is it possible to quantify the amount of uncertainty associated with the observable x? To answer this question, Shannon and Weaver [38] introduced the celebrated *Shannon entropy* $H(x)$,

$$H(x) = -\sum_{i=1}^{N} P(x_i) \log_2 P(x_i) \tag{11.21}$$

to measure the "information" content that one is missing in order to predict the value of the observable x. Here the observable x can take values x_i with $i = 1, \ldots, N$ (e.g., $N = 6$ and $x_i = i$ in the dice's case). The value of $H(x)$ falls between zero and $\log_2 N$, i.e., $0 \leq H(x) \leq \log_2 N$.

In order to get some intuition how the Shannon entropy serves as a measure of uncertainty associated with the observable x, let us consider the case in which the probability is uniform, $P(x_i) = 1/N$ for all x_i (e.g., a fair dice). Since a uniform probability means that there is no preference for any value of x, one can expect that this is the most difficult case to predict the overcome of the observable x. In this case, the Shannon entropy obtains its maximum value ($H(x) = \log_2 N$, i.e., maximum uncertainty). On the other hand, in the other extreme that the probability of observing a particular value of x (let us say x_1) is one and zero for all the others (i.e., $P(x_i) = 0$ for $i = 2, \ldots, N$), it is clear that we are sure (no uncertainty) that the outcome must be x_1. The Shannon entropy in this case indeed obtains its minimum value, $H(x) = 0$.

The Shannon entropy can be extended easily to the case of multi-observables. For example, suppose we have two observables x and y with joint probability $P(x, y)$. The "joint" Shannon entropy is then given by

$$H(x, y) = -\sum_{i=1}^{N_x} \sum_{j=1}^{N_y} P(x_i, y_j) \log_2 P(x_i, y_j), \tag{11.22}$$

where x and y can take N_x and N_y different values, respectively. If the two observables are statistically independent, i.e., $P(x_i, y_j) = P(x_i)P(y_j)$, then $H(x, y)$ becomes additive with respect to each observable, $H(x, y) = -\sum_{i,j} P(x_i)\, P(y_j) \log_2 P(x_i)\, P(y_j) = -\sum_i P(x_i) \log_2 P(x_i) - \sum_j P(x_j) \log_2 P(x_j) = H(x) + H(y)$.

A1.2 Conditional Entropy

On the other hand, if the two observables are statistically dependent, i.e., $P(x_i, y_j) = P(x_i \mid y_j) P(y_j) \neq P(x_i) P(y_j)$ ($P(x_i \mid y_j)$ denotes the conditional probability), then the joint Shannon entropy can be expressed as

$$H(x,y) = -\sum_{i,j} P(x_i, y_j) \log_2 (P(x_i|y_j)P(y_j)) = H(y) + H(x|y), \tag{11.23}$$

with $H(x \mid y) \equiv -\sum_{i,j} P(x_i, y_j) \log_2 P(x_i \mid y_j)$. $H(x \mid y)$ is called the *conditional entropy*. $H(x \mid y)$ provides a measure of uncertainty in knowing the outcome of x if the value of y is given. This interpretation is evident from Eq. 11.23, namely, the total uncertainty in knowing the outcome of both x and y (i.e., $H(x, y)$) is equal to the uncertainty associated with y (i.e., $H(y)$) and the uncertainty associated with x given y (i.e., $H(x \mid y)$).

A1.3 Mutual Information: The Measure of Shared Information

The relationship among the Shannon entropy (Eq. 11.21), joint Shannon entropy (Eq. 11.22) and the condition entropy can be explicitly visualized with the help of a Venn diagram as shown in Fig. 11.13. For example, the relation $H(x, y) = H(y) + H(x \mid y) = H(x) + H(y \mid x)$ can be easily verified in the figure. The Venn diagram also provides us with another important information measure, the *mutual information* $I(x, y)$, corresponding to the intersection area between $H(x)$ and $H(y)$. The explicit form of $I(x, y)$ and its relation with other information measures can be easily read out from Fig. 11.13:

$$\begin{aligned} I(x,y) &= H(x) + H(y) - H(x,y), \\ &= H(y) - H(y|x) = H(x) - H(x|y), \\ &= \sum_{i,j} P(x_i, y_j) \log_2 \frac{P(x_i, y_j)}{P(x_i)P(y_j)}. \end{aligned} \tag{11.24}$$

The intuitive meaning of $I(x, y)$ can be seen from the second line of Eq. 11.24: it measures the amount of uncertainty in an observable reduced by knowing the

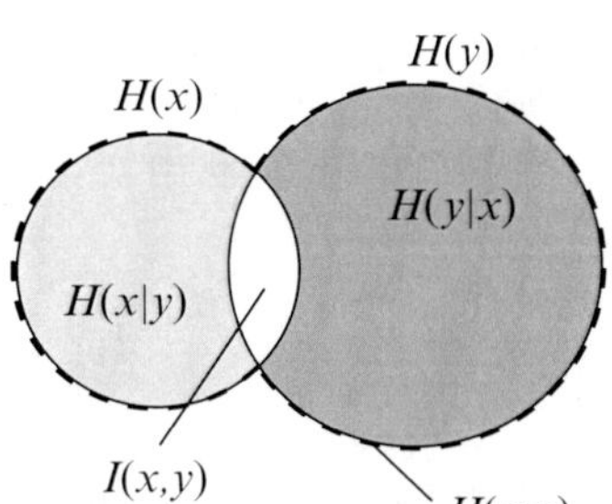

Fig. 11.13 A venn diagram representation showing the relationship among different information measures. The values of these measures are represented by the areas enclosed in different regions. $H(x)$ (and $H(y)$) corresponds to the area enclosed by the *solid circle*. $H(x, y)$ corresponds to the area enclosed by the dash line. $H(x \mid y)$ and $H(y \mid x)$ are denoted by the light and dark gray areas, respectively. The mutual information $I(x, y)$ corresponds to the intersection area between the two *solid circles*.

other. In other words, it is the information shared between the two observables. The last line of Eq. 11.24 also tells us that $I(x, y) = 0$ if and only if x and y are statistically independent (i.e., $P(x_i, y_j) = P(x_i)P(y_j)$).

It should be noted that the Shannon entropy as a measure of uncertainty is, in general, different from the entropy that one considers in thermodynamics and statistical mechanics. A rational connection between the Shannon entropy and the thermodynamic entropy, and therefore between information theory and statistical mechanics, has been established by Jaynes in 1957. Although we will not discuss these important topics in the present article, readers interested in this fundamental connection can refer to the literature (see e.g., [17, 18]).

Summary

1. *Shannon Entropy* $H(x) = -\sum_i P(x_i)\log_2 P(x_i)$: a measure of uncertainty in x
2. *Joint Shannon Entropy* $H(x, y) = -\sum_{i,j} P(x_i, y_j)\log_2 P(x_i, y_j)$: a measure of uncertainty in x and y
3. *Conditional Entropy* $H(x \mid y) = -\sum_{i,j} P(x_i, y_j)\log_2 P(x_i \mid y_j)$: a measure of uncertainty in x when y is known
4. *Mutual Information* $I(x, y) = \sum_{i,j} P(x_i, y_j)\log_2 \frac{P(x_i, y_j)}{P(x_i)P(y_j)}$: a measure of shared information between x and y

A2 Choice of Symbolization Scheme

In practice, the choice of a suitable symbolization scheme and the number of symbols depends on the nature of the time series, including whether the time series has continuous (e.g., single molecule electron transfer experiment), stepwise (e.g., single ion-channel current measurement) or spiked (e.g., single enzymatic turnover measurements) nature, whether the observable x is linear (e.g., single molecule FRET measurement) or angular (e.g., rotation of the F_1_ATPase) variable, the experimental resolution in x and the signal to noise ratio, and so forth.

In the case of time series with discrete nature, a simple and common discretization scheme is to first presume the number of symbols and partition the values of the observable x by thresholding. Figure 11.14a,b show an example of the application of a thresholding method to discretize a two-level intensity trajectory. It is evident from Fig. 11.14a,b that the assignment of symbols is problematic when the noise level is comparable to the difference between the two intensity levels. The existence of the fast artificial transitions can lead to a wrong kinetic scheme represented by the state-space network.

One possible solution to this problem is the use of symbolization schemes based on change point detection. Change points in the time series are the time instants at which the statistical properties, such as the average, variance, correlation, etc., are different before and after the change points. Once all the change points in the time series are assigned, one can replace all time series segments between two

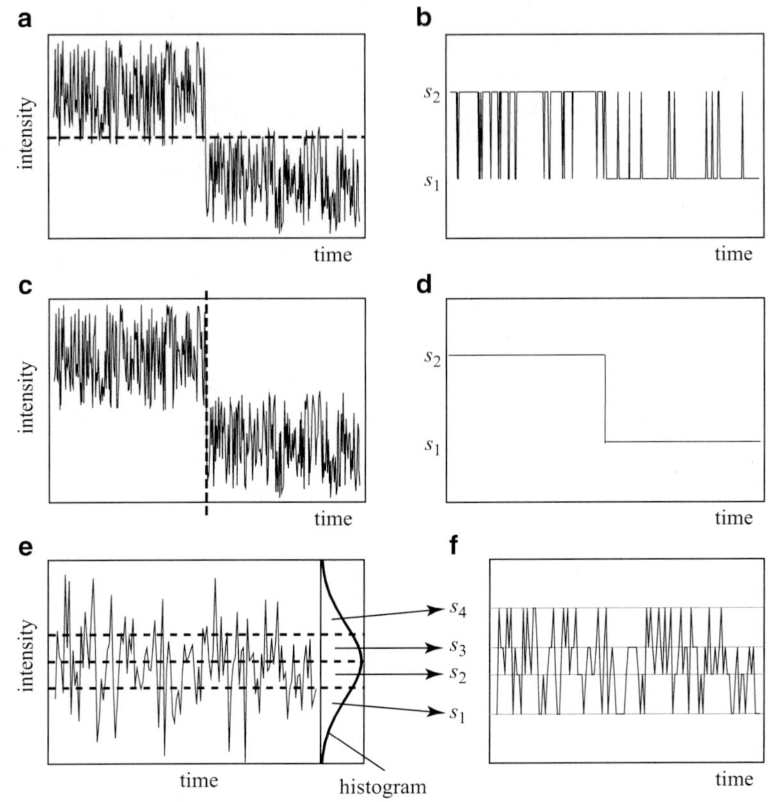

Fig. 11.14 Thresholding and change point detection as symbolization schemes. (**a**) shows the discretization of the two-level intensity trace (*solid line*) with a threshold (*dash line*). The resulting symbolic time series shown in (**b**) contains many artificial fast transitions between the two symbols s_1 and s_2. (**c**) shows the identification of the change point (*dash line*) in the intensity trace which separates the two segments in the time series with different statistical properties. The resulting symbolization by taking the average of the each segments before and after the change points reproduces the correct two-level transitions. (**e**) shows the equal probability partition (or thresholding) of a non-stepwise time series with four symbols in which each symbol has the same occurrence probability. The corresponding symbolized time series is shown in (**f**).

consecutive change points by their mean value, resulting in a clean stepwise trajectory with a few distinguishable steps (see Fig. 11.14d) from which the symbolization is much easier compared with the original noisy time series (see Fig. 11.14c). In this way, change point detection can also be thought as a scheme for "de-noising." Various change point algorithms have been developed recently in the application to single molecule experimental time series, such as in single molecule spectroscopy [45, 49], and single molecule motor protein movement [19].

It should be noted, however, that the symbolization schemes based on change point detection are only suitable for time series with stepwise changes.

For continuous (or non-stepwise) time series, the statistical properties usually change smoothly in time, which means that each time instant can be thought as a change point. In this case, the change point detection may not provide much help in identifying a good discretization compared to dealing directly with the original noisy time series. A simple way to discretize a time series with non-stepwise nature is to first fix the number of symbols and then apply an equal probability partition to the time series in which each symbol has the same occurrence probability (see an example in Fig. 11.14e,f). By increasing the number of symbols from two, to three and so on, the most suitable number of symbols can be found when the structural properties of the state-space networks (see Section A5) no longer change significantly as the number of symbols increases.

A3 An Illustrative Example for the Construction of SSN: Deterministic Property and the Optimal L_{past}

We recall from the procedure of SSN construction that the length of the past subsequence L_{past} is an undetermined parameter which needs to be fixed. In practice, different SSNs are constructed for $L_{\text{past}} = 0, 1, 2, \ldots$. The *optimal* L_{past} is chosen as the minimum number of L_{past} above which the structural properties, which will be discussed in Section A5, do not change (i.e., converge) even when L_{past} increases further. We illustrate in Fig. 11.15 the choice of the "optimal" L_{past} and its implications by constructing the SSN for a two-symbol ($s_1 = 0$ and $s_2 = 1$) time series "…100100100…".

Let us first suppose some properties of the time series before we start to construct the SSN: in order to know what symbol (0 or 1) will be produced in the next time step, one needs to know the past subsequence with a length at least two, e.g., if we see the past subsequence 00, we know that the next symbol to be produced must be 1, but if we only see the past subsequence 0, then the next symbol can be either 0 or 1. On the other hand, knowing the past subsequences with $L_{\text{past}} > 2$ does not give us further information in predicting the future. It is therefore expected that the "optimal" L_{past} should be two. In the language of statistical physics, such a process is called a Markovian process of order two (or Markovian process with memory two), in other words, the future symbol depends only on the past two symbols. We will see below how the optimal $L_{\text{past}} = 2$ is obtained from the SSN construction.

We start with $L_{\text{past}} = 0$ which corresponds to a null sequence, i.e., no information in the past is used to predict the future symbol. The transition probabilities from the null sequence to the two future symbols (0 and 1) are shown in the table of Fig. 11.15a. Since there is only one possible past subsequence – the null sequence, the SSN for $L_{\text{past}} = 0$ has only one state (indicated by the gray area in the table) and the SSN with $L_{\text{past}} = 0$ is completed by combining the transitions according to the transition probabilities as shown in Fig. 11.15a.

We proceed next to the case of $L_{\text{past}} = 1$ with the two past subsequences (0 and 1) and their transition probabilities to the future symbols shown in the table of Fig. 11.15b. Since the two past subsequences have distinct transition probabilities, two states are obtained. The corresponding SSN with $L_{\text{past}} = 1$ is shown in Fig. 11.15b. It is clear that the SSN with $L_{\text{past}} = 1$ is different from the SSN with

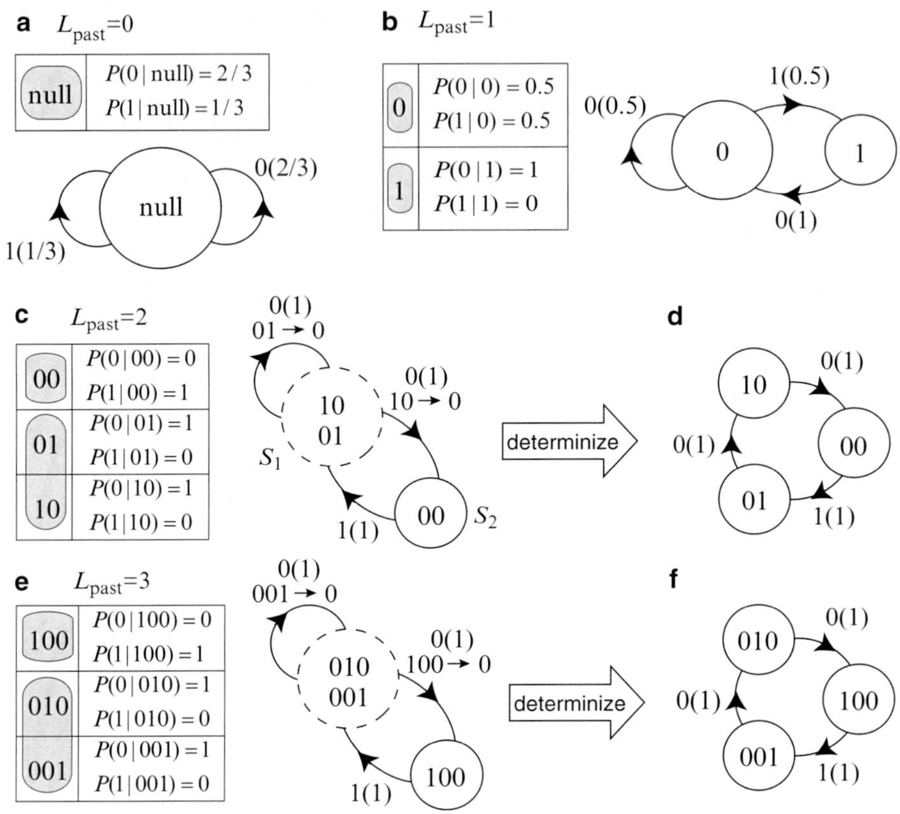

Fig. 11.15 An illustrative example of state-space construction for the time series "...100100100..." (see text for detail). Transition probabilities for transitions from past subsequences to the future symbol, and the corresponding SSNs for (**a**) $L_{past} = 0$, (**b**) $L_{past} = 1$, (**c**) $L_{past} = 2$ and (**e**) $L_{past} = 3$. The SSN at $L_{past} = 2$ in (**c**) and at $L_{past} = 3$ in (**d**) are nondeterministic with the nondeterministic state shown by *dash circles*. The deterministic SSNs for (**d**) $L_{past} = 2$ and (**f**) $L_{past} = 3$ are obtained by splitting the nondeterministic states in (**c**) and (**e**), respectively. The optimal L_{past} is two, as the SSN is not changed by increasing $L_{past} = 2$ to $L_{past} = 3$.

$L_{past} = 0$. This means that the SSN has not yet converged at $L_{past} = 0$. In order to check if $L_{past} = 1$ is the optimal value, we proceed to the construction for $L_{past} = 2$. For $L_{past} = 2$, there are three possible past subsequences, namely 00, 01 and 10, in the time series with the corresponding transition probabilities shown in the table of Fig. 11.15c. Since the transition probabilities of the past subsequences 10 and 01 are the same, they are grouped into the same state (denoted by S_1), whereas the past subsequence 00 belongs to one state by itself (denoted by S_2). This grouping is again indicated by the gray areas in the table of Fig. 11.15c. The resulting two-state network with the transitions among the states is shown in Fig. 11.15c.

We note in Fig. 11.15c that there are two arrows (transitions), both producing the same symbol "0" with probability equal to one, coming out from the state indicated by the dash circle. This situation is called nondeterministic in the sense of automation theory and the state in the dash circle is called a nondeterministic state. On the contrary, a SSN is called deterministic when there exists a unique successor state, given the current state and the next symbol. Depending on which past subsequence (10 or 01) is visited, the nondeterministic state in Fig. 11.15c can make transitions to different successor states. For example, the transition $01 \rightarrow 0$ *connects* the nondeterministic state *with* itself which contains the past subsequence 10, while the transition $10 \rightarrow 0$ *connects* the nondeterministic state *with* the state containing 00.

Such a nondeterministic situation poses two problems in using the SSN as a mathematical machinery to describe the dynamics from a time series. The first problem is that one cannot define the state transition probabilities of the nondeterministic states in the SSN so as to be consistent with probability theory. For example in Fig. 11.15c, the total transition probability from the nondeterministic state is equal to $P(S_2 \mid S_1) + P(S_1 \mid S_1) = 2$, which violates the basic requirement in probability theory that the total transition probability must be one (i.e., $\sum_J P(S_J \mid S_1) = 1$). The second problem is that one needs to include the information about which past subsequences in the nondeterministic state are visited in the time trace in order to specify the transition, as shown by the second line in the labels of the arrows coming from the dash circle in Fig. 11.15c. The need to know which past subsequences are to be visited is equivalent to asking which pathway in the network is taken in arriving at the current state. Thus, this means that grouping together the past subsequences in a nondeterministic state does not have any advantage. Here, we adopt a simple procedure to split the nondeterministic states until they become deterministic, in order to overcome the above problems. Indeed one can easily check that the state transition probabilities of the deterministic SSN in Fig. 11.15d obtained by splitting the nondeterministic state in Fig. 11.15c is consistent with probability theory.

The more complex structure of the SSN with $L_{\text{past}} = 2$ in Fig. 11.15d compared to the SSN with $L_{\text{past}} = 1$ in Fig. 11.15b tells us that $L_{\text{past}} = 1$ is not the optimal value yet since more features from the time series are captured by the SSN when increasing $L_{\text{past}} = 1$ to $L_{\text{past}} = 2$. Therefore, we proceed to the case of $L_{\text{past}} = 3$ and construct the transition probabilities for the three possible past subsequences as shown in Fig. 11.15e. The corresponding SSN is nondeterministic, as those in Fig. 11.15c. The resulting deterministic SSN for $L_{\text{past}} = 3$ in Fig. 11.15f, obtained again by the simple procedure of splitting the nondeterministic state, is topologically the same as that obtained at $L_{\text{past}} = 2$. Therefore $L_{\text{past}} = 2$ is chosen as the optimal L_{past} to capture all the non-Markovian properties of this time series, resulting in the minimal but most predictive network (as described below). We also see that the optimal L_{past} from the converged SSN corresponds to the memory length of the process as discussed above.

A4 What Is So Special About the State-Space Network?

The procedure in the previous section provides us with a straightforward scheme to construct the SSN directly from experimental time series. However, one may simply ask what makes the constructed SSN so special in modeling the underlying dynamics from time series data. In this section, we will look at the SSN from a different viewpoint to understand what advantages the above SSN construction scheme has. Within the limited space, the discussion in this section will be focused on the general scope and implication of the SSNs instead of giving the detail mathematical account of the subject. Readers who are interested in the rigorous mathematical theorems and derivations can look at the corresponding references (e.g., [7, 37]).

A4.1 The Minimal and Best Predictive Model

One of the main goals in constructing a mathematical model to capture features of the underlying dynamics is to grasp the pattern of time evolution hidden in the time series and to predict the future by using the present and past information, i.e., the set of past sequences. Suppose that we have a time series made from experimental data such as shown in Fig. 11.16a with three symbols. Let us first ask what is the best predictive model one can construct from the set of past sequences $\mathbf{s}^{\text{past}}$ with $L_{\text{past}} = 2$ as shown in Fig. 11.16b. It is evident that one can obtain the highest predictability if the full set of nine possible past sequences are all used. However, one expects that all the detailed information of the nine possible past sequences may not be required to be used in order to achieve the same predictive power. Figure 11.16c shows how one can "coarse-grain" the information in the past sequences by taking different partitions of the set of nine possible past sequences. For example, the leftmost panel of Fig. 11.16c corresponds to the most coarse description of the past sequences in which one does not distinguish different past sequences at all. On the other hand, the rightmost panel of Fig. 11.16c corresponds to the finest description in which one utilizes all information of all the nine possible past sequences. Partitions which are intermediate between these two extremes ignore some of details of the set of past sequences, e.g., the second panel from the right does not distinguish the two sequences s_1s_1 and s_1s_2 when seeing them in the time series.

Before we proceed, let us first quantify the concept of predictability. In terms of information theory, a natural measure of predictability for a given partition scheme can be defined using the conditional entropy discussed in Section A1 as follows:

$$\begin{aligned}\text{predictability} &= -H(\text{ future symbol}|\text{ partition of past sequences}) \\ &= \sum_{i,J} P(s_i,\ \text{partition}_J)\log_2 P(s_i|\ \text{partition}_J)\end{aligned} \tag{11.25}$$

where s_i with $i = 1, 2, 3$ is the future symbol and partition$_J$ is the J-th partition of a given partition scheme in Fig. 11.16c. From the discussion of Section A1, the quantity (-1 ×predictability) defined in Eq. 11.25 has the meaning of uncertainty remaining in the prediction of the future symbol, once given a coarse-grained description (i.e., partition scheme) of the set of past sequences.

Among the set of all possible partition schemes (or coarse-grained descriptions of the past) in Fig. 11.16c, one may then ask which of them is(are) as predictive as the finest-grained one, i.e., the rightmost panel in Fig. 11.16c. Suppose that a few partition schemes, including the finest-grained one in Fig. 11.16c, are found to give the best predictive power (i.e., the value of predictability is maximum), as shown in

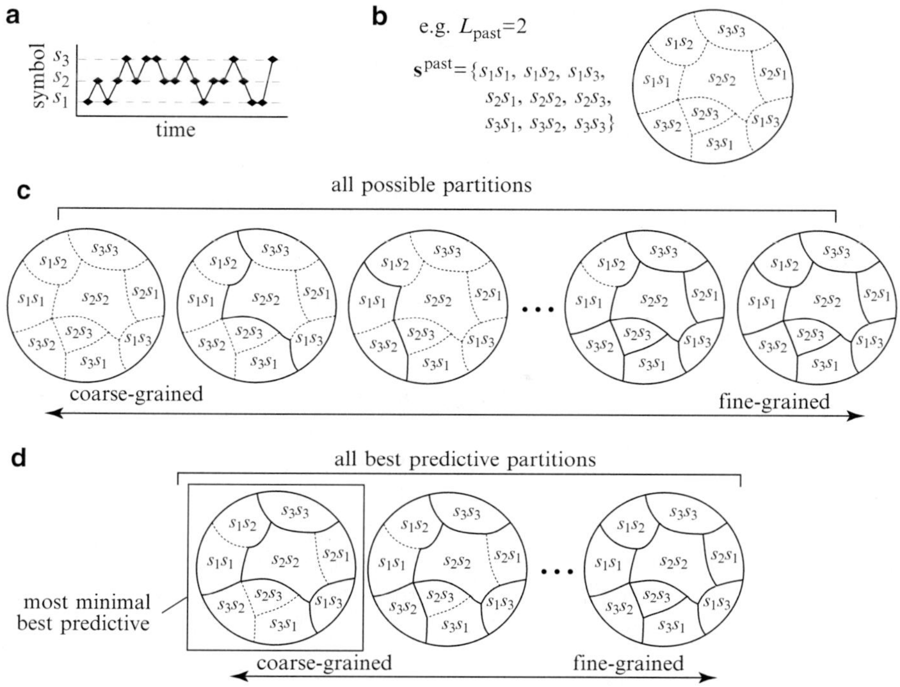

Fig. 11.16 A schematic representation of the constructed SSN as the minimal but best predictive model. Given a symbolized time series (**a**) with three symbols, (**b**) shows all the possible past subsequences with $L_{past} = 2$. Also shown in (**b**) is the representation of the past subsequences as the "elements" of the set of all past subsequences, denoted by the *big circle*. The area of the element corresponds to the occurrence probability of the past subsequence. (**c**) shows all possible partitions of the elements of the set of all past subsequences, ranging from the coarsest (*leftmost*) to the finest (*rightmost*) partition. (**d**) shows the partitions that are as predictive as those keeping all nine possible past subsequences (*the rightmost one*). The states in the SSN correspond to the simplest (i.e., minimal) description that is however as predictive as the best predictive description knowing all possible past subsequences. The minimal but best predictive partition is enclosed by the square in (**d**).

Fig. 11.16d. It is interesting that the simplest (or minimal) partition scheme, shown in the leftmost panel of Fig. 11.16c, corresponds exactly to the SSN discussed in this review article and each partition in this minimal, but best predictive description is a causal state in the SSN. Here, a partition scheme is regarded as minimal if its Shannon entropy, $H(\text{partition}) = -\sum_J P(\text{partition}_J) \log_2 P(\text{partition}_J)$, is the smallest among all the best predictive partition schemes in Fig. 11.16d.

We therefore see that the SSN constructed by the scheme discussed in Section 11.4.1 is the minimal and best predictive model that best captures the pattern of time evolution from the time series. The above discussion also suggests that the SSN construction can be formulated as a variational problem – finding the best predictive partition which has the minimal structure [40]. However, the detailed mathematical discussion of the variational construction of SSN is beyond the scope of this article.

A5 Quantification of the Structure of the State-Space Network

In this section, we will introduce some natural measures in terms of information theory to quantify the complexity of the SSNs. These measures not only allow us to compare the network structure between SSNs (e.g., checking if the structure of SSN converges when determining the optimal L_{past}), but also establish the connections between the structural features of the SSN and their dynamical consequences (e.g., stochasticity of the state transition, correlations and memory contents).

A5.1 Quantifying the State Complexity in the SSN

Two information-theoretic measures are commonly used to quantify how complex the SSN is based on topological and topographical features of its states. The first one is the *topological complexity*, denoted by C_{top},

$$C_{\text{top}} = \log_2 N_{\text{CS}}, \tag{11.26}$$

where N_{CS} is the number of causal states in the SSN. C_{top} is a simple measure of how complex the SSN is in terms of the number of causal states.

It is expected that two SSNs having the same number of causal states can be different if the resident probabilities are different. Therefore, it is useful to come up with a measure that takes into account the resident probabilities $P(S_I)$ of the states in the SSN. A second measure has been defined in terms of the Shannon entropy

$$C_\mu = -\sum_{I=1}^{N_{\text{CS}}} P(S_I)\ \log_2 P(S_I). \tag{11.27}$$

C_μ is called the *statistical complexity* in the literature. For two SSNs with the same number of states N_{CS}, the properties of the Shannon entropy (see Section A1) tell us that the SSN with uniform resident probabilities (i.e., $P(S_I) = 1\ N_{\text{CS}}$ for all states S_I) has the highest statistical complexity. In particular, one can easily see that the highest value of the statistical complexity is equal to the topological complexity by simply substituting $P(S_I) = 1\ N_{\text{CS}}$ into Eq. 11.27. Therefore, the topological complexity is actually not an independent quantity. In the SSN construction, the statistical complexity C_μ is usually used to examine the convergence of the topological features of the SSN to define the optimal L_{past}. On the other hand, since the Shannon entropy is a measure of uncertainty (see Section A1), the statistical complexity C_μ also measures the amount of uncertainty associated with the states in the SSN: for a given N_{CS} it is more difficult to tell which state the system visits preferentially when C_μ is larger. We demonstrate schematically in Fig. 11.17a,b two SSNs with the same number of states (i.e., a same C_{top}) but with different resident probabilities. The statistical complexity of the SSNs in Fig. 11.17a with uniform resident probability of the states is larger than those in Fig. 11.17b.

In addition, the statistical complexity C_μ carries another important meaning as follows: Let us consider the mutual information $I(\mathbf{S}, \mathbf{s}^{\text{past}})$ between the states of the SSN and all the past subsequences with the same length L_{past}, i.e.,

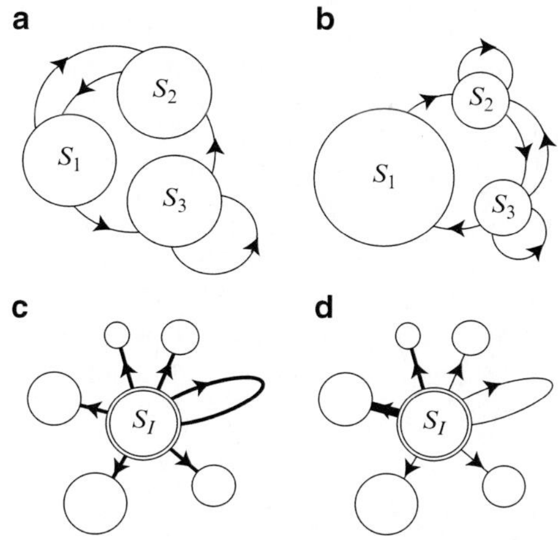

Fig. 11.17 Illustrative examples of the statistical complexity Eq. 11.27 and the transition entropy Eq. 11.29. (**a**) and (**b**) show two SSNs with the same number of states but with different statistical complexity C_μ. The area of the state in the SSNs is proportional to the resident probability of the state. The C_μ of the SSN in (**a**) is bigger than that in (**b**), indicating that the resident probabilities are more uniform in (**a**). (**c**) and (**d**) show two cases of a state S_I (*doubled circle*) having the same degree $k_I = 6$ but different transition entropy $H_{\text{tran}}(S_I)$. In (**c**) and (**d**), the thickness of the arrow is proportional to the transition probability between the state S_I and the corresponding target state. The state S_I in (**c**) with almost uniform transition probability to all the other states has a larger value of $H_{\text{tran}}(S_I)$ compared to the one in (**d**) with a more directional transition.

$$
\begin{aligned}
I(\mathbf{S}, \mathbf{s}^{\text{past}}) &= \sum\nolimits_{I,j} P(S_I, s_j^{\text{past}}) \log_2 \frac{P(S_I, s_j^{\text{past}})}{P(S_I) P(s_j^{\text{past}})} \\
&= \sum\nolimits_{I,j} P(S_I, s_j^{\text{past}}) \log_2 P(S_I | s_j^{\text{past}}) + C_\mu \\
&= C_\mu
\end{aligned}
\tag{11.28}
$$

in which the term $\sum_{I,j} P(S_I, s_j^{\text{past}}) \log_2 P(S_I | s_j^{\text{past}})$ vanishes. This is because we have $P(S_I, s_j^{\text{past}}) = P(S_I | s_j^{\text{past}}) P(s_j^{\text{past}})$ and the conditional probability $P(S_I | s_j^{\text{past}})$ of finding a particular state S_I associated with a past sub-sequence s_j^{past} is either one or zero. We recall from Section A1 that the mutual information $I(x, y)$ measures the information shared by the variables x and y. Therefore the statistical complexity C_μ measures the amount of information carried by the states $\{S_I\}$ in the SSN from the set of all possible past subsequences $\{s_j^{\text{past}}\}$ with length L_{past}, leading to another meaning of C_μ as the average amount of information (in bits) in the past, that is, memory content, which is relevant to predict the future.

A5.2 Quantifying the Transition Complexity in the SSN

The two measures, the topological complexity C_{top} and the statistical complexity C_μ, quantify the structural complexity of the SSN only based on the state properties.

To quantify the transition (connectivity) features of SSN, we introduce a natural measure in terms of the Shannon entropy of transition probability:

$$H_{\mathrm{tran}}(S_I) = -\sum_{J=1}^{N_{\mathrm{CS}}} P(S'_J|S_I)\ \log_2 P(S'_J|S_I), \tag{11.29}$$

In Eq. 11.29, $P(S'_J\,|\,S_I)$ is the transition probability of visiting the state S_I followed by S'_J and N_{CS} is the number of states in the SSN. The measure $H_{\mathrm{tran}}(S_I)$ is termed the *transition entropy* of the state S_I for a reason that will be clear shortly. We first note that the expectation value of $H_{\mathrm{tran}}(S_I)$ over all the states in the SSN, denoted here by $C_{\mathrm{tran}}(S', S)$, which is given by

$$\begin{aligned} C_{\mathrm{tran}}(S', S) &= \sum_{I=1}^{N_{\mathrm{CS}}} P(S_I) H_{\mathrm{tran}}(S_I) \\ &= \sum_{I=1}^{N_{\mathrm{CS}}} P(S_I) \times [-\sum_{J=1}^{N_{\mathrm{CS}}} P(S'_J|S_I)\ \log_2 P(S'_J|S_I)], \\ &= -\sum_{I,J}^{N_{\mathrm{CS}},N_{\mathrm{CS}}} P(S'_J, S_I)\ \log_2 P(S'_J|S_I), \end{aligned} \tag{11.30}$$

corresponds to the conditional entropy in Eq. 11.23 (one can see by simply setting $y = S$ (the state being visited) and $x = S'$ (the state being visited after S)) and, hence, due to Eq. 11.23, $C_{\mathrm{tran}}(S', S)$ can be related to the statistical complexity and the "two-state" statistical complexity of the SSN as

$$C_{(2)}(S', S) = C_\mu(S) + C_{\mathrm{tran}}(S', S), \tag{11.31}$$

where $C_{(2)}(S', S) = -\sum_{I,J=1}^{N_{\mathrm{CS}}} P(S'_J, S_I)\ \log_2 P(S'_J, S_I)$ is the *two-state statistical complexity* of the SSN and $P(S'_J, S_I)$ is the joint probability of visiting the state S_I followed by the state S'_J in the SSN (i.e., the joint Shannon entropy between S and S'). Since the statistical complexity $C_\mu(S)$ depends only on the resident probability of the states, the relation Eq. 11.31 shows that the average transition entropy $C_{\mathrm{tran}}(S', S)$ corresponds to the *transition* part of the two-state statistical complexity $C_{(2)}(S', S)$.

In fact, the Markovian property of the SSN implies that $C_\mu(S)$ and $C_{\mathrm{tran}}(S', S)$ are the only two independent complexity measures that can be obtained from the "multi-(m)-state" statistical complexity $C_{(m)}(S^{(m)}, \ldots, S^{(2)}, S^{(1)})$ defined by the joint probability $P(S_{I_m}^{(m)}, \ldots, S_{I_2}^{(2)}, S_{I_1}^{(1)})$ of visiting the states $S_{I_1}^{(1)}, S_{I_2}^{(2)}, \ldots, S_{I_m}^{(m)}$ successively in the SSN. By using the chain rule of joint probability with Markovian property, $P(S_{I_m}^{(m)}, \ldots, S_{I_2}^{(2)}, S_{I_1}^{(1)}) = P(S_{I_m}^{(m)}|S_{I_{m-1}}^{(m-1)}) \times \cdots \times P(S_{I_3}^{(3)}|S_{I_2}^{(2)}) P(S_{I_2}^{(2)}|S_{I_1}^{(1)}) P(S_{I_1}^{(1)})$, one can easily obtain

$$\begin{aligned} C_{(m)}(S^{(m)},\ldots,S^{(2)},S^{(1)}) &= -\sum_{I_1,I_2,\ldots,I_m=1}^{N_{CS}} P(S^{(m)}_{I_m},\ldots,S^{(2)}_{I_2},S^{(1)}_{I_1}) \\ &\quad\times \log_2 P(S^{(m)}_{I_m},\ldots,S^{(2)}_{I_2},S^{(1)}_{I_1}) \\ &= C_\mu(S) + (m-1)C_{\text{tran}}(S',S) \end{aligned} \tag{11.32}$$

where the properties $\sum_y P(y\,|\,z)=1$ and $\sum_y P(x\,|\,y)P(y\,|\,z)=P(x\,|\,z)$ are used. As a natural extension of the statistical complexity C_μ, we simply call $C_{\text{tran}}(S',S)$ the *transition complexity*.

For a given state S_I with k_I links emanating from it, $H_{\text{tran}}(S_I)$ has the property that it is minimum ($H_{\text{tran}}(S_I)=0$) if S_I can make transition to only one state (i.e, $P(S'_J\,|\,S_I)=1$ for only one S'_J and equal to zero otherwise), and it is maximum ($H_{\text{tran}}(S_I)=\log_2 k_I$) if all links emanated from the state have the same transition probability (i.e., $P(S'_J\,|\,S_I)=1/k_I$ for all connected S'_J). Therefore, $H_{\text{tran}}(S_I)$ measures the stochasticity of transitions from a state, namely, the smaller the $H_{\text{tran}}(S_I)$, the more directional the transition from S_I. The property of the transition entropy $H_{\text{tran}}(S_I)$ as a measure of the dynamical heterogeneity of state transition is shown schematically in Fig. 11.17c and d.

Moreover, in order to differentiate the property of transition heterogeneity from the complementary degree k_I of the state (i.e., the number of transitions from the state), it is convenient in practice to consider the normalized transition entropy

$$\tilde{H}_{\text{tran}}(S_I) \equiv H_{\text{tran}}(S_I)/\log_2 k_I, \tag{11.33}$$

with $0 \le \tilde{H}_{\text{tran}}(S_I) \le 1$, to. Fig. 11.7d,e,f demonstrate how the normalized transition entropy can capture the dynamical heterogeneity of state transitions at different timescales in the single molecule mesurement.

Acknowledgements We thank Prof. Haw Yang for his continuous valuable contributions to our project from his experimentalist's viewpoint. We also thank Profs. Satoshi Takahashi and Mikito Toda for their valuable discussions. We acknowledge financial support from JSPS, JST/CREST, Grant-in-Aid for Research on Priority Areas 'Systems Genomics,' 'Real Molecular Theory', and 'Innovative nano-science,' MEXT.

References

1. Albert R, Barabási AL (2002) Statistical mechanics of complex networks. Rev Mod Phys 74:47–97
2. Baba A, Komatsuzaki T (2007) Construction of effective free energy landscape from single-molecule time series. Proc Natl Acad Sci USA 104(49):19,297–19,302
3. Baba A, Komatsuzaki T (2010) Multidimensional energy landscapes in single molecule biophysics. Adv Chem Phys 145: in press

4. Ball KD, Berry RS, Kunz RE, Li FY, Proykova A, Wales DJ (1996) From Topographies to dynamics on multidimensional potential energy surfaces of atomic clusters. Science 271:963
5. Barabási AL, Albert R (1999) Emergence of scaling in random networks. Science 286:509–512
6. Barkai E, Jung Y, Silbey R (2004) Protein conformational dynamics probed by single-molecule electron transfer. Annu Rev Phys Chem 55:457–507
7. Crutchfield JP, Young K (1989) Inferring statistical complexity. Phys Rev Lett 63:105
8. Daubechies I (1992) Ten lectures on wavelets. (Soc Indust Appl Math, New York)
9. Debnath P, Min W, Xie XS, Cherayil BJ (2005) Multiple time scale dynamics of distance fluctuations in a semiflexible polymer: a one-dimensional generalized langevin equation treatment. J Chem Phys 123:204,903
10. Evans DA, Wales DJ (2003) Free energy landscapes of model peptides and proteins. J Chem Phys 118(8):3891–3897
11. Frauenfelder H, Sligar SG, Wolynes PG (1991) The energy landscapes and motions of proteins. Science 254:1598–1603
12. Gallos LK, Song C, Havlin S, Makse HA (2007) Scaling theory of transport in complex biological networks. Proc Natl Acad Sci USA 104:7746–7751
13. Gfeller D, Rios PDL, Caflisch A, Rao F (2007) Complex network analysis of free-energy landscapes. Proc Natl Acad Sci USA 104:1817–1822
14. Gopich IV, Szabo A (2003) Single-macromolecule fluorescence resonance energy transfer and free-energy profiles. J Phys Chem B 107:5058–5063
15. Grote RF, Hynes JT (1980) The stable states picture of chemical reactions. II. rate constants for condensed and gas phase reaction models. J Chem Phys 73:2715–2732
16. Huang NE, Shen Z, Long SR, Wu MC, Shih HH, Zheng Q, Yen NC, Tung CC, Liu HH (1998) The empirical mode decomposition and the Hilbert spectrum for nonlinear and non-stationary time series analysis. Proc R Soc Lond A 454:908
17. Jaynes ET (1957) Information theory and statistical mechanics. Phys Rev 106:620
18. Jaynes ET (1957) Information theory and statistical mechanics II. Phys Rev 108:171
19. Kalafut B, Visscher K (2008) An objective, model-independent method for detection of non-uniform steps in noisy signals. Comp Phys Commun 179:716
20. Kinoshita M, Kamagata K, Maeda M, Goto Y, Komatsuzaki T, Takahashi S (2007) Development of a technique for the investigation of folding dynamics of single proteins for extended time periods. Proc Natl Acad Sci USA 104:10,453
21 Klimov DK, Thirumalai D (1997) Viscosity dependence of the folding rates of proteins. Phys Rev Lett 79:317–320
22. Komatsuzaki T, Baba A, Kawai S, Toda M, Straub JE, Berry RS (2010) Ergodic problems for real complex systems in chemical physics. Adv Chem Phys 145: in press
23. Kramers HA (1940) Brownian motion in a field of force and the diffusion model of chemical reactions. Physica 7:284–304
24. Krivov SV, Karplus M (2004) Hidden complexity of free energy surfaces for peptide (protein) folding. Proc Natl Acad Sci USA 101:14,766–14,770
25. Krzanowski WJ (2003) Non-parametric estimation of distance between groups. J App Stat 30 (7):743–750
26. Li CB, Yang H, Komatsuzaki T (2008) Complex network of protein conformational fluctuation buried in single molecule time series. Proc Natl Acad Sci USA 105:536–541
27. Li CB, Yang H, Komatsuzaki T (2009) New quantification of local transition heterogeneity of multiscale complex networks constructed from single-molecule time series. J Phys Chem B 113:14,732–14,741
28. Lippitz M, Kulzer F, Orrit M (2005) Statistical evaluation of single nano-object fluorescence. ChemPhysChem 6:770–789
29. Luo G, Andricioaei I, Xie XS, Karplus M (2006) Dynamic distance disorder in proteins is caused by trapping. J Phys Chem B 110:9363–9367

30. Michalet X, Weiss S, Jager M (2006) Single-molecule fluorescence studies of protein folding and conformational dynamics. Chem Rev 106(5):1785–1813
31. Min W, Luo G, Cherayil BJ, Kou SC, Xie XS (2005) Observation of a power-law memory Kernel for fluctuations within a single protein molecule. Phys Rev Lett 94: 198,302
32. Moerner WE, Fromm DP (2003) Methods of single-molecule fluorescence spectroscopy and microscopy. Rev Sci Inst 74(8):3597–3619
33. Moser CC, Keske JM, Warncke K, Farid RS, Dutton PL (1992) Nature of biological electron-transfer. Nature 355(6363):796–802
34. Rao F, Caflisch A (2004) The protein folding network. J Mol Biol 342:299–306
35. Rhoades E, Gussakovsky E, Haran G (2003) Watching proteins fold one molecule at a time. Proc Natl Acad Sci USA 100(6):3197–3202
36. Schuler B, Lipman EA, Eaton EA (2002) Probing the free-energy surface for protein folding with single-molecule fluorescence spectroscopy. Nature 419:743–747
37. Shalizi CR, Crutchfield JP (2001) Computational mechanics: pattern and prediction, structure and simplicity. J Stat Phys 104:819–881
38. Shannon C, Weaver W (1948) A mathematical theory of communication. University of Illinois Press, Urbana, IL
39. Socci ND, Onuchic JN, Wolynes PG (1996) Diffusive dynamics of the reaction coordinate for protein folding funnels. J Chem Phys 104:5860–5868
40. Still S, Cutchfield JP, Ellison CJ (2007) Optimal causal inference. http://lanl.arxiv.org/abs/0708.1580
41. Stillinger FH (1995) A topographic view of supercooled liquids and glass formation. Science 267:1935–1939
42. Talaga DS, Lau WL, Roder H, Tang JY, Jia YW, DeGrado WF, Hochstrasser RM (2000) Dynamics and folding of single two-stranded coiled-coil peptides studied by fluorescent energy transfer confocal microscopy. Proc Natl Acad Sci USA 97(24):13,021–13,026
43. Tang J, Marcus RA (2006) Chain dynamics and Power-law distance fluctuations of single-molecule systems. Phys Rev E 73:022,102
44. Wales DJ (2003) Energy landscapes. Cambridge University Press, Cambridge
45. Watkins LP, Yang H (2005) Detection of intensity change points in time-resolved single-molecule measurements. J Phys Chem B 109(1):617–628
46. Xie XS, Trautman JK (1998) Optical studies of single molecules at room temperature. Annu Rev Phys Chem 49:441–480
47. Yanagida T, Ishii Y (2009) Single molecule dynamics in life science. Wiley, Weinheim
48. Yang H, Luo G, Karnchanaphanurach P, Louie TM, Rech I, Cova S, Xun L, Xie XS (2003) Protein conformational dynamics probed by single-molecule electron transfer. Science 302:262–266
49. Zhang K, Chang H, Fu A, Alivisatos AP, Yang H (2006) Continuous distribution of emission intensity and its non-linear correlation to luminescence decay rates from single cdse/zns quantum dots. Nano Lett 6:843–847

Chapter 12
Statistical Analysis of Lateral Diffusion and Reaction Kinetics of Single Molecules on the Membranes of Living Cells

Satomi Matsuoka

Abstract Single-molecule imaging has made it possible to directly observe the behavior of signaling molecules functioning on the membrane of living cells, revealing multiple subpopulations which can be characterized by their lateral diffusion coefficients on the membrane and or kinetics of dissociation from the membrane. The transition kinetics between these functional states is a central problem for understanding bio-signaling mechanisms. Here I propose a novel method to simultaneously analyze lateral diffusion coefficient and reaction kinetics from single-molecule trajectories. Based on the probability density function of displacement derived from a diffusion equation with appropriate reaction terms, the temporal development of diffusive mobility can be analyzed in a quantitative manner. I discuss simple diffusion models for a molecule that exhibits one or two states with different diffusion coefficients in the absence or presence of state transitions and/or membrane dissociation. I use numerical simulation based on my model to generate single-molecule trajectories to demonstrate the practice of this method with special emphasis on revealing reaction schemes.

Keywords Diffusion · Membrane · Single-molecule imaging · Single particle tracking · Reaction kinetics · State transition · Membrane association/dissociation · Signal transduction

---

S. Matsuoka (✉)
Laboratories for Nanobiology, Graduate School of Frontier Biosciences, Osaka University, 1-3 Yamadaoka, Suita, Osaka 565-0871, Japan
and
Japan Science and Technology Agency (JST), CREST, 1-3 Yamadaoka, Suita, Osaka 565-0871, Japan
e-mail: matsuoka@fbs.osaka-u.ac.jp

Y. Sako and M. Ueda (eds.), *Cell Signaling Reactions: Single-Molecular Kinetic Analysis*, 265
DOI 10.1007/978-90-481-9864-1_12,

12.1 Introduction

Cells are able to adjust to fluctuations in their environment by discerning subtle and vital changes hidden inside noise. To detect and transmit a signal inside the cytosol, a cascade of reactions such as intermolecular interactions or enzymatic reactions is required. However, these reactions are stochastic in nature, meaning they inevitably contain uncertainty that gives rise to randomness in the concentration of the signaling molecules along the cell membrane, even under constant and uniform conditions [9,18]. Yet, despite this, cellular responses occur in a highly organized manner in time and space [7]. After a step-wise stimulation, an ensemble of certain molecules shows a transient response such as membrane localization or phosphorylation, finally leading to the decision of the cell's fate. Upon stimulation with a chemical gradient, spatially localized activation occurs on a membrane to determine the direction of cell migration or extension of a growth cone [5,10]. Clarifying how the stochastic reactions are organized into a signaling system is fundamental to understanding how living organisms respond to changes in their environment. To do this, essential processes or molecules in a given signaling system should be extracted from the whole and the dynamics on the membrane should be quantified. One method for such a task is single-molecule imaging.

Direct observation of signaling molecules' real-time behavior on the membranes of living cells has revealed the heterogeneity in diffusive behavior [6,17]. Diffusion characteristics can serve as an index of the molecular functional states. Molecules can adopt multiple states between which transitions occur without any other signals and/or under the regulation by upstream signals. For membrane proteins, spontaneous fluctuations in their conformation may change the characteristics of the interface between them and the membrane lipids, which in turn may change the diffusion coefficient. For proteins that transmit signals by intermolecular interactions, binding with other signaling proteins has significant effects on the diffusion coefficient. Additionally, the diffusion coefficient can be changed by environmental conditions surrounding the given molecule. In fact, recent studies have revealed that the diffusion coefficient of a given molecule correlates with nearby synapses or microdomain structures like caveolae [2,12]. Since multiple diffusion coefficients for a molecule reflect possible states in its signaling process, temporal changes in diffusion coefficient offer kinetic information on the signaling reaction.

Kinetic analysis requires temporal information about how the diffusion coefficient changes depending on time, which cannot be acquired with the conventional analysis of the mean square displacement (MSD) [8]. One usually calculates MSD in a time-averaged manner from a single-molecule trajectory, $(x(t), y(t))$, as,

$$MSD(\Delta t) = \left\langle \{(x(t+\Delta t)) - (x(t))\}^2 + \{(y(t+\Delta t)) - (y(t))\}^2 \right\rangle,$$

MSD is a function of the time interval, Δt, and calculated for individual molecules [13,15]. Since displacements during Δt are measured at any point in a single trajectory irrespective of time, t, the temporal information is lost when obtaining the mean of the squared values. Another problem with this method lies in the dynamic nature of signaling molecules shuttling between the membrane and cytosol. In order to examine the reaction kinetics that describes a molecule's membrane association, its interaction with other signaling molecules, and its return to the cytosol, determining the temporal change of the diffusive mobility after the onset of membrane association is essential. Thus, understanding whole reaction kinetics of only a single kind of molecule requires integrating information describing the transition described in the diffusion coefficient and the length of time of the membrane association. However, conventional analysis divides the spatial (diffusion mobility) and temporal (lifetime on the membrane) information, although a given trajectory on the membrane contains both. Spatiotemporal information should be analyzed simultaneously, meaning a novel method is required.

In this chapter, I propose such a method. This method is based on the diffusion equation in which the Brownian movement of a molecule is described with the state transitions and membrane dissociation considered. Both diffusion coefficients and reaction rate constants can be estimated from a molecule's displacement during an arbitrary time interval by fitting the distribution to the probability density function. I concentrate on six models to describe variations of simple diffusion (Fig. 12.1). Models 1 through 3 consider membrane-integrated molecules, where the diffusive mobility is characterized by a single diffusion coefficient (model 1) or two different diffusion coefficients in the absence (model 2) or presence (model 3) of state transitions. Each of these models is further analyzed for the effect of membrane dissociation (models 4 through 6). These models can be tested objectively by using statistics that describe the trajectory length, molecular instantaneous displacement and temporal correlation analysis of the molecular movement. Using this method, important kinetic parameters for a signaling process can be quantified directly in living cells.

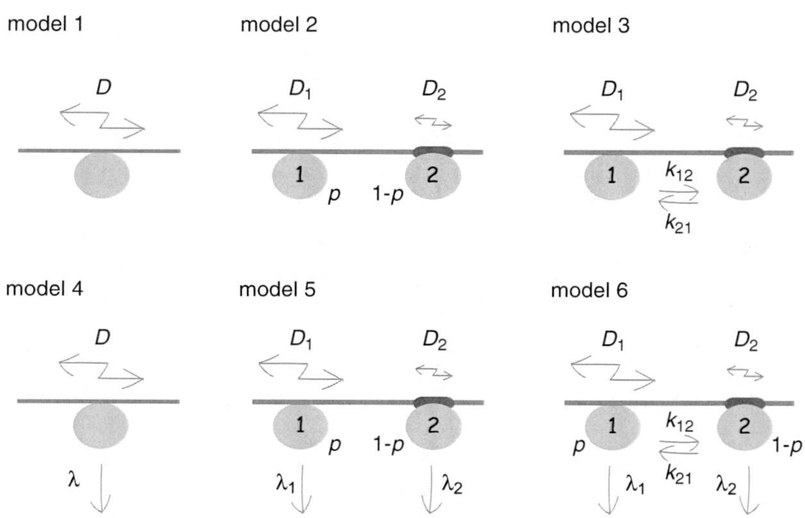

Fig. 12.1 Schematic view of six models.

12.2 Models of Diffusion

12.2.1 Membrane-Integrated Molecules

12.2.1.1 Simple Diffusion (Model 1)

First, I explain the diffusion process of a molecule showing simple diffusion on a two-dimensional plane. In order to describe the theoretical analysis method that will be introduced in the following section, both the probability density function of molecular displacement and an autocorrelation function of squared displacement are described. This method can then be extended to incorporate state transitions and/or membrane dissociation (Models 2 through 6).

Diffusion Coefficient

Diffusion is a well known phenomena that can be observed without any special equipment. When an aliquot of ink is placed on a surface of water, it gradually disperses into the surrounding area until finally no spatial difference in the intensity is observed. We can intuitively decide what will happen when the same ink is placed on a more viscous liquid; slower dispersion but eventually the same result. Therefore, for a given molecule, how fast it diffuses depends on the environmental conditions. This is because the diffusion is driven by the molecule's Brownian motion, in which countless collisions with the solvent molecules lead to molecular movement in random directions. The index for this diffusive rate is the diffusion coefficient, which for a molecule freely diffusing in a three-dimensional liquid is given as,

$$D = \frac{k_b T}{6\pi\eta a},$$

where k_b, T, η and a are Boltzmann's constant, the absolute temperature, the viscosity of the fluid solvent and the radius of the molecule, respectively. It is apparent that the diffusion coefficient is also dependent on the molecular size. The larger the molecule is, the slower the diffusive movement becomes. The dependence of the diffusion coefficient on the environmental viscosity and molecular size applies to molecules in the membrane plane [4,11,14].

Diffusion Equation

Irrespective of the molecular species or environmental conditions, diffusion occurs in the same manner. Molecules that initially exist as a point source disperse radially with the highest concentration of molecules maintained at the original point source until a spatially uniform distribution is achieved. So long the molecule is in a homogeneous environment, the diffusion process is described by a single partial differential equation. The diffusion equation for a molecule exhibiting lateral diffusion on a two-dimensional plane with diffusion coefficient D is,

$$\frac{\partial P(x,y,t)}{\partial t} = D\left(\frac{\partial^2}{\partial x^2} + \frac{\partial^2}{\partial y^2}\right)P(x,y,t), \tag{12.1}$$

where $P(x,y,t)$ represents a probability density function (PDF) at position x, y and time t. PDF explains the probability density that a molecule will be found in an infinitely small area around position (x, y), which gives the probability written as $P(x,y,t)dxdy$. Equation 12.1 tells us the temporal development of the spatial distribution. Assuming that the probability density is 0 except at the origin at time 0 ($P(x, y,0) = \delta(x,y)$), the equation can be solved as,

$$P(x,y,t) = \frac{1}{4\pi Dt} e^{-\frac{x^2+y^2}{4Dt}}, \tag{12.2}$$

This function shares the same profile as a Gaussian distribution with a mean 0 and a variance $4Dt$. The variance corresponds to the MSD. Changing the diffusion coefficient changes the width of the PDF under constant t. The distribution along the x and y axes are the same as long as there are no anisotropies on the membrane plane which may perturb molecular movement such that a bias for a specific direction occurs. For molecules showing two-dimensional diffusion, the PDF along the x axis is written as,

$$P(x,t) = \frac{1}{\sqrt{4\pi Dt}} e^{-\frac{x^2}{4Dt}}, \tag{12.3}$$

Displacement Distribution

Based on the PDF, displacement statistics provides a powerful tool for estimating the diffusion coefficient. By using single-molecule imaging, one can directly measure a molecule's displacement over an arbitrary time interval, Δt. The PDF of the displacement, $\Delta r = \text{sqrt}(\Delta x^2 + \Delta y^2)$, can be obtained by transforming variables x and y into r in Eq. 12.2, which gives rise to

$$P(\Delta r, \Delta t) = \frac{\Delta r}{2D\Delta t} e^{-\frac{\Delta r^2}{4D\Delta t}}. \tag{12.4}$$

The plot of the PDF exhibits a distribution with a single positive peak (Fig. 12.3a). Over time, the peak shifts rightwards and the distribution becomes wider. Even for the trajectory of a single molecule freely diffusing in a temporally and spatially homogeneous environment, the displacement naturally shows random values that distribute in accordance with the PDF. By fitting the displacement distribution with the PDF, it is possible to estimate the diffusion coefficient of the molecule. When molecule-to-molecule differences can be neglected, the same holds when estimating the diffusion coefficient of the ensemble. A lack of data due to short sized or a small number of trajectories only affects the precision of the estimate, not the PDF itself [8].

The discussion above is based on theoretical descriptions that assume no experimental artifacts like measurement errors. Actual application of this analysis to empirical data, however, cannot ignore these errors, especially regarding the error in the molecule's position. In single-molecule imaging, the fluorescence emission of a fluorophore like green fluorescent protein (GFP) or tetramethylrhodamine (TMR) can be regarded as a point light source since the size of the fluorophore molecule is less than the wavelength. The fluorescence collected via an objective lens exhibits a radially decaying intensity distribution known as a point-spread function or Airy disk, which describes a diffraction pattern of light passed through a circular diaphragm. The center of the intensity profile corresponds to the position of the molecule. In practice, the profile can be affected by several factors. When considering an immobilized molecule, the time duration of the image acquisition affects the photon number detectable in a single image. A longer acquisition time causes a finer intensity distribution with a trade-off between a signal-to-noise ratio and temporal resolution. When considering a diffusing molecule, the position of the molecule is inevitably blurred due to the movement. In most experiments, an effective way to estimate position is to fit the obtained intensity distribution to a two-dimensional Gaussian function [3].

Estimating position by using some sort of fit like the two-dimensional Gaussian leads to statistical errors. Consider that the *estimated* position, (x', y'), distributes around the *actual* position, (x, y), with a fluctuation following the Gaussian distribution. Then, the conditional probability that the estimated position is (x', y') when the actual position is (x, y) is written as,

$$P((x',y')|(x,y)) = \frac{1}{2\pi\varepsilon^2} e^{-\frac{(x'-x)^2+(y'-y)^2}{2\varepsilon^2}},$$

where ε represents the standard deviation (SD) of the measurement error. In the single-molecule trajectory, the error is independent of time and imposed on the estimated position at every time point. Considering the displacement between time t and $t + \Delta t$ along the trajectory, the displacement measures the distance between two *estimated* positions. The position in the x-direction, x', and the displacement during Δt, $\Delta r'$, fluctuate around the actual values, x and Δr, respectively, with a variance of $2\varepsilon^2$. The PDFs for x' and $\Delta r'$ respectively follow,

$$P(x',t) = \frac{1}{\sqrt{4\pi Dt + 4\pi\varepsilon^2}} e^{-\frac{x'^2}{4Dt+4\varepsilon^2}}, \tag{12.5}$$

and

$$P(\Delta r', \Delta t) = \frac{\Delta r'}{2D\Delta t + 2\varepsilon^2} e^{-\frac{\Delta r'^2}{4D\Delta t + 4\varepsilon^2}}. \quad (12.6)$$

In the presence of a measurement error, the distribution becomes broader than that of the error-free case (Fig. 12.2a). The effect of the error becomes more significant when shortening the time interval of the displacement measurement, i.e. increasing the temporal resolution. Assuming that the diffusion coefficient is 0.01 $\mu m^2/s$ and the SD of the error is 40 nm, the theoretical curves of the PDF in the absence and presence of the error are quite different at $t = 0.033$ s (Fig. 12.2a). If the measurement data are fitted to Eq. 12.3, the diffusion coefficient is overestimated by ε^2/t. Assuming $\varepsilon = 0.04$ and $t = 0.033$ s, the estimated diffusion coefficient will be larger than the actual value by 0.048 $\mu m^2/s$.

The SD of the error can be quantified from MSD. MSD in the presence of a measurement error follows $MSD(\Delta t) = 4D\Delta t + 4\varepsilon^2$ (Eq. 12.5). When MSD is calculated from the trajectories of diffusing molecules, the SD can be estimated from the y-intercept of the $MSD(\Delta t)$-Δt plot (Fig. 12.2b) [13]. When calculated from those of immobile molecules ($D = 0$), MSD is expected to be $4\varepsilon^2$ independent of Δt. Regarding immobilization, single molecules are visualized in fixed cells or on a glass surface. In our typical experiments, when visualizing TMR molecules under an EB-CCD camera equipped with an image intensifier, the SD is 40 nm.

Autocorrelation Function of the Squared Displacements

Diffusion is driven by molecular collisions. A difference in the diffusion coefficient comes from a difference in the frequency of collisions, reflecting the molecular size or solvent viscosity. Since the time between intermolecular collisions ($<10^{-12}$ s) is much shorter than the time resolution of observation (10^{-6}~10^{-3} s at least), observed molecular movement is temporally uncorrelated as long as the diffusion coefficient is constant. A change in the diffusion coefficient during the observation time is accompanied by a temporal correlation in the amplitude of the random molecular movement. The amplitude at time t can be measured from the instantaneous velocity, $v(t) = (\Delta x(t)^2 + \Delta y(t)^2)^{1/2}/\Delta t$, or by squaring the instantaneous displacement, $\Delta x(t)^2$ or $\Delta y(t)^2$. An instantaneous displacement along the x-axis is given by $\Delta x(t) = x(t + \Delta t) - x(t)$. Since $\Delta x(t)$ shows fluctuations around 0, it has no temporal correlation even if the amplitude of the fluctuation varies depending on the diffusion coefficient. For convenience, we use Δx^2 below to examine the temporal correlation.

In the case of simple diffusion, the squared displacement during an instant time interval, Δt, fluctuates in time (Fig. 12.6c, inset) in a temporally uncorrelated manner. The autocorrelation function when assuming no error is,

$$\left\langle (\Delta x(t))^2 (\Delta x(t'))^2 \right\rangle = \begin{cases} 4D^2\Delta t^2, & t \neq t', \\ 12D^2\Delta t^2, & t = t', \end{cases}$$

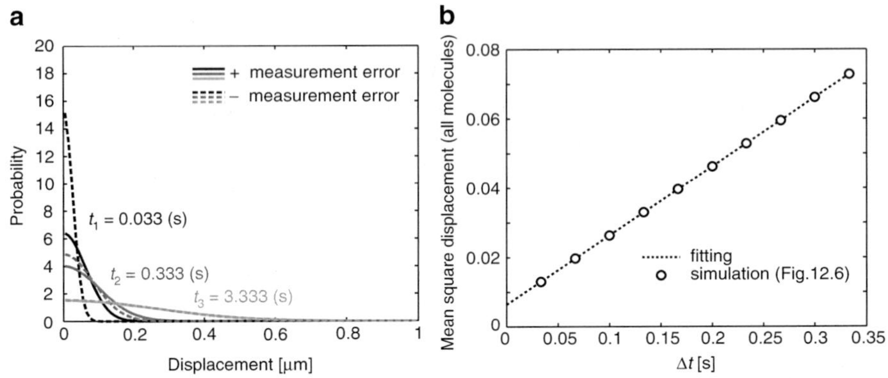

Fig. 12.2 Probability density function of molecular position. (**a**) The PDF of the simple diffusion model (model 1) before (*dotted lines*) and after (*solid lines*) incorporation of the measurement error. Theoretical curves at $t = 0.033$, 0.333 and 3.333 s are shown for $D = 0.01\ \mu m^2/s$ and standard deviation (SD) of the error, $\varepsilon = 0.04\ \mu m$. (**b**) An estimation of the SD from MSD-Δt plot. The SD was estimated by fitting an ensemble average of MSD plotted against Δt to $4D\Delta t + 4\varepsilon^2$. Trajectories generated by the numerical simulations seen in Fig. 12.6 were used for the analysis.

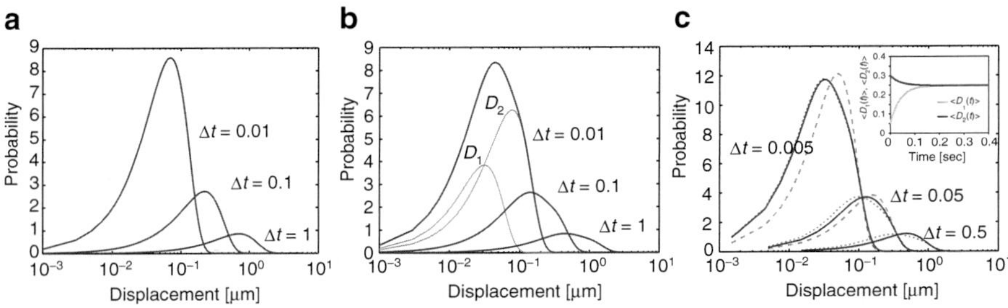

Fig. 12.3 Diffusion without membrane dissociation. (**a**) Model 1. Theoretical PDF curves of displacements at $\Delta t = 0.01$, 0.1 and 1 are shown for $D = 0.25$. (**b**) Model 2. Theoretical PDF curves of displacements at $\Delta t = 0.01$, 0.1 and 1 are shown for $D_1 = 0.05$, $D_2 = 0.3$ and $p = 0.2$ (*dark gray lines*). Each curve is a superposition of two PDFs from model 1 as seen for $\Delta t = 0.01$ (*pale gray lines*). (**c**) Model 3. Theoretical PDF curves of displacements at $\Delta t = 0.005$, 0.05 and 0.5 are shown for $D_1 = 0.05$, $D_2 = 0.3$, $k_{12} = 20$ and $k_{21} = 5$. The theoretical curves from model 1 with $D = 0.25$ (*large dotted lines*) and model 2 with $D_1 = 0.05$, $D_2 = 0.3$, and $p = 0.2$ (*small dotted lines*) show a similar profile at $\Delta t = 0.5$ and $\Delta t = 0.005$, respectively. (*inset*) Apparent diffusion coefficients of the molecules adopting state 1 or 2 at $D = 0$. D, $\mu m^2/s$; Δt, s; k, 1 s^{-1}.

which shows delta-correlated behavior (Fig. 12.6c). The nature of the autocorrelation function is insensitive to the measurement error. Upon incorporating the error, the function can be rewritten as,

$$\left\langle \Delta x'^2(t)\Delta x'^2(t')\right\rangle = \begin{cases} 12D^2\Delta t^2 + 24D\Delta t\varepsilon^2 + 12\varepsilon^4, & t = t', \\ 4D^2\Delta t^2 + 8D\Delta t\varepsilon^2 + 6\varepsilon^4, & t = t' + \Delta t, \\ 4D^2\Delta t^2 + 8D\Delta t\varepsilon^2 + 4\varepsilon^4, & t \neq t', t \neq t' + \Delta t. \end{cases}$$

In this case, the autocorrelation function decays in a delta-correlated manner, although the value at $t = t'+\Delta t$ is larger than the base line value of the autocorrelation function.

12.2.1.2 Multiple States with Different Diffusive Mobility (Model 2)

When a membrane-integrated molecule is located in two different environments on the membrane, it is possible that the molecule has two different diffusion coefficients. Assuming that the molecule does not enter the different environment during an observation, it is considered that the molecule has two states with different D but no state transitions between them. In this case, the diffusion processes of the two subpopulations with respective diffusion coefficients D_1 and D_2 are mutually independent. Letting the PDF of the molecule in state 1 and state 2 at position (x, y) at time t be represented by P_1 and P_2, respectively, the diffusion processes follow,

$$\begin{cases} \frac{\partial P_1(x,y,t)}{\partial t} = D_1\left(\frac{\partial^2}{\partial x^2}+\frac{\partial^2}{\partial y^2}\right)P_1(x,y,t), \\ \frac{\partial P_2(x,y,t)}{\partial t} = D_2\left(\frac{\partial^2}{\partial x^2}+\frac{\partial^2}{\partial y^2}\right)P_2(x,y,t). \end{cases}$$

When the subpopulations of D_1 and D_2 are p and $1-p$, respectively, the PDF of the ensemble, $P(x,y,t) = pP_1(x,y,t) + (1-p)P_2(x,y,t)$, is described as,

$$P(x',t) = p\frac{1}{\sqrt{4\pi D_1 t + 4\pi\varepsilon^2}}e^{-\frac{x'^2}{4D_1 t+4\varepsilon^2}} + (1-p)\frac{1}{\sqrt{4\pi D_2 t + 4\pi\varepsilon^2}}e^{-\frac{x'^2}{4D_2 t+4\varepsilon^2}},$$

which is a summation of the PDFs for two simple diffusion processes with different diffusion coefficients. After transforming the spatial coordinates, the PDF becomes,

$$P(\Delta r',\Delta t) = p\frac{\Delta r'}{2D_1\Delta t + 2\varepsilon^2}e^{-\frac{\Delta r'^2}{4D_1\Delta t+4\varepsilon^2}} + (1-p)\frac{\Delta r'}{2D_2\Delta t + 2\varepsilon^2}e^{-\frac{\Delta r'^2}{4D_2\Delta t+4\varepsilon^2}}. \quad (12.7)$$

It is apparent in the plot at a given time that the PDF is composed of two components (Fig. 12.3b). Separating the two subpopulations is possible at any given Δt.

Because each molecule displays simple diffusion while retaining a constant diffusion coefficient, the resulting autocorrelation for individual molecular displacement is uncorrelated. When the average is taken over the ensemble, the autocorrelation function theoretically follows,

$$\langle \Delta x'^2(t)\Delta x'^2(t')\rangle = \begin{cases} 12\{pD_1{}^2+(1-p)D_2{}^2\}\Delta t^2+24\{pD_1+(1-p)D_2\}\Delta t\varepsilon^2+12\varepsilon^4, & t=t', \\ 4\{pD_1{}^2+(1-p)D_2{}^2\}\Delta t^2+8\{pD_1+(1-p)D_2\}\Delta t\varepsilon^2+6\varepsilon^4, & t=t'+\Delta t, \\ 4\{pD_1{}^2+(1-p)D_2{}^2\}\Delta t^2+8\{pD_1+(1-p)D_2\}\Delta t\varepsilon^2+4\varepsilon^4, & t\neq t', t\neq t'+\Delta t. \end{cases}$$

For a time series of the squared displacements obtained by using numerical simulations, the amplitude of the fluctuations vary molecule to molecule depending on the diffusion coefficient (Fig. 12.7c).

12.2.1.3 State Transitions (Model 3)

When a membrane-integrated molecule changes its conformation spontaneously, for example, it possibly causes switching behavior of its diffusion coefficient. It is considered that the molecule displays transition between two states with different diffusion coefficients. In this section, I specifically concentrate on those that are integrated into the membrane, like receptors, and assume that the cells are in a resting state. Thus, the state transitions are in equilibrium, and the ratio of the number of molecules in state 1 to those in state 2 is constant at $k_{21}/(k_{12} + k_{21})$ irrespective of time, with k_{12} and k_{21} representing the rate constants for the transitions from state 1 to state 2 and vice versa, respectively. The diffusion equations are simultaneous partial differential equations written as,

$$\begin{cases} \frac{\partial P_1(x,y,t)}{\partial t} = D_1\left(\frac{\partial^2}{\partial x^2} + \frac{\partial^2}{\partial y^2}\right)P_1(x,y,t) - k_{12}P_1(x,y,t) + k_{21}P_2(x,y,t), \\ \frac{\partial P_2(x,y,t)}{\partial t} = D_2\left(\frac{\partial^2}{\partial x^2} + \frac{\partial^2}{\partial y^2}\right)P_2(x,y,t) + k_{12}P_1(x,y,t) - k_{21}P_2(x,y,t), \end{cases} \tag{12.8}$$

with $\int\int (P_1(x,y,t) + P_2(x,y,t))dxdy = 1$. The Fourier-Bessel transform of $P(r, t)$ is given as,

$$p(v) = \frac{(B+k)k + (D_2 - D_1)(k_{21} - k_{12})v^2}{4Bk\pi} e^{\frac{A+B}{2}t} + \frac{(B-k)k + (D_2 - D_1)(k_{12} - k_{21})v^2}{4Bk\pi} e^{\frac{A-B}{2}t},$$

where

$$k = k_{12} + k_{21},$$
$$A = -\{v^2(D_1 + D_2) + k\},$$
$$B = \sqrt{\{v^2(D_1 + D_2) + k\}^2 - 4\{(v^2 D_1 + k_{12})(v^2 D_2 + k_{21}) - k_{12}k_{21}\}}.$$

$P(r)$ is obtained by an inverse transformation performed by numerical integration. The PDF of displacement r' with the measurement error incorporated is obtained as follows,

$$P(r') = \int P(r'|r)P(r)dr = 2\pi \int p(v)e^{-\frac{\varepsilon^2}{2}v^2} J_0(vr')vdv,$$

where $J_0(x)$ is the Bessel function. For a sufficiently short time, the PDF is approximately the same as that from Eq. 12.7 with the same diffusion coefficients, D_1 and D_2, and $p = k_{21}/(k_{12} + k_{21})$ (Fig. 12.3c). On the other hand, for a sufficiently long time interval, the PDF approaches the same profile as Eq. 12.6 with a single effective diffusion coefficient, $D_{\text{eff}} = (D_1k_{21} + D_2k_{12})/(k_{12} + k_{21})$ [8].

The transition between two states with different diffusion coefficients leads to a temporal correlation in the amplitude of the molecular random movement. The two diffusion coefficients can be discerned in a time series of squared displacements as two different amplitudes (Fig. 12.8c, inset). The time duration of each state is characterized by inversing the rate constant of the transition for the other state. The autocorrelation function follows,

$$\langle \Delta x'^2(t)\Delta x'^2(t')\rangle$$
$$= \begin{cases} 4\frac{k_{12}k_{21}}{k^4}(D_1 - D_2)^2(e^{-k\Delta t} + e^{k\Delta t} - 2) + 4D_{eff}{}^2\Delta t^2 \\ +8\frac{D_1{}^2k_{21}+D_2{}^2k_{12}}{k}\Delta t^2 + 24D_{eff}\Delta t\varepsilon^2 + 12\varepsilon^4, & t = t', \\ 4\frac{k_{12}k_{21}}{k^4}(D_1 - D_2)^2e^{k\Delta t}(e^{-k\Delta t} + e^{k\Delta t} - 2) \\ \quad +4D_{eff}{}^2\Delta t^2 + 8D_{eff}\Delta t\varepsilon^2 + 6\varepsilon^4, & t = t' + \Delta t, \\ 4\frac{k_{12}k_{21}}{k^4}(D_1 - D_2)^2e^{-k(t'-t)}(e^{-k\Delta t} + e^{k\Delta t} - 2) \\ +4D_{eff}{}^2\Delta t^2 + 8D_{eff}\Delta t\varepsilon^2 + 4\varepsilon^4, & t \neq t',\, t \neq t' + \Delta t, \end{cases}$$

where k represents a sum of the two rate constants ($k = k_{12} + k_{21}$). The autocorrelation function displays an exponential decay with a rate constant k instead of delta-correlated behavior, independent of the measurement error (Fig. 12.8c). Even if the difference in the fluctuation amplitude is not obvious in the time trajectory, the autocorrelation has the potential to reveal the state transition.

12.2.2 *Membrane-Associating Molecules*

12.2.2.1 Simple Diffusion with Membrane Dissociation (Model 4)

Next, the diffusion process of a molecule that dissociates from the membrane is discussed. Let all molecules be located on the same position in an ideal homogeneous plane at $t = 0$, and allow them to diffuse with the diffusion coefficient D. To incorporate the dissociation process, it is assumed that the molecule disappears from the plane with a rate constant λ, which is equivalent to the dissociation rate constant. As time passes, the position of the molecules disperses around the original. Furthermore, because molecules are dissociating from the membrane, their actual count will decrease at the same time. If dissociation occurs independently of the diffusion process, the PDF is written as,

$$\frac{\partial P(x,y,t)}{\partial t} = D\left(\frac{\partial^2}{\partial x^2} + \frac{\partial^2}{\partial y^2}\right)P(x,y,t) - \lambda P(x,y,t),$$

where the term for membrane dissociation is added to the diffusion equation that describes simple diffusion (Eq. 12.1). In the case of membrane-associating molecules, displacements, $\Delta r'$, cannot be used when considering both diffusion and the reaction kinetics simultaneously. Instead, the position of the molecules at time t should be considered. One-dimensionally this means,

$$P(x,t) = \frac{e^{-\lambda t}}{\sqrt{4\pi D t}} e^{-\frac{x^2}{4Dt}}.$$

The PDF becomes broader with increasing t while the mean remains at the origin. When the measurement error is included, the PDF is rewritten as,

$$P(x',t) = \frac{e^{-\lambda t}}{\sqrt{4\pi D t + 4\pi\varepsilon^2}} e^{-\frac{x^2}{4Dt+4\varepsilon^2}}. \tag{12.9}$$

Compared to the PDF for simple diffusion (Eq. 12.5), the PDF in Eq. 12.9 depends on time in an exponential manner. This means for a given t, the probability density is $\exp(-\lambda t)$-fold smaller than that calculated from Eq. 12.5 over all x values (Fig. 12.4a). In the absence of membrane dissociation, the Riemann integral over x equals 1 independent of time t. In the presence of dissociation, the integral of Eq. 12.9 shows an exponential decay with rate λ,

$$R(t) = e^{-\lambda t}.$$

$R(t)$ describes the probability the molecule remains bound to the membrane, and is hereafter called the "membrane residence probability" (Fig. 12.4a, inset). The PDF from Eq. 12.9 indicates all molecules irrespective of position have the same probability of dissociation, which agrees with the assumptions.

12.2.2.2 Multiple States with Different Diffusion Coefficients and Dissociation Rates (Model 5)

When a molecule has two binding sites on the membrane, for example, two complexes often show different diffusion coefficients and dissociation rate constants. In case that the molecule does not exchange one binding site for the other and vice versa on the membrane, model 5 is applicable. The analysis becomes more complicated than that used for model 2 because the subpopulations vary with t. The PDFs for subpopulations with D_1 and D_2, respectively, follow,

$$\begin{cases} \frac{\partial P_1(x,y,t)}{\partial t} = D_1\left(\frac{\partial^2}{\partial x^2} + \frac{\partial^2}{\partial y^2}\right)P_1(x,y,t) - \lambda_1 P_1(x,y,t), \\ \frac{\partial P_2(x,y,t)}{\partial t} = D_2\left(\frac{\partial^2}{\partial x^2} + \frac{\partial^2}{\partial y^2}\right)P_2(x,y,t) - \lambda_2 P_2(x,y,t). \end{cases}$$

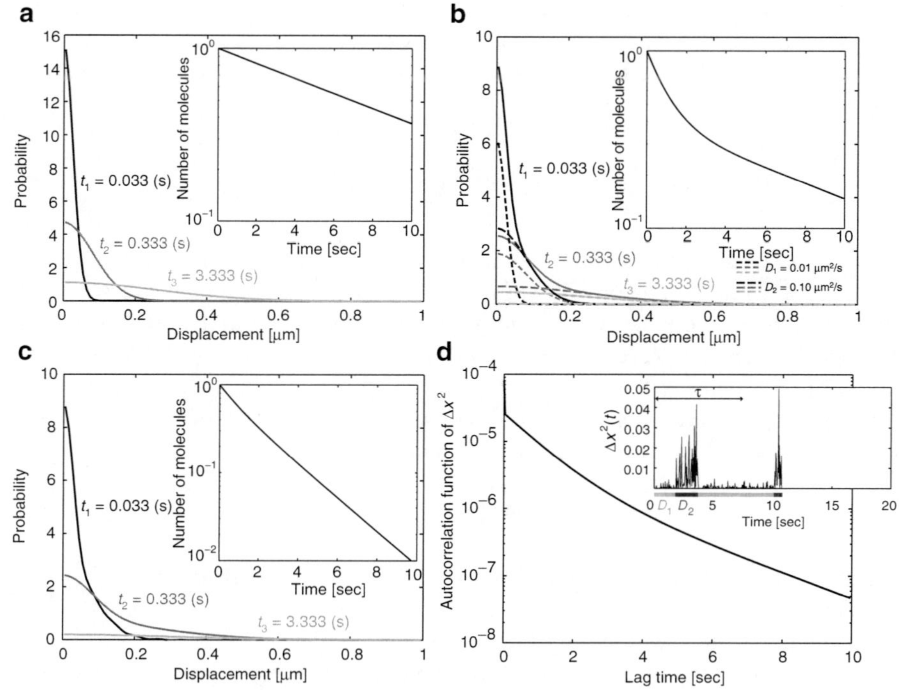

Fig. 12.4 Diffusion with membrane dissociation. (**a**) Model 4. Theoretical curves of the position PDF at $t = 0.033$, 0.333 and 3.333 are shown for $D = 0.01$ and $\lambda = 0.1$. (*inset*) Release curve. (**b**) Model 5. Theoretical curves of the position PDF at $t = 0.033$, 0.333 and 3.333 are shown for $D_1 = 0.01, D_2 = 0.1, \lambda_1 = 0.1, \lambda_2 = 1$ and $p = 0.4$ (*solid lines*). Each curve is a superposition of two subpopulations adopting state 1 (*small dotted lines*) or state 2 (*large dotted lines*). (*inset*) Release curve. (**c**) Model 6. Theoretical curves of the position PDF at $t = 0.033$, 0.333 and 3.333 are shown for $D_1 = 0.01$, $D_2 = 0.1$, $\lambda_1 = 0.1$, $\lambda_2 = 1$, $k_{12} = 0.4$, $k_{21} = 0.1$ and $p = 0.4$. (*inset*) Release curve. (d) Theoretical curve for the autocorrelation function of the squared displacements, $<\Delta x^2(0)\Delta x^2(\tau)>$. An ensemble average of $\Delta x^2(0)\Delta x^2(\tau)$ was taken assuming $\Delta x^2(\tau) = 0$ after membrane dissociation. (*inset*) Time series of squared displacements. D, μm²/s; t, s; k, 1 s^{-1}; λ, 1 s^{-1}.

Assuming that the proportion of molecules in state 1 is p at $t = 0$, the PDF of the ensemble is $P(x,y,t) = pP_1(x,y,t) + (1 - p)P_2(x,y,t)$, which can be obtained as,

$$P(x', t) = p\frac{e^{-\lambda_1 t}}{\sqrt{4\pi D_1 t + 4\pi\varepsilon^2}} e^{-\frac{x^2}{4D_1 t + 4\varepsilon^2}} + (1 - p)\frac{e^{-\lambda_2 t}}{\sqrt{4\pi D_2 t + 4\pi\varepsilon^2}} e^{-\frac{x^2}{4D_2 t + 4\varepsilon^2}}, \quad (12.10)$$

in the presence of the measurement error (Fig. 12.4b). When taking the Riemann integral, the membrane residence probability is obtained as,

$$R(t) = pe^{-\lambda_1 t} + (1 - p)e^{-\lambda_2 t},$$

which contains two exponential decays and indicates the subpopulation with D_1 decays with the rate constant λ_1 and the subpopulation with D_2 with the rate constant λ_2 (Fig. 12.4b, inset). The content ratios, $Q_1(t)$ and $Q_2(t)$, of the two subpopulations are written as,

$$Q_1(t) = \frac{pe^{-\lambda_1 t}}{pe^{-\lambda_1 t}+(1-p)e^{-\lambda_2 t}}, \quad Q_2(t) = \frac{(1-p)e^{-\lambda_2 t}}{pe^{-\lambda_1 t}+(1-p)e^{-\lambda_2 t}}.$$

Initially, the ratios equal p and $1 - p$, respectively, by definition. At infinity, they approximate 0 and 1, respectively, if $\lambda_1 >> \lambda_2$. Thus, depending on the dissociation rate constants, the content ratio temporally varies from the initial values. The PDF plot demonstrates these characteristics (Fig. 12.4b). Since the subpopulations exhibit no mutual exchange, the PDF is regarded as the sum of two plots corresponding to the subpopulations. However, their contribution to the PDF changes over time, a different result from model 2.

12.2.2.3 State Transitions and Membrane Dissociation (Model 6)

When a molecule has two binding sites and exchanges one binding site for the other and vice versa on the membrane, it is possible that transitions between two complexes with different diffusion coefficients and dissociation rate constants occur. For a molecule showing both state transitions and membrane dissociation, it is difficult to understand the PDF intuitively. However, it can be obtained from the diffusion equation into which the reactions of state transitions and dissociation are incorporated. The PDFs for those adopting state 1 and state 2, P_1 and P_2, respectively, are,

$$\begin{cases} \frac{\partial P_1(x,y,t)}{\partial t} = D_1\left(\frac{\partial^2}{\partial x^2} + \frac{\partial^2}{\partial y^2}\right)P_1(x,y,t) - (k_{12}+\lambda_1)P_1(x,y,t) + k_{21}P_2(x,y,t), \\ \frac{\partial P_2(x,y,t)}{\partial t} = D_2\left(\frac{\partial^2}{\partial x^2} + \frac{\partial^2}{\partial y^2}\right)P_2(x,y,t) + k_{12}P_1(x,y,t) - (k_{21}+\lambda_2)P_2(x,y,t). \end{cases} \tag{12.11}$$

At $t = 0$, it is assumed molecules that adopt state 1 make up fraction p of the total population. Solving analytically, the Fourier transform of the PDF for all molecules becomes,

$$\begin{aligned} p(k_x,k_y,t) = &\left[\left\{\left(k_x^2+k_y^2\right)(D_1-D_2)+\lambda_1-\lambda_2\right\}(1-2p)+k_{12}+k_{21}+B\right]e^{\frac{A+B}{2}t}/4B\pi \\ &-\left[\left\{\left(k_x^2+k_y^2\right)(D_1-D_2)+\lambda_1-\lambda_2\right\}(1-2p)+k_{12}+k_{21}-B\right]e^{\frac{A-B}{2}t}/4B\pi, \end{aligned}$$

where

$$A = -\left\{\left(k_x^2 + k_y^2\right)(D_1 + D_2) + k_{12} + k_{21} + \lambda_1 + \lambda_2\right\},$$

$$B = \sqrt{\left\{\left(k_x^2 + k_y^2\right)(D_1 + D_2) + k_{12} + k_{21} + \lambda_1 + \lambda_2\right\}^2} \times \sqrt{-4\left[\left\{\left(k_x^2 + k_y^2\right)D_1 + k_{12} + \lambda_1\right\}\left\{\left(k_x^2 + k_y^2\right)D_2 + k_{21} + \lambda_2\right\} - k_{12}k_{21}\right]}.$$

The inverse transformation is performed to obtain $P(x,y,t)$ by numerical integration (Fig. 12.4c). The Riemann integral of the PDF corresponds to the membrane residence probability written as,

$$R(t) = (0.5 - C)e^{s_1 t} + (0.5 + C)e^{s_2 t}, \tag{12.12}$$

where

$$s_1 = -0.5(k_{12} + k_{21} + \lambda_1 + \lambda_2 + \sqrt{(k_{12} + k_{21} + \lambda_1 + \lambda_2)^2 - 4(k_{12}\lambda_2 + k_{21}\lambda_1 + \lambda_1\lambda_2)}),$$

$$s_2 = -0.5(k_{12} + k_{21} + \lambda_1 + \lambda_2 - \sqrt{(k_{12} + k_{21} + \lambda_1 + \lambda_2)^2 - 4(k_{12}\lambda_2 + k_{21}\lambda_1 + \lambda_1\lambda_2)}),$$

$$C = \frac{(k_{12} + k_{21} - \lambda_1 + \lambda_2)p + (k_{12} + k_{21} + \lambda_1 - \lambda_2)(1 - p)}{2\sqrt{(k_{12} + k_{21} + \lambda_1 + \lambda_2)^2 - 4(k_{12}\lambda_2 + k_{21}\lambda_1 + \lambda_1\lambda_2)}}.$$

The function contains two exponentials derived from the two states (Fig. 12.4c, inset). Depending on the parameter values, the function can take either a concave or convex profile; a convex profile meaning there exists certain rate-limiting process.

The autocorrelation of the squared displacements obtained from a subpopulation of molecules bound to the membrane for a relatively long period exhibits an exponential decay. The exponential decay is an indication of state transitions (Fig. 12.11b). However, the rate constant of the exponential decay does not coincide with the sum of two rate constants for state transitions, which is different from model 3. The formulation of this type of autocorrelation function is currently unknown. On the other hand, the autocorrelation function can be calculated from all molecules as,

$$\left\langle \Delta x^2(0)\Delta x^2(\tau)\right\rangle = \frac{1}{X}\sum_{i=1}^{X} \Delta x_i{}^2(0)\Delta x_i{}^2(\tau),$$

in which the squared displacement of i-th molecule, Δx_i^2, is 0 when the molecule detaches from the membrane. The theoretical autocorrelation function is,

$$\langle \Delta x^2(0)\Delta x^2(\tau)\rangle = \begin{cases} 2\Delta t^2\{(D+E)e^{s_1\tau} + (D-E)e^{s_2\tau}\}, & \tau \neq 0, \\ 12D\Delta t^2, & \tau = 0, \end{cases} \tag{12.13}$$

where,

$$D = D_1{}^2 p + D_2{}^2(1-p),$$
$$E = \frac{(k_{12} - k_{21} + \lambda_1 - \lambda_2)\{D_1{}^2 p - D_2{}^2(1-p)\} - 2\{k_{12}p + k_{21}(1-p)\}D_1D_2}{\sqrt{(k_{12} + k_{21} + \lambda_1 + \lambda_2)^2 - 4(k_{12}\lambda_2 + k_{21}\lambda_1 + \lambda_1\lambda_2)}}.$$

Incorporation of the measurement error is described in the Appendix. The autocorrelation function is described by the sum of two exponentials with rate constants equal to the residence probability function $R(t)$ (Fig. 12.4d). By fitting the autocorrelation function and the release curve to Eqs. 12.13 and 12.12, respectively, values for s_1, s_2, C, D and E can be acquired, which serves as constraints limiting possible parameter values as described below.

12.3 Method of Model Selection

Diffusion coefficients and reaction rate constants determine the PDF of displacement. Regarding the six models, the questions to be clarified are whether the molecule is membrane-integrated or not, how many states with different diffusive mobility the molecule can adopt, and whether the molecule shows state transition or not, in order to reveal the reaction scheme.

12.3.1 Membrane Dissociation

At first, the presence or absence of membrane dissociation can be discriminated by comparing the disappearance rate of the fluorescence signal and photobleaching rate of the fluorophore (Fig. 12.5a). Single-molecular fluorescence intensities are visualized during their membrane association under excitation by an evanescent field. Generally, continuous excitation of a fluorophore leads to photobleaching, which is an irreversible process rendering fluorophores unable to emit fluorescence. It inevitably occurs in all fluorescent molecules. Even if the fluorophore-conjugated

Fig. 12.5 Criteria for model selection. (**a**) Distinction between the presence and absence of membrane dissociation. A faster *apparent* rate of fluorescence disappearance ("observation") than the rate of photo-bleaching ("photo-bleaching") indicates that the molecule is not membrane-integrated. The decay rates are $\lambda = 0.1\ \mathrm{s}^{-1}$ and $k_b = 0.06\ \mathrm{s}^{-1}$. (**b**) Log likelihood function used for AIC calculation to estimate the number of states with different D. (**c**) Dependence of the autocorrelation function of the squared displacements on the temporal resolution. By changing the time interval Δt, theoretical curves using the parameter values from Fig. 12.3c were plotted against lag time/Δt. (**d**) A diagram describing a series of analyses required to hypothesize the reaction schema.

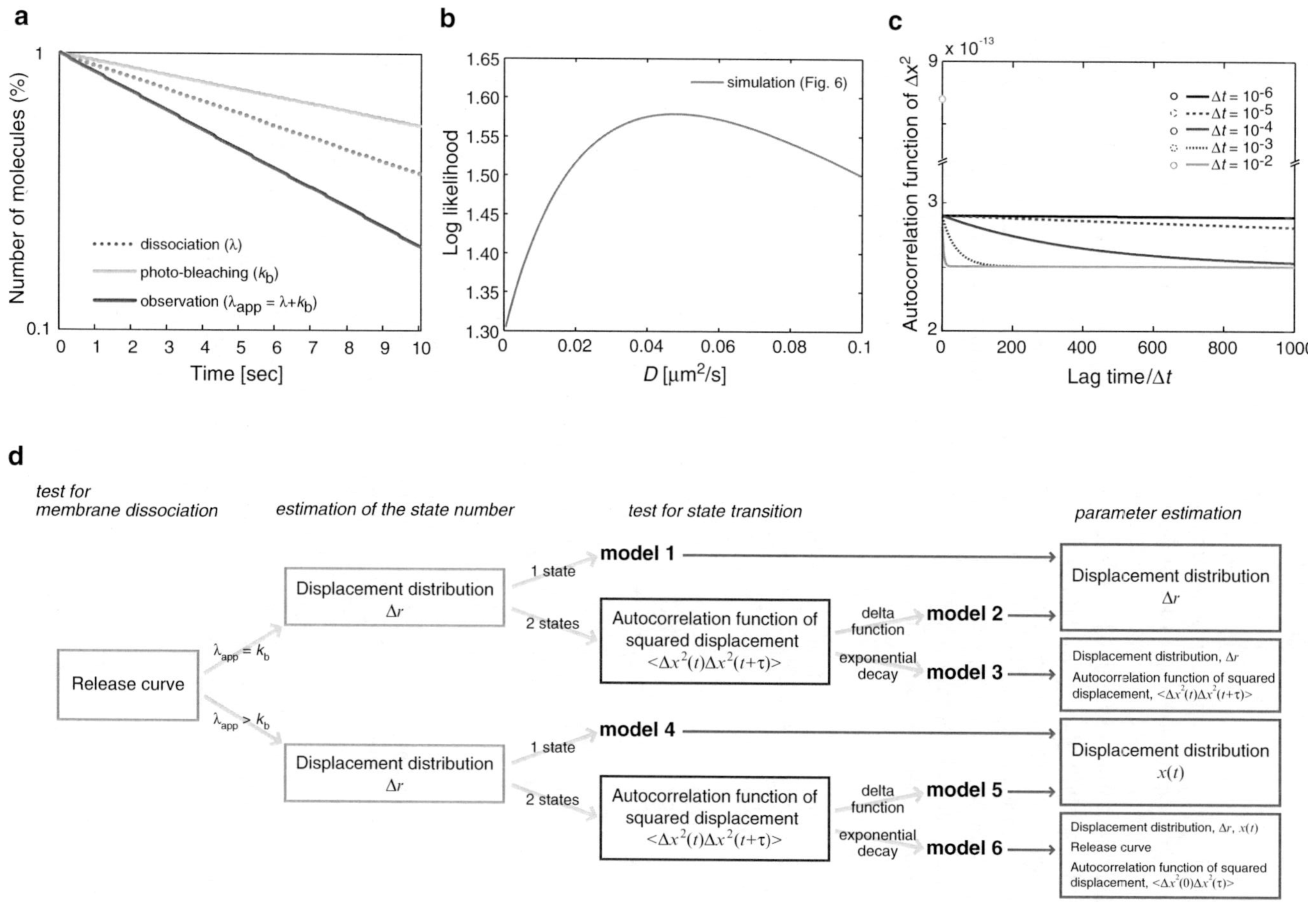

Fig. 12.5 (continued)

molecule is integrated into the membrane, the trajectory has a finite length that varies molecule to molecule. For membrane-integrated molecules, photobleaching is the only way for their fluorescence intensities to diminish. Therefore, in these cases, the fluorescence disappearance rate is equal to the photobleaching rate. On the other hand, for those molecules shuttling between the membrane and cytosol, both photobleaching and membrane dissociation causes the fluorescence to disappear. Since these processes are thought to be mutually independent, the observed fluorescence disappearance rate is greater than the actual dissociation rate because of photobleaching. Therefore, when the fluorescence disappears faster than the rate of photobleaching, the molecule is not membrane-integrated.

These two rates are estimated from the lifetime of the fluorescent signal measured for each molecule: the fluorescence disappearance rate from molecules on the membrane of living cells; the photobleaching rate from molecules immobilized in fixed cells or on a glass surface. Assuming all molecules are visible at $t = 0$, when the number of molecules is plotted against time after the onset of signal detection, an exponential decay is seen (Fig. 12.5a). The actual dissociation rate constant can be obtained by subtracting the photobleaching rate constant, k_{b}, from the apparent rate constant, λ_{app}, such that $\lambda = \lambda_{\mathrm{app}} - k_{\mathrm{b}}$. I shall refer to the experimentally obtained decay curve as the "release curve".

12.3.2 Number of States

The number of states with different diffusion coefficients is related to the number of parameters in the PDF to be fitted. While increasing the parameter number results in a better fit, it also complicates the model to the point that it is difficult to decipher functional meaning. As an objective criterion for finding a compromise, we apply Akaike Information Criterion (AIC) [1]. Here, the goal is to count the number of states with different diffusion coefficients that the molecule can adopt. For this purpose, the time resolution of the displacement measurement is required to be sufficiently high so not to overlook any states with short lifetimes, meaning the time interval of the image acquisition should be as short as possible. In addition, the timing of the state transitions can be neglected such that the displacement can be measured from the whole trajectory. When the time interval is infinitely short, the displacement distribution obeys a sum of PDFs that each describes simple diffusion,

$$P_i(\Delta r, \Delta t) = \sum_{j=1}^{i} p_j \frac{\Delta r}{2D_j \Delta t + 2\varepsilon^2} e^{-\frac{\Delta r^2}{4D_j \Delta t + 4\varepsilon^2}},$$

where

$$\sum_{j=1}^{i} p_j = 1$$

and i indicates the number of states with different diffusion coefficients. By maximum likelihood estimation (MLE), how well the statistic data set is fitted to the model PDF is measured as a log likelihood, which is a function of the parameter vector $\boldsymbol{\theta}$ calculated from data containing n displacements,

$$l_i(\tilde{\theta}) = \sum_{m=1}^{n} \log P_i(r_m|\theta).$$

The greater log likelihood is obtained when differences between the data set and the model become smaller. In MLE, the parameter value $\tilde{\theta}$ is sought so that the maximum of the log likelihood is returned (Fig. 12.5b). By increasing the parameter number, the log likelihood generally grows. To distinguish which model is most plausible, a penalty for increasing the parameter number is introduced, which gives rise to AIC,

$$AIC_i = -2l_i(\tilde{\theta}) + 2k_i,$$

where k_i denotes the number of parameters used. The most likely model is predicted as the model that returns the minimum AIC.

12.3.3 State Transitions

The state transition can be revealed from single-molecule stochastic trajectories when the diffusion coefficient changes depending on the state. If the molecule changes the diffusion coefficient temporally between D_1 and D_2, the molecular movement will change the amplitude depending on the diffusion coefficient. The difference will appear in the time trajectory of the squared displacement, giving rise to a fluctuation with two different amplitudes. Whether the state transition occurs or not can be discerned by taking the autocorrelation. This analysis is valid even in the presence of membrane dissociation as described above.

The temporal resolution of a single-molecule image sequence affects the potential to detect the correlation (Fig. 12.5c). In order to detect a slow transition with a small k value, high temporal resolution is not required. Instead, the correlation over a sufficiently long lag time must be examined. Otherwise, state transitions rarely occur within the analyzed time scale, leading to a delta-correlated behavior in the autocorrelation function, which suggests that the molecules behave as if they can adopt two states without any transitions. On the other hand, detecting a fast transition with a large k value requires high temporal resolution of the trajectory. If the time interval of the displacement measurement is much longer than the characteristic time, the state transitions reach equilibrium during the measurement interval and the molecule behaves as if it obeys a simple diffusion process with a single diffusion coefficient, D. Therefore, it is required to adjust the temporal resolution of the image acquisition accordingly so as not to overlook any state transitions.

12.4 Analysis of Single-Molecule Trajectories

In this section, the overall analysis is described starting from the model selection and ending with the estimation of the parameter values. I use trajectories generated from numerical simulations based on the six models. The analysis proceeds as follows: hypothesize a schema of reactions in which the molecule is involved; construct diffusion equation(s) according to the schema; obtain the PDF from the diffusion equation(s); and estimate the parameter values from the molecular displacements based on the PDF. Hypothesizing the reaction schema requires examining single-molecule trajectories (Fig. 12.5d). At first, a release curve is examined to distinguish whether the molecule is integrated into the membrane or not. As stated above, if the curve exhibits decay at the same rate as fluorescent photobleaching, the molecule is integrated into the membrane (models 1 to 3). If it decays faster, the molecule should be thought to transiently associate to the membrane (models 4 to 6). The next process is AIC analysis in order to estimate the number of states with different diffusion coefficients, which is performed in the same manner for both membrane-integrated and membrane-associating molecules. A measurement error estimated by using a MSD plot is used here. MSD averaged over all molecules can be used to estimate a single SD value typical for a given experimental conditions. When the state number is estimated to be 1, models 1 or 4 is appropriate. When it is estimated to be 2, the autocorrelation of the squared displacements is next analyzed to distinguish whether the molecule shows state transitions or not. If it is delta-correlated, state transitions need not to be considered (model 2 or 5). On the other hand, an exponential decay indicates a transition between the two states (model 3 or 6). Cases with more than two states will not be discussed here. In addition, it is assumed that a molecule shows simple diffusion without any corrals or directional flows on the membrane which may cause anomalous diffusion. In the case of the anomalous diffusion, the MSD plot displays some deviations from $4D\Delta t$, which serves as the indication [6].

Trajectories were made by using numerical simulations according to a method described previously [8]. Briefly, the Langevin equations $\frac{dx(t)}{dt} = \eta_x(t)$and $\frac{dy(t)}{dt} = \eta_y(t)$ are solved by the Euler scheme with a time step of 1/300 s. ($x(t)$, $y(t)$) is the position on the membrane, while $\eta_x(t)$ and $\eta_y(t)$ are Gaussian white noise satisfying $<\eta_i(t)> = 0$ and $<\eta_i(t)\eta_j(t')> = 2D\delta_{\{i,j\}}\delta(t - t')$ with $i,j = x$ or y and D being the diffusion coefficient of a given molecular state. A time series of positions starting from the origin $(x, y) = (0, 0)$ was generated for $t = 0$–60 s consisting of 18,000 time steps. In models 2, 3, 5 and 6, an initial state of a molecule was determined randomly to be one of two states with probability p for state 1 and $1 - p$ for state 2. A trajectory was composed of 1,800 time steps extracted from the time series at a unit time interval, 1/30 s. In the trajectories, a Gaussian error with a 40 nm variance was added to the molecular position at every time point. In the following analyses, 100 (models 1 to 3) or 3,000 (models 4 to 6) trajectories were used. Photo-bleaching was not included in the simulations. Values used as diffusion coefficients are typical for membrane proteins [16].

12.4.1 Model 1

The simulation was performed using $D = 0.05$ μm^2/s. From the trajectories, MSD was calculated and plotted against the time interval of the displacement measurement, Δt, where the SD of the measurement error was estimated to be 40 nm from the y-intercept. Using this value, the number of states with different diffusion coefficients was estimated by AIC analysis. Displacement during the shortest time interval, $\Delta t = 1/30$ s, was measured irrespective of time along the trajectory. A distribution was obtained from all trajectories (Fig. 12.6a). AIC showed a single state with a diffusion coefficient of 0.049 μm^2/s (Fig. 12.6b). Consistent with this, no time correlation was detected in the squared displacements (Fig. 12.6c). Thus, model 1 is most appropriate for explaining the molecular behavior.

12.4.2 Model 2

The simulation was performed using $D_1 = 0.01$ μm^2/s (75%) and $D_2 = 0.05$ μm^2/s (25%). From MSD averaged over all trajectories, the SD of the error was estimated to be 40 nm. Two states with different D were predicted by the AIC analysis (Fig. 12.7b). The two diffusion coefficients and the proportion of the smaller D subpopulation were estimated to be 0.010, 0.054 μm^2/s and 0.849, respectively (Fig. 12.7a). Transitions between the two states were not detected from the autocorrelation of the squared displacements, which followed a delta function (Fig. 12.7c). Thus, model 2 is most appropriate. The estimated parameter values explained all the distributions of the displacements examined at different time intervals (data not shown). In general, accurate parameter estimates depend on the number of analyzed trajectories, although this applies more to the ratio of the two subpopulations than to the diffusion coefficients. Despite estimating diffusion coefficient from displacement data containing $1{,}800 \times 100$ samples, the ratio estimate effectively uses only 100 individual samples.

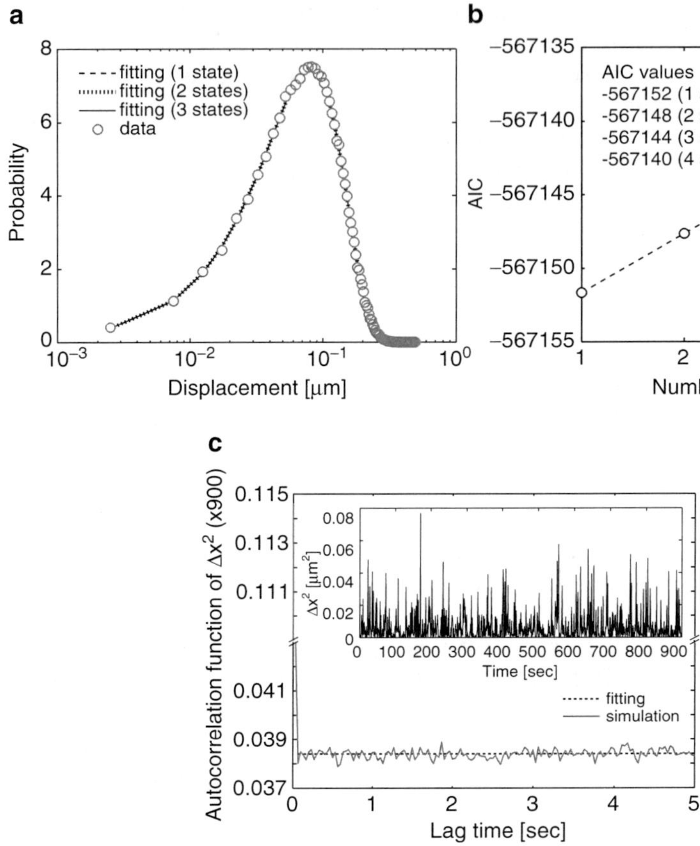

Fig. 12.6 Estimation of the diffusion coefficient for model 1. (**a**) A histogram of displacements measured from 100 simulated trajectories, each containing 1,800 time steps, generated assuming simple diffusion with $D = 0.05$ and $\varepsilon = 0.04$. Displacement during a time interval of 0.033 s was measured at any point in the trajectories. The AIC analysis showed a one-state model is most appropriate (**b**). The estimated diffusion coefficient was 0.049. (**b**) The result of the AIC analysis. (**c**) The autocorrelation function of squared displacement calculated from all simulated trajectories, showing no state transitions. (*inset*) Representative time series of squared displacements, $\Delta x^2(t)$, during $\Delta t = 0.033$ calculated from a single trajectory. D, μm^2/s; ε, μm; t, s.

12.4.3 Model 3

Numerical simulations were performed based on model 3 using $D_1 = 0.01$ μm^2/s, $D_2 = 0.05$ μm^2/s, $k_{12} = 1$ s^{-1} and $k_{21} = 3$ s^{-1}. The SD of the measurement error was 40 nm, as estimated from MSD averaged over all trajectories. Using displacement data during the shortest time interval in the trajectories, AIC analysis based on the PDF with the SD value incorporated displayed a minimum when hypothesizing two states with different D (Fig. 12.8b). Consistent with this, the histogram of displacement was better fitted with the PDF of the two-state case (Fig. 12.8a). By analyzing

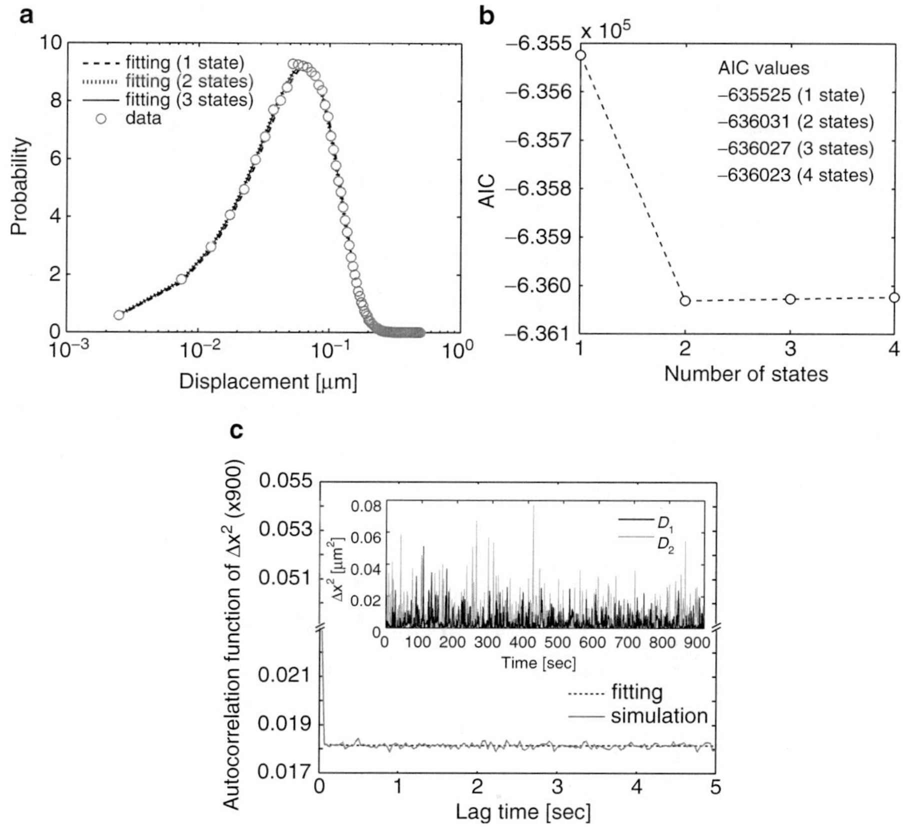

Fig. 12.7 Estimation of parameters for model 2. (**a**) A histogram of displacements measured from 100 simulated trajectories, each containing 1,800 time steps, generated assuming simple diffusion with $D_1 = 0.01$, $D_2 = 0.05$, and $p = 0.75$. The histogram of displacements was well explained by assuming two states. The parameters estimated were $D_1 = 0.010$, $D_2 = 0.054$, and $p = 0.849$. (**b**) The result of the AIC analysis. (**c**) The autocorrelation function of squared displacement calculated from all simulated trajectories, showing no state transitions. (*inset*) Representative time series of squared displacements, $\Delta x^2(t)$, during $\Delta t = 0.033$ calculated from two trajectories with D_1 and D_2. D, $\mu m^2/s$; t, s.

the autocorrelation of squared displacements, it became clear that state transitions occur and that the exponential decay of the function has a time constant of $k = 3.962\ s^{-1}$ (Fig. 12.8c). Therefore, model 3 is best and that Eq. 12.8 should be used to obtain the PDF for parameter estimation. Using k to constrain k_{12} and k_{21}, the diffusion coefficients and transition rates were estimated to be $D_1 = 0.011\ \mu m^2/s$, $D_2 = 0.062\ \mu m^2/s$, $k_{12} = 0.752\ s^{-1}$ and $k_{21} = 3.211\ s^{-1}$ by MLE (Fig. 12.8a). Since the PDF obtained from Eq. 12.8 with these parameters coincided with the histograms of the displacements during time intervals ranging from 0.001 to 0.1 s, the analysis was confirmed to be reasonable (data not shown).

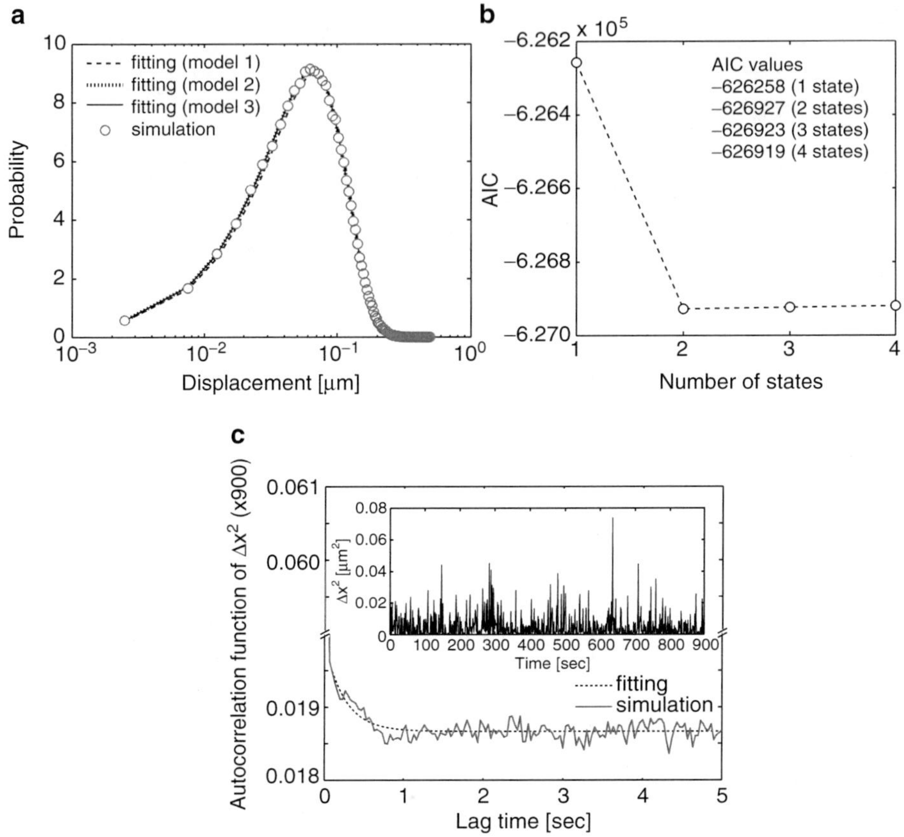

Fig. 12.8 Estimation of parameters for model 3. (**a**) A histogram of displacements obtained by numerical simulations in model 3 assuming 100 molecules, each containing 1,800 time steps, using $D_1 = 0.01, D_2 = 0.05, k_{12} = 1$ and $k_{21} = 3$. The histogram of displacements during a time interval of 0.033 was better fitted using the probability density function from Eqs. 12.7 and 12.8 than that from Eq. 12.6, indicating the molecule has two states with different diffusion coefficients. The parameters estimated were $D_1 = 0.011$, $D_2 = 0.062$, $k_{12} = 0.752$ and $k_{21} = 3.211$. (**b**) The result of the AIC analysis. (**c**) The autocorrelation function of squared displacement calculated from all simulated trajectories, showing state transitions. (*inset*) Representative time series of squared displacements, $\Delta x^2(t)$, during $\Delta t = 0.033$ calculated from a single trajectory. D, μm^2/s; k, 1 s^{-1}; t, s^{-1}.

12.4.4 Model 4

Simulations were performed using $D = 0.01$ μm^2/s and $\lambda = 0.1$ s^{-1} to obtain the trajectories. The SD of the measurement error was set to 40 nm. The length of the trajectories varied from trajectory to trajectory. The release curve, the number of diffusing molecules plotted against time, showed an exponential decay (Fig. 12.9d, inset). Since it is assumed that the molecule dissociates from the membrane, models 4, 5 and 6 are all candidates for explaining the molecular behavior. Displacement

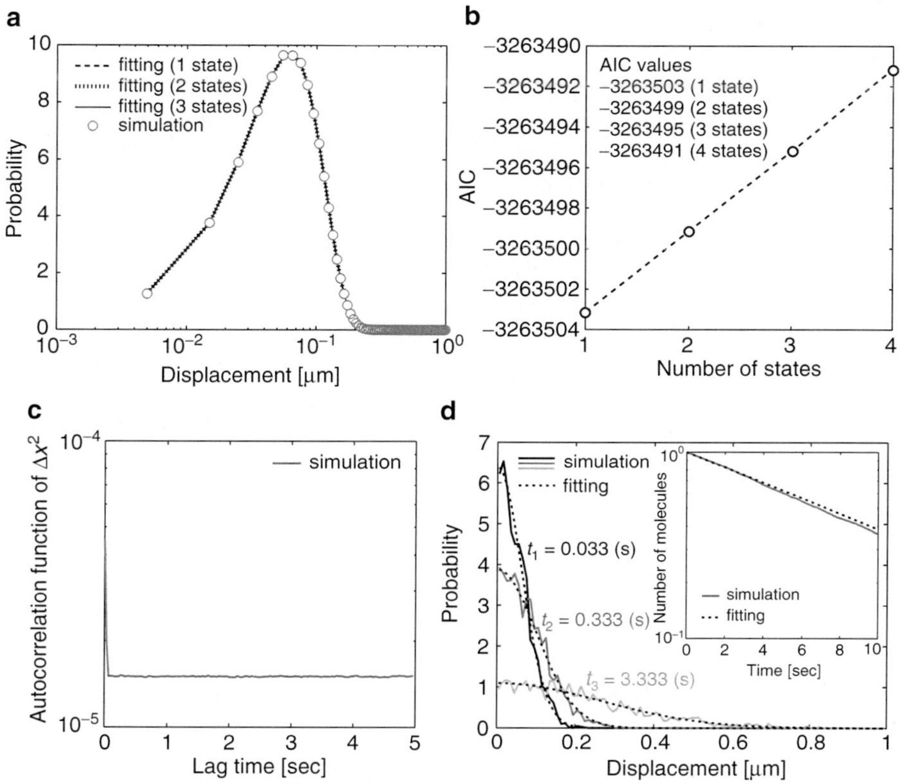

Fig. 12.9 Estimation of parameter values for model 4. (**a**) A histogram of displacements, Δr, measured from simulated trajectories. Numerical simulations were performed assuming 3,000 molecules, each containing at most 1,800 time steps, showing simple diffusion with $D = 0.01$ and a dissociation rate constant of $\lambda = 0.1$. Displacement during a time interval of $\Delta t = 0.033$ was measured from the trajectories containing an error with a SD of $\varepsilon = 0.04$. The histogram was well fitted when assuming at least one state. (**b**) The result of AIC analysis showing one-state model is the most likely. (**c**) The autocorrelation function of squared displacements calculated from the simulated trajectories, which were all longer than 10 s. (**d**) Histograms of position, x, at $t = 0.033$, 0.333 and 3.333. The estimated diffusion coefficient and dissociation rate constant were 0.010 and 0.095, respectively. (*inset*) The release curve obtained from the simulated trajectories. D, $\mu m^2/s$; λ, $1\ s^{-1}$; t, s; ε, μm.

during a minimum time interval in the trajectories was used for MLE based on four hypothetical models with one to four states with different D. A minimum AIC was achieved when assuming only one state (Fig. 12.9a, b). Model 4 is most consistent with this, and a time series of squared displacements calculated from 1,093 trajectories longer than 10 s showed no temporal correlation in the autocorrelation analysis (Fig. 12.9c). By fitting three histograms of molecular positions after three different time intervals to Eq. 12.9, D and λ were estimated to be 0.010 $\mu m^2/s$ and 0.095 s^{-1}, respectively (Fig. 12.9d). The estimated dissociation rate constant accurately described the release curve (Fig. 12.9d, inset).

12.4.5 Model 5

Simulations were performed using $D_1 = 0.01$ μm²/s, $D_2 = 0.10$ μm²/s, $\lambda_1 = 0.1$ s^{-1} and $\lambda_2 = 1.0$ s^{-1} with the subpopulation of D_1 being 0.4. AIC was minimal when assuming the molecule has two states with different D (Fig. 12.10a, b). The autocorrelation of the squared displacements, which was calculated from 472 trajectories longer than 10 s, showed a delta-correlated function, indicating no state transitions (Fig. 12.10c). The histograms of the molecular positions for $t = 0.033$, 0.333 and

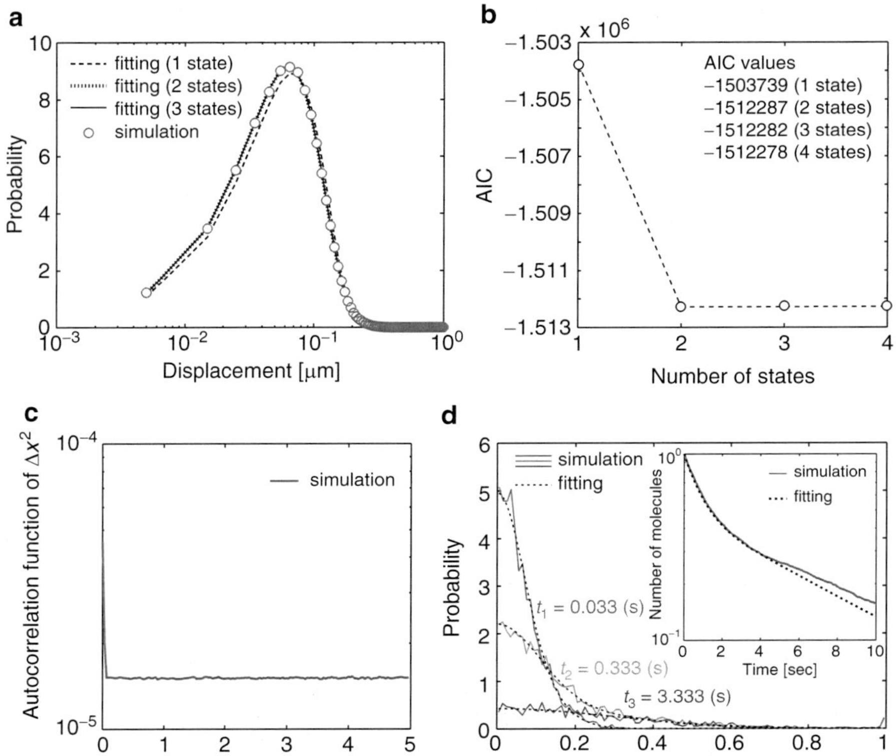

Fig. 12.10 Estimation of parameter values for model 5. (**a**) A histogram of displacements, Δr, measured from simulated trajectories. Numerical simulations were performed assuming 3,000 molecules, each containing at most 1,800 time steps, showing simple diffusion with $D_1 = 0.01$ and $D_2 = 0.10$ and dissociation rate constants of $\lambda_1 = 0.1$ and $\lambda_2 = 1.0$. Molecules that adopt state 1 were set to $p = 0.4$. Displacement during a time interval of $\Delta t = 0.033$ was measured from the trajectories containing an error with a SD of $\varepsilon = 0.04$. The histogram was well fitted by assuming at least two states. (**b**) AIC analysis concluded a two-state model is best. (**c**) The autocorrelation function of squared displacements calculated from the simulated trajectories, which were all longer than 10 s. (**d**) Histograms of position, x, at $t = 0.033$, 0.333 and 3.333. The estimated parameter values were $D_1 = 0.013$, $D_2 = 0.103$, $\lambda_1 = 0.127$, $\lambda_2 = 1.216$ and $p = 0.474$. (*inset*) The release curve obtained from the simulated trajectories. D, μm²/s; λ, 1 s^{-1}; t, s; ε, μm.

3.333 s were fitted to Eq. 12.10 from model 5. A single set of estimated parameter values was obtained as $D_1 = 0.013$ μm^2/s, $D_2 = 0.103$ μm^2/s, $\lambda_1 = 0.127$ s^{-1}, $\lambda_2 = 1.216$ s^{-1} and a D_1 subpopulation ratio of 0.474 (Fig. 12.10d). The estimated parameter set was consistent with the release curve (Fig. 12.10d, inset).

12.4.6 *Model 6*

Simulations were performed using $D_1 = 0.01$ μm^2/s, $D_2 = 0.10$ μm^2/s, $\lambda_1 = 0.1$ s^{-1}, $\lambda_2 = 1.0$ s^{-1}, $k_{12} = 0.4$ s^{-1}, $k_{21} = 0.1$ s^{-1}, and an initial ratio of the slower diffusing subpopulation, $p = 0.4$. According to AIC, the molecule contains two states with different D, $D_1 = 0.01$ μm^2/s, $D_2 = 0.10$ μm^2/s (Fig. 12.11a). The autocorrelation of the squared displacements calculated from 25 trajectories, each longer than 10 s, showed an exponential decay, indicating that the molecule exhibits transitions between the two states (Fig. 12.11b). The results above suggest that model 6 is most appropriate. The number of parameters is large to be estimated at the same time by fitting the distribution. Then, as the diffusion coefficients, those estimated in AIC analysis were exploited. Further, constraints on possible parameter values were obtained from the release curve (Fig. 12.11d, inset) and an autocorrelation of squared displacement calculated from all molecules (Fig. 12.11c). They were fitted to Eqs. 12.12 and 12.13, respectively, giving rise to four constraints: $k_{12} + k_{21} + \lambda_1 + \lambda_2 = -(s_1 + s_2) = 1.4692$, $k_{12}\lambda_2 + k_{21}\lambda_1 + \lambda_1\lambda_2 = s_1 s_2 = 0.4509$, $(k_{12} + k_{21} - \lambda_1 + \lambda_2)p + (k_{12} + k_{21} + \lambda_1 - \lambda_2)(1 - p) = 2C(s_2 - s_1) = 0.2361$, and $D_1{}^2 p + D_2{}^2(1 - p) = D = 0.0061$. From the estimated D_1 and D_2 values, p was calculated to be 0.3967. Under these constraints, the three histograms of position were fitted and a single set of estimated parameter values were obtained: $\lambda_1 = 0.004$ s^{-1}, $\lambda_2 = 1.019$ s^{-1}, $k_{12} = 0.438$ s^{-1}, and $k_{21} = 0.008$ s^{-1} (Fig. 12.11d). The major reactions in this simulation are transitions from state 1 to state 2 and dissociations from state 2. The rate constants for these two reactions, k_{12} and λ_2, can be estimated with high precision.

12.5 Concluding Remarks

The transition kinetics between multiple states with different diffusion coefficient for a signaling molecule provides important clues about the spatiotemporal properties of the signaling process. The method proposed here expands our ability to quantify the kinetics and diffusion of such processes by analyzing the single-molecule trajectory on the membrane. Principally based on a displacement distribution, this method estimates important dynamic parameters such as diffusion coefficients, the composition of molecules in different states, transition rates between the states, and dissociation rates from the membrane. The PDF of the displacement is derived from a diffusion equation that describes molecular

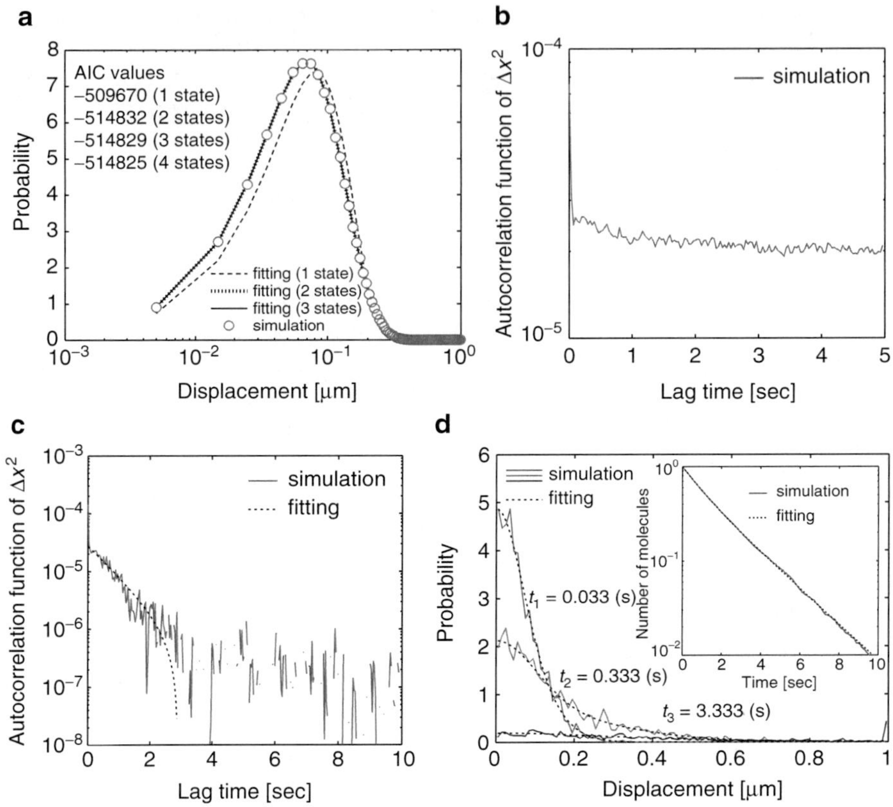

Fig. 12.11 Estimation of parameter values for model 6. (**a**) A histogram of displacements, Δr, measured from simulated trajectories. Numerical simulations were performed assuming 3,000 molecules, each containing at most 1,800 time steps, showing simple diffusion with $D_1 = 0.01$ and $D_2 = 0.10$ and dissociation rate constants of $\lambda_1 = 0.1$ and $\lambda_2 = 1.0$. Molecules that adopt state 1 were set to $p = 0.4$ for an initial condition. State transitions had rate constants of $k_{12} = 0.4$ and $k_{21} = 0.1$. Displacement during a time interval of $\Delta t = 0.033$ was measured from the trajectories containing an error with a SD of $\varepsilon = 0.04$. The histogram was well fitted by assuming at least two states. AIC analysis showed two-state model is best. (**b**) The autocorrelation function of squared displacements calculated from the simulated trajectories, which were all longer than 10 s. (**c**) The autocorrelation function of squared displacements calculated from all simulated trajectories. Due to the membrane dissociation, the curve decays exponentially. (**d**) Histograms of position, x, at $t = 0.033$, 0.333 and 3.333. The estimated parameter values were $D_1 = 0.010$, $D_2 = 0.100$, $\lambda_1 = 0.004$, $\lambda_2 = 1.019$, $k_{12} = 0.438$, $k_{21} = 0.008$ and $p = 0.397$. (*inset*) The release curve obtained from the simulated trajectories. D, μm²/s; λ, 1 s^{-1}; k, 1 s^{-1}; t, s; ε, μm.

diffusion and reactions on a membrane, which enables one to overcome limitations in current MSD analysis techniques. The proposed method can be applied to molecules involved in more complicated reactions as long as they show simple diffusion. On the other hand, in its current form this technique is not capable of handling anomalous diffusion, a phenomenon that probably requires the diffusion equation to consider the effects of directed flow or confinements. Further study is also needed to reflect the revealed multiple states in each trajectory to visualize the relationship between the molecular states and the molecule's location on the membrane. Various applications of this technique are possible by focusing on the displacement distribution and the temporal development, which together provide significant information concerning the spatiotemporal properties of signaling molecules in living cells.

Acknowledgement The author would like to thank Masahiro Ueda and Tatsuo Shibata for helpful discussion, Hiroaki Takagi, Yuichi Togashi, Masatoshi Nishikawa and members of Stochastic biocomputing group in Osaka University for generous suggestions and Peter Karagiannis for critical reading of the manuscript. This work is supported by JST, CREST.

Appendix

When calculated from experimental trajectories, the autocorrelation function of the squared displacements in model 6 contains an error term dependent on the lag time, τ. Let us assume the trajectory consists of an *estimated* molecular position at time t, $x'(t)$, that distributes around the *actual* position, $x(t)$, with fluctuations, $\eta(t)$, such that the variance is ε^2 and $<\eta(t)> = 0$. The displacement along the x-axis during time t and $t + \Delta t$, $\Delta x'(t)$, is written as,

$$\Delta x'(t) = \Delta x(t) + \Delta\eta(t),$$

where

$$\Delta x(t) = x(t + \Delta t) - x(t),$$
$$\Delta\eta(t) = \eta(t + \Delta t) - \eta(t).$$

The variance of $\Delta\eta(t)$ is $2\varepsilon^2$. Then, an autocorrelation function calculated from the trajectories of all molecules theoretically follows,

$$\begin{aligned}\left\langle \Delta x'^2(0)\Delta x'^2(\tau)\right\rangle = {} & \langle \Delta x^2(0)\Delta x^2(\tau)\rangle + \langle \Delta x^2(0)\rangle\langle \Delta\eta^2(\tau)\rangle \\ & + 4\langle \Delta x(0)\Delta x(\tau)\rangle\langle \Delta\eta(0)\Delta\eta(\tau)\rangle + 2\langle \Delta x(0)\rangle\langle \Delta\eta(0)\Delta\eta^2(\tau)\rangle \\ & + \langle \Delta x^2(\tau)\rangle\langle \Delta\eta^2(0)\rangle + 2\langle \Delta x(\tau)\rangle\langle \Delta\eta^2(0)\Delta\eta(\tau)\rangle \\ & + \langle \Delta\eta^2(0)\Delta\eta^2(\tau)\rangle.\end{aligned}$$

When $\tau \neq 0$and $\tau \neq \Delta t$, it is,

$$\left\langle \Delta x'^2(0)\Delta x'^2(\tau)\right\rangle = \langle \Delta x^2(0)\Delta x^2(\tau)\rangle + \langle \Delta x^2(0)\rangle\{\langle \eta^2(\tau+\Delta t)\rangle + \langle \eta^2(\tau)\rangle\} + \langle \Delta x^2(\tau)\rangle\{\langle \eta^2(\Delta t)\rangle + \langle \eta^2(0)\rangle\} + \langle \eta^2(\Delta t)\rangle\langle \eta^2(\tau+\Delta t)\rangle + \langle \eta^2(\Delta t)\rangle\langle \eta^2(\tau)\rangle + \langle \eta^2(0)\rangle\langle \eta^2(\tau+\Delta t)\rangle + \langle \eta^2(0)\rangle\langle \eta^2(\tau)\rangle, \tag{12.14}$$

which is composed of three terms: an autocorrelation of the squared displacements in the absence of the error, $<\Delta x^2(0)\Delta x^2(\tau)>$, an ensemble average of the error, $<\eta^2(t)>$, and an actual value of the ensemble-averaged squared displacements, $<\Delta x^2(t)>$. $<\Delta x^2(0)\Delta x^2(\tau)>$ is explained by Eq. 12.13.

If the molecule does not exhibit membrane dissociation, the ensemble average of the error is equal to ε^2 irrespective of time t. However, in the presence of dissociation, the number of molecules decreases depending on t, leading to a concomitant decrease in the ensemble average. The ensemble average of the error imposed on molecules that dissociate at $t=t_r$ is,

$$\langle \eta^2(t)|t_r\rangle = \varepsilon^2 H(t_r - t) = \begin{cases} \varepsilon^2, & t \leqslant t_r, \\ 0, & t \geqslant t_r, \end{cases}$$

where $H(t)$ is a Heaviside function. Taking the integral for t_r from 0 to plus infinity, the ensemble average is,

$$\begin{aligned}\langle \eta^2(t)\rangle &= \int_0^\infty \langle \eta^2(t)|t_r\rangle P(t_r)dt_r \\ &= \varepsilon^2 \int_0^\infty H(t_r - t)P(t_r)dt_r \\ &= \varepsilon^2 \int_t^\infty P(t_r)dt_r \\ &= \varepsilon^2\left(1 - \int_0^t P(t_r)dt_r\right)\end{aligned}$$

where $P(t_r)$ represents the probability that a molecule dissociates at $t=t_r$. Theoretically, the membrane residence probability, $R(t)$, is equivalent to $1 - \int_0^t P(t_r)dt_r$. Experimentally, the release curve can be used as an estimate of $R(t)$. Thus, the ensemble average of the error can be calculated from the experimental data.

The actual value of the ensemble-averaged squared displacements is calculated from the trajectories as follows. Since the estimated value is,

$$\begin{aligned}\langle \Delta x'^2(t)\rangle &= \left\langle (\Delta x(t) + \Delta\eta(t))^2\right\rangle \\ &= \langle \Delta x^2(t)\rangle + \langle \Delta\eta^2(t)\rangle \\ &= \langle \Delta x^2(t)\rangle + \langle \eta^2(t+\Delta t)\rangle + \langle \eta^2(t)\rangle,\end{aligned}$$

the actual value of the ensemble-averaged squared displacements is,

$$\left\langle \Delta x^2(t) \right\rangle = \left\langle \Delta x'^2(t) \right\rangle - \left\langle \eta^2(t+\Delta t) \right\rangle - \left\langle \eta^2(t) \right\rangle.$$

The ensemble average of the estimated squared displacements is obtained from the time series of squared displacements of i-th molecule,

$$\left\langle \Delta x'^2(t) \right\rangle = \frac{1}{X} \sum_{i=1}^{X} \Delta x'^2_i(t).$$

From Eq. 12.14 and calculating $<\eta^2(t)>$ and $<\Delta x^2(t)>$, the autocorrelation function in the absence of the measurement error, $<\Delta x^2(0)\Delta x^2(\tau)>$, is,

$$\begin{aligned}\left\langle \Delta x^2(0)\Delta x^2(\tau) \right\rangle = & \left\langle \Delta x'^2(0)\Delta x'^2(\tau) \right\rangle \\ & - \varepsilon^2 \left[\left\langle \Delta x'^2(0) \right\rangle \{R'(\tau+\Delta t) + R'(\tau)\} + \left\langle \Delta x'^2(\tau) \right\rangle \{R'(\Delta t) + 1\} \right] \\ & + \varepsilon^4 \{R'(\Delta t) + 1\}\{R'(\tau+\Delta t) + R'(\tau)\},\end{aligned}$$

where $R'(t)$ represent the release curve. The calculated autocorrelation function is fitted to Eq. 12.13 to obtain values of s_1, s_2, D and E.

References

1. Akaike H (1974) A new look at the statistical model identification. IEEE Trans Automat Contr 19:716–723
2. Bannai H, Lévi S, Schweizer C, Dahan M, Triller A (2006) Imaging the lateral diffusion of membrane molecules with quantum dots. Nat Protoc 1:2628–2634
3. Cheezum MK, Walker WF, Guilford WH (2001) Quantitative comparison of algorithms for tracking single fluorescent particles. Biophys J 81(4):2378–2388
4. Gambin Y, Lopez-Esparza R, Reffay M, Sierecki E, Gov NS, Genest M, Hodges RS, Urbach W (2006) Lateral mobility of proteins in liquid membranes revisited. Proc Natl Acad Sci USA 103:2098–2102
5. Jin T, Xu X, Hereld D (2008) Chemotaxis, chemokine receptors and human disease. Cytokine 44:1–8
6. Kusumi A, Sako Y, Yamamoto M (1993) Confined lateral diffusion of membrane receptors as studied by single particle tracking (nanovid microscopy). Effects of calcium-induced differentiation in cultured epithelial cells. Biophys J 65:2021–2040
7. Matsuoka S, Iijima M, Watanabe TM, Kuwayama H, Yanagida T, Devreotes PN, Ueda M (2006) Single-molecule analysis of chemoattractant-stimulated membrane recruitment of a PH-domain-containing protein. J Cell Sci 119:1071–1079
8. Matsuoka S, Shibata T, Ueda M (2009) Statistical analysis of lateral diffusion and multistate kinetics in single-molecule imaging. Biophys J 97(4):1115–1124

9. Miyanaga Y, Matsuoka S, Yanagida T, Ueda M (2007) Stochastic signal inputs for chemotactic response in Dictystelium cells revealed by single molecule imaging techniques. Biosystems 88(3):251–260
10. Mortimer D, Fothergill T, Pujic Z, Richards LJ, Goodhill GJ (2008) Growth cone chemotaxis. Trends Neurosci 31(2):90–98
11. Petrov EP, Schwille P (2008) Translational diffusion in lipid membranes beyond Saffman-Delbruck approximation. Biophys J 94(5):L41–L43
12. Pinaud F, Michalet X, Iyer G, Margeat E, Moore H-P, Weiss S (2009) Dynamic partitioning of a GPI-anchored protein in glycosphingolipid-rich microdomains imaged by single-quantum dot tracking. Traffic 10(6):691–712
13. Qian H, Sheetz MP, Elson EL (1991) Single particle tracking. Analysis of diffusion and flow in two-dimensional systems. Biophys J 60:910–921
14. Saffman PG, Delbrück M (1975) Brownian motion in biological membranes. Proc Natl Acad Sci USA 72:3111–3113
15. Saxton MJ (1997) Single-particle tracking: the distribution of diffusion coefficients. Biophys J 72:1744–1753
16. Saxton MJ, Jacobson K (1997) Single-particle tracking: applications to membrane dynamics. Annu Rev Biophys Biomol Struct 26:373–399
17. Shibata SC, Hibino K, Mashimo T, Yanagida T, Sako Y (2006) Formation of signal transduction complexes during immobile phase of NGFR movements. Biochem Biophys Res Commun 342:316–322
18. Ueda M, Sako Y, Tanaka T, Devreotes P, Yanagida T (2001) Single-molecule analysis of chemotactic signaling in Dictyostelium cells. Science 294(5543):864–867

Chapter 13
Noisy Signal Transduction in Cellular Systems

Tatsuo Shibata

Abstract Stochastic fluctuations of chemical reactions are particularly prominent in small systems such as cells. Such fluctuations of cellular signal processes have been observed directly by single-molecule imaging. Recent theoretical studies have also revealed that stochastic cellular reactions generate fluctuations in the number of a molecule, which can be related to their functioning such as amplification of signals. Here, we study how large each signal reaction generates and amplifies the stochastic fluctuations. The general framework of fluctuation response relation in non-equilibrium physics throws light upon this problem. The result has been applied to the chemotactic signal processing of eukaryotic cell, such as *Dictyostelium* cell, revealing that the accuracy of chemotaxis is determined by the signal to noise ratio in the signal of the reaction between G-protein and G-protein coupled receptor.

Keywords Fluctuations · Noise · Gain · Single molecule imaging · Gain fluctuation relation · Intrinsic noise · Extrinsic noise · Master equation · Fokker–Planck equation · Langevin equation · Gaussian white noise · Linear noise approximation · Power spectrum · Poisson process · Markov process · cAMP · cAR1 · Dictyostelium push–pull reaction · Michaelis–Menten · Fokker–Planck operator · Response-fluctuation relation · Linear response · Chemotaxis · PTEN · PtdIns(3,4,5)P3 · G-protein · Signal to noise ratio (SNR) · MAPK · Cascade

---

T. Shibata (✉)
RIKEN Center for developmental biology, 2-2-3 Minatojima-minamimachi, chuo-ku, Kobe 650-0047, Japan
and
Japan Science and Technology Agency, CREST, 1-3 Yamadaoka, Suita, Osaka 565-0871, Japan
e-mail: tatsuoshibata@cdb-riken.jp

Y. Sako and M. Ueda (eds.), *Cell Signaling Reactions: Single-Molecular Kinetic Analysis*, DOI 10.1007/978-90-481-9864-1_13,

13.1 Introduction

Cellular systems consist of a variety of processes of molecular machines, which are typically proteins, whose stochastic dynamics are essential for their functioning. Since the cell is such a small system, stochastic chemical reactions inside the cell result in its stochastic fluctuating behaviors. The recent development in experimental methods of quantitative measurement of cellular processes and the accumulation of knowledge of molecular biology enable us to study the stochastic cellular behaviors and the variability of properties between cells in connection with the molecular processes inside cells. For instance, motile cells often exhibit motions in random directions without external cues. To produce a motion, a cell has to produce a motile activity in a particular direction, by breaking uniformity spontaneously. Stochastic chemical reactions may play roles to produce random motile activities. Random motilities have been observed in many types of cellular systems, in both prokaryotic and eukaryotic cells. Berg and his colleague have been shown that bacterial motion without chemoattractant gradient is a random motion [2]. The flagella motors, which produce the cellular motility, are responsible for the random motility. The switching in the direction of their rotational motions takes place in a stochastic way. More recently, stochastic activations of the chemotactic signaling system have been shown to be amplified and contributing to produce a variability of the bacterial random motility [15]. In the case of paramecia, Oosawa discussed that the intracellular noise is hierarchically organized from thermal fluctuations to spike-like large fluctuations, which produce spontaneous signals to change the behavior of the swimming cell [19, 20]. In the molecular scale, stochastic behaviors have been shown experimentally by single-molecule imaging [25, 30].

In this chapter, the stochastic properties of signal transduction reactions are studied. Here, the stochastic fluctuations in the number of molecules is referred as "molecular noise". In Section 13.2, the stochastic process of a chemical reaction is introduced. Mathematically, the probability distribution of the number of molecules can be described by a chemital master equation. When the molecular noise is relatively small, a stochastic differential equation can describe the process, which is called the chemical Langevin equation. As the increase of the probability having a reaction event in an interval, the molecular noise can be negligible. In such a case, the reaction process can be described by a kinetic equation without effects of noise. In Section 13.3, some parameters to describe reaction processes can be obtained by the single molecule imaging. A single molecule imaging data of cAMP receptor cAR1 of chemotactic cell *Dictyostelium* is analyzed to obtained several kinetic parameters.

One essential feature of biological signal transduction systems is to amplify small changes in input signals [16]. However, the molecular noise in signal transduction has been discussed, suggesting that a large amplification results in a generation of strong stochastic fluctuations in the output signal [3, 5]. It has been shown theoretically that how the abrupt response of ultrasensitive signal transduction reactions results in both generation of large inherent noise and high amplification of input noise [27]. Section 13.4 is devoted to the gain-fluctuation relation, which connect the noise generated by a reaction (intrinsic noise) to the gain which is the signal amplification rate. The relation is derived mathematically in Section 13.5, based on the more general response-fluctuation relation. The molecular noise generated at a reaction affect another reactions in the signal transduction network. High gain reactions, which amplify a small change in the input signal, also amplify noise in the input. The relation between the gain and the amplified noise, which is called extrinsic noise, is shown in Section 13.6. The extrinsic noise is distinguished from the noise inherent in its own reaction (intrinsic noise). Depending on the intensity of gain, there are two cases: intrinsic noise is dominant and extrinsic noise is dominant. As an example, the signal transduction reaction of chemotactic eukaryotic cell is studied in Section 13.7. The accuracy of chemotaxis is shown to be well explained by the signal to noise ratio at the signal of G-protein and G-protein coupled receptor cAR1. In Section 13.8, propagation of molecular noise in a signal transduction cascade is studied.

13.2 The Hierarchy of Molecular Noise

Consider a cell containing a mixture of huge kinds of molecules. The state of molecule can be described by discrete states, such as active state and inactive one. We suppose positions of individual molecules can be ignored. Thus, the cell is considered to be well stirred. In such a case, the state of the cell is described by the numbers of these molecules. When the information about a particular reaction event disappear fast enough due to the huge number of events, such as collision among molecules, the two successive reactions of a given type can be considered statistically independent. This means that the probability of occurring two successive reactions at particular times t_1 and t_2 is given just by the product of the two probabilities of occurring the reaction at time t_1 and occurring the reaction at time t_2. The time interval of successive reactions obeys an exponential distribution. In such a process, the evolution in the number of a molecule can be described by Poisson processes.

Let us consider a reaction taking place between molecules Y and S producing X; $\mathsf{Y} + \mathsf{S} \rightarrow \mathsf{X}$. For instance, in the case of reaction between receptor and its ligand, this reaction is the association reaction between receptor Y and ligand S, producing receptor–ligand complex X. When a cell with volume V can be considered as a well-stirred system, the probability that a reaction taking place is proportional to the numbers of molecules Y and S. Then, the probability in time interval dt is given by

$$k_a \frac{N_{\mathsf{Y}}}{V} \frac{N_{\mathsf{S}}}{V} V dt \tag{13.1}$$

where k_a is the rate constant of the reaction, and N_{Y} and N_{S} are the numbers of molecules Y and S, respectively, at a given time [14]. Hereafter, we suppose that the number of S is constant, or we consider N_S/V can be replaced by an average number s. Then, the probability is given by $k_a N_Y s dt$. For a particular molecule of Y, the probability of having a reaction in a time interval dt within the cell is $k_a\, sdt$. Thus, the probability $q_1(t)$ that one particular molecule of Y to survive without reaction at t is described by the equation,

$$\frac{dq_1(t)}{dt} = -k_a s q_1(t), \tag{13.2}$$

which can be solved as

$$q_1(t) = \exp(-k_a st) \tag{13.3}$$

For the ensemble containing N molecules of Y, when there is no correlation and statistically independent between molecules, the process is simply a combination of mutually independent reaction processes of individual molecule Y. The reaction probability within a unit time conditional upon existing n molecules is given by $k_a s \times n$, where k_a and s are the reaction rate constant and concentration of signal S, respectively. Thus, with the probability of having n molecules at t, $q(n,t)$, the reaction probability rate at time t is given by $k_a\, s\, n\, q(n,t)$. Therefore, the probability for n molecules of Y to survive at t is described by

$$\frac{dq(n,t)}{dt} = k_a\, s\,(n+1)q(n+1,t) - k_a\, s\, n\, q(n,t) \tag{13.4}$$

where the first term is the probability current that the number is reduced from $n+1$ to n, i.e., the probability rate at time t that the number of molecule is $n+1$ and a reaction takes place so that the number is reduced to be n, and the second term is from n to $n-1$. With $q(N,0) = 1$, meaning that all molecules occupy the state Y, the equation is solved by

$$q(n,t) = \binom{N}{n} \exp(-k_a st)^n (1 - \exp(-k_a st))^{N-n} \tag{13.5}$$

$$= \binom{N}{n} q_1(t)^n (1 - q_1(t))^{N-n} \tag{13.6}$$

Therefore, the probability is described by a binominal distribution with parameters N and $q_1(t)$ given by Eq. 13.3.

We also consider the inverse reaction, X $\rightarrow$ Y + S, with the reaction rate k_d. In the case of receptor and its ligand, this reaction is the dissociation reaction. Then, the probability $p_1(t)$ of one molecule of X to survive without reaction at a given time t follows the equation that is similar to Eq. 13.2, and the probability is given by

$$p_1(t) = \exp(-k_d t). \tag{13.7}$$

This exponential distribution of time interval until the next reaction has been observed typically by single molecule imaging in vitro and in vivo [8, 25]. For instance, in the case of receptor, the distribution of association duration is typically follows an exponential distribution. Thus, the parameter k_d can be obtained by the single molecule imaging (see below).

By both forward and inverse reactions, the molecule switches repeatedly between Y and X. When the distribution of time interval from the state X to Y is independent of the previous history of transition times between X and Y, i.e., each single molecule has no memory about the previous histories, the state of the molecule can be described by a Markov process. Then, the probability $P_1(t)$ that the molecule is in the state X can be described by the equation with the probability $Q_1 = 1 - P_1$ of molecule being in the state Y

$$\frac{dP_1(t)}{dt} = -k_d P_1(t) + k_a s Q_1(t), \tag{13.8}$$

$$= k_a s - \left(k_a s + k_d\right) P_1(t) \tag{13.9}$$

When the molecule is in the state Y at time $t = 0$, i.e., $P_1(0) = 0$ the probability $P_1(t)$ is obtained by solving this equation as

$$P_1(t) = \frac{s}{K_d + s}\left(1 - e^{-(k_a s + k_d)t}\right) \tag{13.10}$$

where K_d is the dissociation constant given by $K_d = k_d / k_a$. For the ensemble of N molecules, when the molecules are mutually independent with each other, the time series of number n of molecule X can also be described by a Markov process. The probability $P(n, t)$ of having n molecules of X at t is described by the equation,

$$\begin{aligned}\frac{dP(n,t)}{dt} &= k_a s\Big((N-n+1)P(N-n+1,t) - (N-n)P(N-n,t)\Big)\\ &+ k_d\Big((n+1)P(n+1,t) - nP(n,t)\Big)\end{aligned} \tag{13.11}$$

where the first term describe the forward reaction, while the last term corresponds to the inverse reaction. In the process described by this equation, the time interval between succeeding two evens follows an exponential distribution. Thus, the process can be numerically produced by generating a series of random numbers that satisfy the exponential time interval distribution. This idea gives the Gillespie's algorithm a numerical scheme to simulate time evolutions of stochastic chemical reactions [9]. For stationary state where $\frac{dP(n,t)}{dt} = 0$, Eq. 13.11 gives the solution given by,

$$P(n,t) = \binom{N}{n} P_1(t)^n (1 - P_1(t))^{N-n} \tag{13.12}$$

Therefore, the number distribution follows a Binomial distribution. The average number of X and its variance, $\overline{N_X}$ and $\overline{(N_X - \overline{N_X})^2}$ are respectively given by $\overline{N_X} = NP_1(t)$ and $\overline{(N_X - \overline{N_X})^2} = NP_1(t)(1 - P_1(t))$. At the steady state after sufficiently long time t, the average number of X and its variance, are respectively approaches to

$$\overline{N_X} = N\frac{s}{K_d + s}, \quad \overline{(N_X - \overline{N_X})^2} = N\frac{K_d s}{(K_d + s)^2}, \tag{13.13}$$

So far, the number of molecule is a discrete integer variable. When the number of molecules are relatively large, the number of molecule is well described by continuous variables. Even in such a case, if stochastic fluctuations can still not be ignored, then we employ stochastic differential equation or the chemical Langevin equation for the temporal change in the number of molecules. Let X and Y be the number of molecules X and Y, respectively. Then, the time evolution of X and Y can be described by the chemical Langevin equation,

$$\frac{dX}{dt} = k_a sY - k_d X + \sqrt{k_a sY + k_d X}\xi(t) \tag{13.14}$$

where the first two terms are deterministic parts of reactions, and the last term describe the stochastic aspect of reactions. Here, $\xi(t)$ is the Gaussian white noise term associated with the stochastic reaction with $\langle\xi(t)\rangle = 0$ and $\langle\xi(t)\xi(t')\rangle = \delta(t - t')$. $\delta(t)$ is the Dirac's delta function. Letting N be the total number of molecules, i.e., $N = X + Y$, the evolution equation of X can be rewritten as

$$\frac{dX}{dt} = k_a sN - (k_a s + k_d)X + \sqrt{k_a sN - k_a sX + k_d X}\xi(t) \tag{13.15}$$

In order to study the fluctuation in the number of bounded receptor X, we consider the temporal evolution of small deviation x from the average number $\bar{X}$

$$\bar{X} = N\frac{s}{K_d + s}, \tag{13.16}$$

with $K_d = k_d / k_a$. The temporal evolution of x can be approximated by the following linearized Langevin equation, given by,

$$\frac{dx}{dt} = -\Gamma x + \sigma\xi(t) \tag{13.17}$$

with

$$\Gamma = k_a s + k_d, \quad \text{and} \quad \sigma^2 = \frac{2Nk_d s}{K_d + s}. \tag{13.18}$$

In this approximation, the noise intensity σ^2 is not time-dependent but is a constant, which is called the linear noise approximation.

The power spectrum density characterizes the frequency content of a stochastic process. The power spectrum density $I(f)$ of $x(t)$ at frequency f is defined by $\overline{\hat{x}(f)\hat{x}^*(f')} = I(f)\delta(f - f')$, where $\hat{x}(f)$ is the Fourier transform of $x(t)$, $\hat{x}(f) = \int_{-\infty}^{\infty} x(t)e^{-2\pi i f t}dt$, and $\hat{x}^*$ is its complex conjugate. The power spectrum density is obtained by solving Eq. 13.17 as

$$I(f) = \frac{\sigma^2}{(2\pi f)^2 + \Gamma^2} \tag{13.19}$$

The noise intensity, which is the variance of the distribution of X, is given by frequency integral of $I(f)$,

$$\overline{(X - \bar{X})^2} = \overline{x^2} = \int_{-\infty}^{\infty} I(f)df = \frac{\sigma^2}{2\Gamma} = N\frac{K_d s}{(K_d + s)^2} \tag{13.20}$$

For the system described by a Langevin equation, we consider the probability density of x, $P(x, t)$, at time t. The evolution equation of $P(x, t)$ is described by the Fokker–Planck equation. Corresponding to the linearized Langevin equation, Eq. 13.17, the Fokker–Planck equation is given by

$$\frac{dP(x,t)}{dt} = \frac{\partial}{\partial x}\left(\Gamma x + \frac{\sigma^2}{2}\frac{\partial}{\partial x}\right)P(x,t) \tag{13.21}$$

The stationary solution of this equation is given by the Gaussian distribution

$$P(x) = \frac{1}{\sqrt{\pi\sigma^2/\Gamma}} e^{-\frac{x^2}{\sigma^2/\Gamma}}. \tag{13.22}$$

In a Poisson process, when the probability of having a reaction within a second is r, the statistical variation of the number of reaction taking place within the same time interval, measured by the standard deviation, is $\sqrt{r}$. Therefore, the relative strength of the statistical variation compared with the mean number of reaction within a second is given by $1/\sqrt{r}$, which decreases as the reaction probability increases. If the probability is large enough, one can neglect the statistical variation in the number of reactions occurring per a second. Then, the evolution of the number is again described by a deterministic dynamical system, such as kinetic equations described by ordinary differential equations. Such descriptions are expected when the number of molecules of a given type is large.

13.3 An Example: cAMP Receptor cAR1 in *Dictyostelium* Cells

Chemotaxis of eukaryotic cells is a typical example of cellular information processing. Chemotaxis plays important roles in diverse functions, such as finding nutrients, forming multicellular structures in protozoa, and tracking bacterial infections in neutrophils. In the case of chemotactic eukaryotic cell *Dictyostelium*, the chemoattractant molecule is cyclic adenosine $3^{'}$, $5^{'}$-monophosphate (cAMP). Shallow chemoattractant gradient of 2–5% in cell length is enough to induce chemotaxis. This indicates that these cells can compare and process extremely small differences in concentrations of the extracellular stimuli . The cell size of this chemotactic cell is typically 10–20 μm, which contains about 80,000 cAMP receptor cAR1 on membrane. The dissociation constant of receptor is $K_d \approx 100$ nM. Base on these numbers, the number of cAMP molecule bound on the membrane surface can be calculated, which is about 16,000 at 25 nM where the cell exhibits chemotaxis the most efficiently [7]. At this concentration, when the chemoattractant gradient is 2% in 10 μm, the difference in the number of bound cAMP between anterior and posterior halves of cell is calculated at about 60, which is the signal that the cell has to detect. At 1nM, about 800 cAMPs are bound on the surface, and about 5 is the difference between two regions under the gradient of 2%. This small system can show chemotaxis for almost six order of magnitude ranging from about 10 pM to 10 μM. The mechanism of how the cells can detect such small signal has not been clarified for far.

By using the single molecule imaging for this single chemotactic cell, binding and unbinding processes between cAR1 and cAMP can be monitored [30]. The binding and unbinding events are stochastic processes, which is typically described by Poisson processes. If the events are statistically independent in time, the binding duration is distributed exponentially. In fact, the binding duration between cAR1 and cAMP has been measured under the condition that fluorescence labeled cAMP, Cy3-cAMP, was added uniformly to *Dictyostelium* cells at 10 nM. In Fig. 13.1a, the probability distribution of bound Cy3-cAMP to survive at time t without unbinding from membrane is plotted, which can be described by a mixture of two exponential distributions

$$P(t) = p_1 e^{-k_1 t} + p_2 e^{-k_2 t} \tag{13.23}$$

with $p_1 + p_2 = 1$, as shown in Fig. 13.1a. The parameter of exponential distribution gives the dissociation rate constants k_d. This mixture distribution can be explained simply by assuming that each receptor is in one of two binding states.

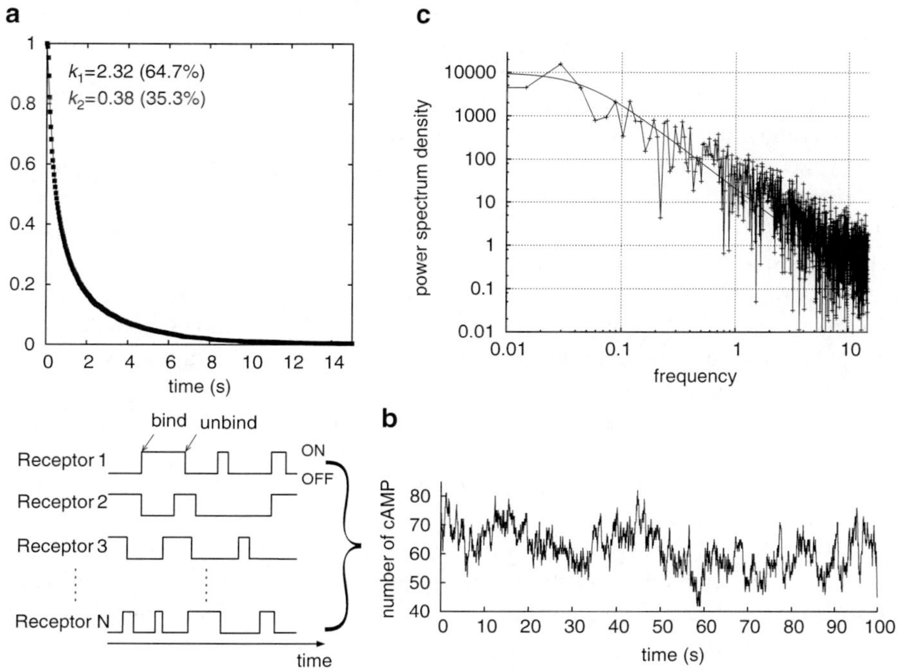

Fig. 13.1 Single molecule to molecular noise. (**a**) Cumulative frequency histogram of lifetime of Cy3-cAMP spots bound on the cell surface. A fluorescent-labeled cAMP (Cy3-cAMP) was added uniformly to a *Dictyostelium* cell at 10 nM. The basal surface of the cells was observed by using total internal reflection fluorescence microscopy (see [31] Courtesy of Prof. Ueda). (**b**) Time series of the number of bound Cy3-cAMP. (**c**) Power spectrum density obtained from the time series.

With the dissociation constant K_d and the ligand concentration, the association rate k_a can also be obtained. In addition to the statistically independent of binding and unbinding events in time, the events between receptors are also expected to be independent (Fig. 13.1b). As a result, the number of ligand on the membrane fluctuates in time. In Fig. 13.1b, the time series of the number of ligand on the membrane is plotted, which in fact exhibits stochastic fluctuations in time. Notice that this fluctuation is not a kind of measurement error, but intrinsic in a reaction of this scale. By performing the discrete Fourier transform for the time series, the power spectrum density can be obtained. Applying Eq. 13.19 with Eq. 13.18 to the spectrum, parameters such as the dissociation constant K_d and the association rate constant k_a can be obtained (Fig. 13.1c).

13.4 The Gain-Fluctuation Relation

Many cellular processes respond quickly to internal and external variations by using chemical reaction networks. For instance, chemotactic amoeba cells such as *Dictyostelium discoideum* responds to move up a shallow chemoattractant gradient within a minutes. This time scale is much faster than the time scale of gene expression responses, which is typically slower than a tens of minutes. Signal transduction networks are responsible for these quick responses, which are typically consists of interactions between proteins.

As has shown in previous section, the chemotactic cells *Dictyostelium* can move up shallow chemoattractant gradients of 2–5% in the cell length. It indicates that they can compare and process extremely small differences in the concentrations of extracellular stimuli. One strategy to respond to such small signals is to adopt switch-like reactions, which can generate abrupt responses from a small change in the input stimuli. There are several reactions that sharpen the responses, such as cooperative reactions in single proteins [16–18], and push–pull antagonistic reactions [11, 29]. In these reactions, the response is switch-like with a threshold in the concentration of stimuli. In a cascade of such switch-like reactions, as observed in mitogen activated protein kinase cascade, the amplification of the whole cascade can be much larger [12]. Is it more appropriate for the cellular systems to have reactions that exhibit steep responses in order to generate all-or-none cellular behaviors?

As we have seen in the previous sections, cellular processes are inherently noisy. Such noisy characteristic of reactions may affect the behaviors of switch-like signal transduction reactions. Here we study how the noise generated by a reaction relates to the amplification of signal [27].

As examples, we study two types of reactions which is always found in signal transduction cascades: one is a simple binding-unbinding reaction, and the other one is a reaction in which a messenger is activated and deactivated cyclically by a pair of opposing enzyme, as observed in a combination of kinase and phosphatase reactions.

The binding-unbinding reaction is the simplest signal transduction reaction that behaves as a molecular switch:

$$\mathsf{Y} + \mathsf{S} \underset{k_d}{\overset{k_a}{\rightleftharpoons}} \mathsf{X} \tag{13.24}$$

where S is the signaling molecule that binds to the inactive state Y so that the protein is switched on to the active state X (Fig. 13.2a). This best-known reaction gives rise to a hyperbolic curve, which is the same as the Michaelis–Menten kinetics (Fig. 13.3a). One typical example is a reaction between a receptor and a ligand, in which Y, S and X are the receptor, ligand, and ligand–receptor complex, respectively. For the chemotaxis of *Dictyostelium* cells, the chemoattractant ligand is cAMP, which binds to the cAMP receptor cAR1.

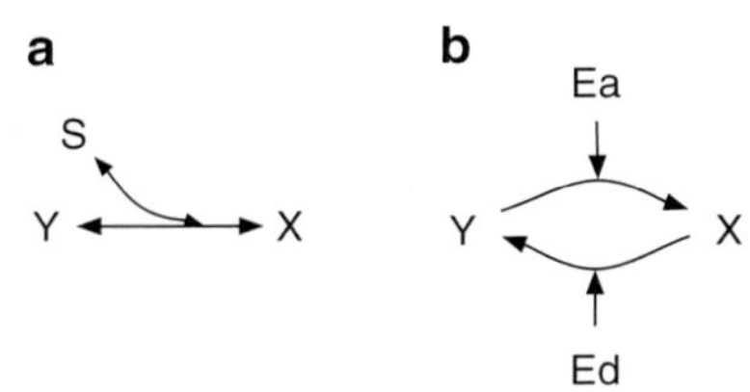

Fig. 13.2 Signal transduction reactions: (**a**) the Michaelis–Menten type reaction. (**b**) The push–pull antagonistic reaction.

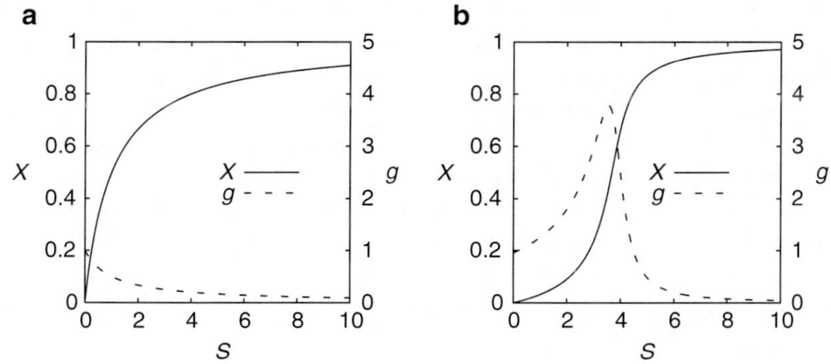

Fig. 13.3 Ultrasensitive responses in signal transduction reactions. The fractional concentration of the output signal X (*left axis*) and the gain g (*right axis*) are plotted as functions of the concentration of signal molecule S. (**a**) The Michaelis–Menten type reaction, (**b**) the push–pull antagonistic reaction.

The push–pull antagonistic reaction is the simplest example of cyclic modification reactions that can show a sharp response [11, 29]. In the push–pull reaction, the signaling molecule, that is an enzyme, switches its substrate protein from an inactive state to an active state, whereas the other enzyme switches the protein off (Fig. 13.2b). Each step is characterized by Michaelis–Menten kinetics:

$$\begin{aligned} \mathsf{Y} + \mathsf{E_a} &\leftrightarrow \mathsf{YE_a} \rightarrow \mathsf{X} + \mathsf{E_a} \\ \mathsf{X} + \mathsf{E_d} &\leftrightarrow \mathsf{XE_d} \rightarrow \mathsf{Y} + \mathsf{E_d} \end{aligned} \tag{13.25}$$

where $\mathsf{E_a}$ is the signaling enzyme which switches inactive state Y to active state X, and $\mathsf{E_d}$ switches X off. Thus, the input signal is the concentration of $\mathsf{E_a}$. If each Michaelis–Menten reaction works near saturation, sharp response is obtained (Fig. 13.3b).

In order to characterize the amplification of small changes in signal, we introduce the gain, which is defined as the ratio between the fractional change in the output signal X and the fractional change in the input signal S,

$$g = \frac{\Delta X/\bar{X}}{\Delta S/S}. \tag{13.26}$$

In this chapter, we consider that there is only a small change in S. Then, the gain is rewritten as $g = \frac{d \log X}{d \log S}$. For the Michaelis–Menten-type reaction and the push–pull reaction, the gain is plotted in Fig. 13.3. Ultrasensitivity is defined as the response of a system that is more sensitive to change in S than is the normal hyperbolic response in Michaelis–Menten kinetics in which the maximum gain g is unity. Thus, the maximum gain g of an ultrasensitive system is larger than unity.

In Fig. 13.4, the gain g is plotted as a function of the variance of X divided by the average number of X. The gain g is calculated by changing the intensity of the signal S, to $S + \Delta S$ and measuring the response ΔX in the output signal X from its stationary value $\bar{X}$. The gain varies by changing the level of S. The noise intensity is calculated under the stationary condition without changing the signal intensity. The variance and its average number are obtained by performing stochastic simulations of scheme (13.25) according to the Gillespie's numerical algorithm [9]. Since the noise is not applied externally, but is intrinsically produced, this noise is called *intrinsic noise*. The numerical result clearly indicates that the gain is linearly proportional to the intrinsic noise intensity, i.e., $g \propto \sigma_{\text{in}}^2/\bar{X}$, where σ_{in}^2 is the variance of fluctuation in the concentration of X, and $\bar{X}$ is the average concentration. This proportional relation is also obtained by analytical calculation and it is written as

$$g = \Theta \frac{\sigma_{\text{in}}^2}{\bar{X}} \tag{13.27}$$

where Θ is a particular constant depending on the reaction [27]. Note that Θ is not a dimensionless parameter but it has a dimension of volume, which depends on the measurement of X. For the derivation, see Ref. [27]. Not only the push–pull reaction, but many types of reactions follows this gain-intrinsic noise relation Eq. 13.27. For instance, the Michaelis–Menten type reactions, cooperative reactions such as allosteric enzyme models [1, 4, 17, 18] and gene expressions. This relation tells us that a high response that is characterized by high gain results in large intrinsic noise, while large intrinsic noise implies high gain. The gain and the intrinsic noise cannot be controlled independently.

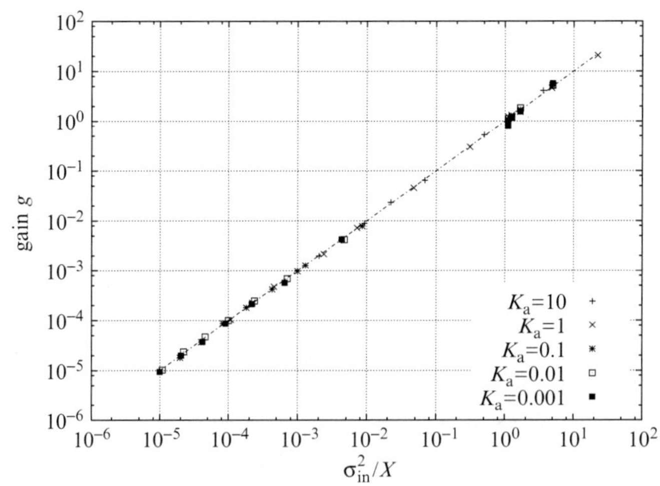

Fig. 13.4 The gain-intrinsic noise relation of signal transduction systems. The gain g is plotted as a function of the variance of X divided by the average number of X. The Push–pull reaction: The Michaelis constant is K_a for the activation reaction as is indicated in the graph, and $K_d = 1$ for the deactivation reaction. The maximum velocities of activation and deactivation reactions are given by $V_a = V_d = 100$, respectively.

13.5 The Response-Fluctuation Relation

Here, we derive the gain-intrinsic noise relation Eq. 13.27 for the case of the push–pull reaction (13.25) based on the response-fluctuation relation. Let X and Y be the number of X and Y, respectively, and $N = X + Y$ be the total number. Then, the chemical Langevin equation [10] for the push–pull reaction is given by

$$\frac{dX}{dt} = \Gamma_{\mathrm{a}}(Y)Y - \Gamma_{\mathrm{d}}(X)X + \sigma_{\xi}\xi(t) \tag{13.28}$$

where $\Gamma_{\mathrm{a}}(Y)$ and $\Gamma_{\mathrm{d}}(X)$ are the activation and deactivation reaction rates respectively, which depend on the numbers X and Y. In the case of the push–pull reaction [5], the reaction rates are given by

$$\Gamma_{\mathrm{a}}(Y) = k_a S \frac{K_a}{K_a + Y},$$

and

$$\Gamma_{\mathrm{d}}(X) = k_d E_d \frac{K_d}{K_d + X},$$

where S is the concentration of the input molecule E_{a}, K_{a} and K_{d} are the Michaelis constants of each enzymatic reaction, k_{a} and k_{d} are the reaction rate constants. The last term is the white Gaussian noise with $\langle \xi(t) \rangle = 0$ and $\langle \xi(t)\xi(t') \rangle = \delta(t - t')$. Since chemical reaction events take place in time as a Poisson process, the noise intensity σ_{ξ}^2 is given by $\sigma_{\xi}^2 = \Gamma_{\mathrm{a}}(Y)Y + \Gamma_{\mathrm{d}}(X)X$ [10]. The stationary solution X_s and Y_s is given by solving the equation

$$\Gamma_{\mathrm{a}}(Y_s)Y_s = \Gamma_{\mathrm{d}}(X_s)X_s \tag{13.29}$$

We consider the linear response of X to a change in S. For the linear response, we study a small deviation x from the stationary solution X_s with the linear noise approximation, in which the noise intensity σ_{ξ}^2 is given by

$$\sigma_{\xi}^2 = \Gamma_a(Y_s)Y_s + \Gamma_d(X_s)X_s = \frac{2\Gamma_a\Gamma_d N}{\Gamma_a + \Gamma_d} \tag{13.30}$$

Thus, the linearized Langevin equation for the small deviations x and s from X_s and S is given by Eq. 13.28,

$$\frac{dx}{dt} = \gamma s - \Gamma x + \sigma_{\xi}\xi(t) \tag{13.31}$$

with the regression coefficient

$$\Gamma = \Gamma_a \frac{K_a}{K_a + Y_s} + \Gamma_d \frac{K_d}{K_d + X_s} \tag{13.32}$$

and the response coefficient

$$\gamma = \frac{\partial \Gamma_a}{\partial S} Y_s = \frac{\Gamma_a \Gamma_d N}{(\Gamma_a + \Gamma_d) S}. \tag{13.33}$$

For the system described by the linearized Langevin equation Eq. 13.31, we consider the probability density of x, $P(x,t)$, at time t. The evolution equation of $P(x,t)$ is described by the Fokker–Planck equation,

$$\frac{dP(x,t)}{dt} = L_{FP}(x,t)P(x,t) \tag{13.34}$$

where $L_{FP}(x,t)$ is the Fokker–Planck operator which can depend on time. The Fokker–Planck operator can be written in the form

$$L_{FP}(x,t) = L_{FP}(x) + L_{ext}(x,t) \tag{13.35}$$

The first term is the time-independent Fokker–Planck operator $L_{\mathrm{FP}}(x)$ that has the stationary solution $P_s(x)$ satisfying

$$L_{FP}(x)P_s(x) = \frac{\partial}{\partial x}\left(\Gamma x + \frac{\sigma_\xi^2}{2}\frac{\partial}{\partial x}\right)P_s(x) = 0. \tag{13.36}$$

Thus, the stationary solution $P_s(x)$ is given by the Gaussian distribution

$$P_s(x) = \frac{1}{\sqrt{\pi\sigma_\xi^2/\Gamma}} e^{-\frac{x^2}{\sigma_\xi^2/\Gamma}}. \tag{13.37}$$

The time-dependent Fokker–Planck operator $L_{\mathrm{ext}}(x,t)$ is the effect of a change in the concentration of S. The time-dependent solution $P(x,t)$ of Fokker–Planck equation Eq. 13.34 is split into the stationary solution $P_s(x)$ and deviation from it, $p(x,t)$, i.e., $P(x,t) = P_s(x) + p(x,t)$. Thus, the Fokker–Planck equation is rewritten as

$$\frac{dp(x,t)}{dt} = (L_{FP}(x) + L_{ext}(x,t))(P_s(x) + p(x,t)) \tag{13.38}$$

If a change in s is small, we may neglect the term $L_{\mathrm{ext}}(x,t)p(x,t)$ and retain only the linear terms. Then, the Fokker–Planck equation is given as

$$\frac{dp(x,t)}{dt} = L_{FP}(x)p(x,t) + L_{ext}(x,t)P_s(x). \tag{13.39}$$

A formal solution of this equation is given by

$$p(x,t) = \int_{-\infty}^{t} e^{L_{FP}(x)(t-t')} L_{ext}(x,t') P_s(x) dt' \tag{13.40}$$

For a step increase of s at $t = 0$, the average response $\overline{x(t)}$ is thus given by

$$\overline{x(t)} = \int xp(x,t)dx = \int_0^t R_x(t-t')dt' \tag{13.41}$$

where $R_x(t)$ is the response function given by

$$R_x(t) = \int x e^{L_{FP}t} L_{ext}(x,t) P_s(x) dx. \tag{13.42}$$

We consider temporal correlation function of x, $C_x(t) = \overline{x(t)x(0)}$ for $t \geqq 0$, given by

$$C_x(t) = \int\int x\, x' P(x,t;x',0) dx dx' = \int x\, e^{L_{FP}t} x\, P_s(x) dx \tag{13.43}$$

where $P(x, t; x', 0)$ is the joint probability of having x at t and x' at $t = 0$. By comparing Eqs.(13.42) and (13.43), when $\alpha x\, P_s(x) = L_{\text{ext}}(t) P_s(x)$ with a constant α, the response function $R_x(t)$ is proportional to the temporal correlation function as the response-fluctuation relation $R_x(t) = \alpha C_x(t)$. For a change in s, the time-dependent Fokker–Planck operator is given by

$$L_{ext}(x,t) = -\gamma s \frac{\partial}{\partial x}. \tag{13.44}$$

Thus, with Eqs. 13.30, 13.32 and 13.37, we have

$$L_{ext}(x,t) P_s(x) = \Gamma \frac{s}{S} x P_s(x), \tag{13.45}$$

which leads to *the response-fluctuation relation*

$$R_x(t) = \Gamma \frac{s}{S} C_x(t). \tag{13.46}$$

From Eq. 13.43, the time derivative of the temporal correlation function $C_x(t)$ is given by

$$\frac{dC_x(t)}{dt} = \int x\, e^{L_{FP}t} L_{FP} x\, P_s(x) dx = -\Gamma \int x\, e^{L_{FP}t} x\, P_s(x) dx \tag{13.47}$$

Substituting Eq. 13.45 into this, it leads to

$$R_x(t) = -\frac{s}{S}\frac{dC_x(t)}{dt} \tag{13.48}$$

Consider the time dependent gain defined by

$$g(t) = \frac{\overline{x(t)}/X_s}{s/S} = \frac{S}{sX_s}\int_0^t R_x(t-t')dt' \tag{13.49}$$

Substituting Eq. 13.48 into this, and noting that $C_x(0)$ is the variance of X, i.e., $C_x(0) = \overline{x^2}$, the time dependent gain is written as

$$g(t) = \frac{\overline{x^2} - C_x(t)}{X_s}, \tag{13.50}$$

For sufficiently long time ($t \rightarrow \infty$), the temporal correlation function $C_x(t)$ becomes vanish and the gain-intrinsic noise relation obtained as

$$g = \frac{\overline{x^2}}{X_s}. \tag{13.51}$$

Note that X and x are the dimensionless number. If these are concentrations, then a coefficient is necessary as shown in Eq. 13.27.

13.6 Propagation of Noise in Reaction Networks

No reaction works alone in cells. Many types of reactions interact with one another to form cascades and networks. If a signal, which is generated by some reaction, is noisy, the reaction regulated by the signal can be affected by the noise in the signal. Therefore, there are two noise sources; one is the noise inherent in its own reaction (intrinsic noise), and the other one is the noise generated due to the noise in the signal (extrinsic noise). Elowitz, et. al. first pointed out and demonstrated this distinction experimentally[6] in gene expression. Theoretically, Paulsson analyzed the propagation of noise in gene network[23]. If a signal transduction reaction amplifies the small changes in the input signal, the noise in the input signal may also be amplified. Here, we show how the amplification of noise is related to the gain[27].

When the signal S fluctuates with standard deviation σ_S, the fluctuation in the concentration X contains the extrinsic noise component. The standard deviation of the extrinsic noise is designated by σ_{ex}. The relative extrinsic noise intensity $\sigma_{ex}/\bar{X}$, is given by

$$\frac{\sigma_{ex}}{\bar{X}} = g\sqrt{\frac{\tau_s}{\tau+\tau_s}}\frac{\sigma_s}{S}. \tag{13.52}$$

where g is the gain, τ_S is the time constant of the noise in the input signal, and τ is the time constant of the signal transduction reaction. For the derivation, see Ref. [27].

This gain-extrinsic noise relation indicates that the amplification rate of the input noise is at most the gain g. When the time constant of the input noise is large (slow noise), i.e., $\tau_s \gg \tau$, the amplification rate of the input noise approaches the gain g. When the time constant of the reaction is much larger than the noise, i.e., $\tau_s \ll \tau$, the noise in the input signal is averaged out and the amplification rate decreases proportionally to $\sqrt{\tau_s}$ as the time constant τ_s decreases. In this way, the amplification rate of the input noise depends on the time constants τ and τ_s as well as the gain g.

The total noise σ_{tot} is made up of the intrinsic noise σ_{in} and the extrinsic noise σ_{ex}, i.e., $\sigma_{tot}^2 = \sigma_{in}^2 + \sigma_{ex}^2$. Therefore, from Eqs. 13.27 and 13.52, the total noise σ_{tot} is written as

$$\frac{\sigma_{tot}^2}{\bar{X}^2} = g\frac{1}{\Theta\bar{X}} + g^2\frac{\tau_s}{\tau + \tau_s}\frac{\sigma_s^2}{S^2} \tag{13.53}$$

where the first term on the right hand side is the intrinsic noise (Eq. 13.27), and the second term is the extrinsic noise (Eq. 13.52).

Since the intrinsic noise σ_{in} is proportional to the square root of g, and the extrinsic noise σ_{ex} linearly depends on the gain g, the dependence of the total noise σ_{tot} on the gain is both square root and linear of g. When g is small, the total noise dependence on g is square root of g, and when g is large, the total noise dependence is linearly proportional to g. Therefore, depending on the gain g, it is expected that there are two regions; the region in which the intrinsic noise is dominant in the total noise, and the region in which the extrinsic noise dominates the total noise.

For the push–pull reaction, the noises were calculated numerically by applying stochastic input signal. In Fig. 13.5, the total noise intensity is plotted as a function of the gain. When g is small, the total noise σ_{tot}^2 linearly depends on the gain g. As the increase of g, the dependence of the total noise σ_{tot}^2 on the gain g approaches to square dependence on g. In this way, it is shown numerically that two kinds of regime exist; one is the intrinsic noise dominant regime and the other one is the extrinsic noise dominant regime.

Signal transduction systems that amplify signals also amplify the noise in the input signals. Consequently, the total noise is made up of the extrinsic noise as well as the intrinsic noise, as the relation Eq. 13.53 indicates. This relation is generalized in cascade reactions. In a cascade, a signal transduction system regulates another downstream signal transduction. Depending on the gain in each reaction, the output signal of such a cascade is dominated by the intrinsic or extrinsic noise. If the cascade consists of reactions with high gain, the extrinsic noise dominates the fluctuation in the output signal.

The amplification of the noise along a cascade implies the convective instability in dynamical systems with flux, typically found in fluid dynamics, traffic systems, and cascade reaction systems [26]. In the systems with convective instability, small disturbances are amplified as they are advected downstream. In the present case, from Eq. 13.52, the amplification rate of noise λ is given by

$$\lambda = \frac{\sigma_{ex}/\bar{X}}{\sigma_s/S}. \tag{13.54}$$

If the amplification rate λ is larger than unity, the system is convectively unstable. In such a case, although the disturbances in a signal transduction are damped out in time, the disturbances are transmitted with the amplification in the downstream reactions [26]. In such convectively unstable systems, the amplified noise often results in the formation of temporal patterns, such as temporal oscillation [26] and spontaneous switching. It would be quite interesting if the extrinsic noise leads to the formation of such noise-sustained temporal structures that could perform functions, which could not be possible by deterministic means.

13.7 Chemotaxis Is Limited by Noise: An Application to Chemotaxis in Eukaryotic Cells

Here, we apply the gain-fluctuation relation to the problem of accuracy of chemotaxis of eukaryotic cells. The mechanism of gradient sensing in eukaryotic cells, such as *Dictyostelium* cell, has not yet been clarified in the molecular level. In the system level, one possibility is the temporal sensing mechanism, in which the

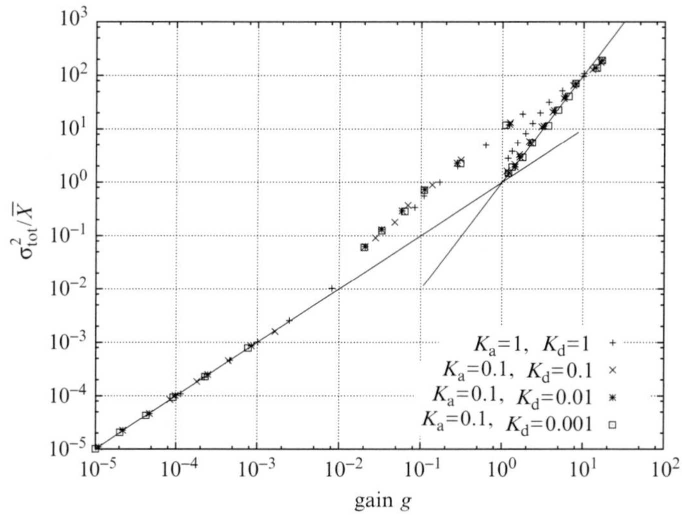

Fig. 13.5 Amplification of noise in signal transduction systems. In order to see the dependence of the total noise intensity σ_{tot} on the gain g, $\sigma_{tot}^2/\bar{X}$ is plotted as a function of the gain g. Changing the average concentration of the input signal, g, σ_{tot}, and $\bar{X}$ were obtained numerically. The numerical calculation was performed using the Gillespie's algorithm [9] as was the case in Fig. 13.4. In the present case, the concentration of the input signal also fluctuates in time, and the average concentration increases under the condition that the relative noise intensity is maintained to be constant. The parameters: $V_a = V_d = 10$, K_a and K_d indicated in the figure. As the increase of gain g, the noise intensity increases with g. For the further increase of g, the noise intensity increases with g^2. The deviation from the linear and square increase of g at the intermediate region is due to the dependence of $\bar{X}$ on the conditions.

system senses a temporal change in the concentration of chemoattractant. For such a system, the motility is essential. When an entire cell or a part of the cell moving up (down) the gradient, the system detects the increase (decrease) in the concentration. By changing the motile behavior depending on the increase or decrease in the concentration sensed, the cells can exhibit chemotaxis. In the case of *Dictyostelium* cell, however, it has been known that the motile activity is not necessary for the gradient information processing. Even when the motile activity is restricted by inhibiting the actin polymerization, some molecules form a gradient inside the cell along the external chemoattractant gradient. For instance, Phosphatidylinositol 3,4,5-trisphosphate (PtdIns(3,4,5)P3) forms a positive gradient, while Phosphatase and tensin homolog (PTEN) forms a negative gradient along the external gradient [21, 22, 32]. This formation of internal gradient indicates that this chemotactic system can detect the spatial difference in the chemoattractant concentration without motile activities through a distribution of the chemoattractant ligand on membrane, i.e., a distribution of the receptor occupancy. Thus, *Dictyostelium* cell adopts a spatial sensing mechanism, which does not necessarily require the temporal sensing mechanism.

Here, we study the signal and noise propagation based on the spatial sensing mechanism, in which the chemotactic signals are the spatial differences in the receptor occupancy of cAR1 and the subsequent activation of G-protein on membrane along cell body [13]. The chemoreceptor is activated upon the binding of chemoattractant ligand, which leads to the production of activated second messenger. The occupied and activated receptor and the activated second messenger are denoted by S and X, respectively (Fig. 13.6a). The respective concentration of occupied receptor and the number of 2nd messenger are given by S and X. For the case of *Dictyostelium* cells, we suppose that X is the activated G-protein on membrane. The external chemoattractant gradient ΔL induces a gradient of receptor occupancy ΔS between the anterior and posterior halves of chemotactic cells, which then lead to the difference in the number of the second messenger X, $\Delta X = X_a - X_p$ where X_a and X_p are the numbers of the two regions (Fig. 13.6b). The difference ΔX formed inside the cell is considered as an internal signal, which induces the motile activity such as actin polymerization. We performed a stochastic numerical simulation of this signaling process for the case of *Dictyostelium* cell to calculate a time series of the difference ΔX, which is plotted in (Fig. 13.6c). The difference sometimes exhibits a negative value, indicating that the spatial signal can be reversed against the external chemoattractant gradient by the stochastic fluctuations in the process of ligand binding and the second messenger activation.

To consider the accuracy of gradient sensing, we first study the stochastic fluctuation in the signal ΔX. For a shallow chemoattractnat gradient, let us consider a small temporal deviation from the average ΔX_s, Δx, i.e., $\Delta X = \Delta X_s + \Delta x$. The linear evolution equation for Δx can be written as

$$\frac{d\Delta x}{dt} = \gamma\Delta s - \Gamma\Delta x + \sigma_{\xi}\Delta\xi(t) \tag{13.55}$$

where $\Delta\xi = \xi_a - \xi_p$ with the noises in the anterior and posterior region respectively. Thus, $\Delta\xi$ is delta-correlated with zero mean and $\overline{\Delta\xi(t)\Delta\xi(t')} = 2\delta(t - t')$. According to the result obtained in the previous sections, the noise intensity $\sigma^2_{\Delta x}$ is approximately given by

$$\sigma^2_{\Delta x} = g\bar{X} + g^2 \frac{\bar{X}^2}{S^2} \frac{\tau_s}{\tau + \tau_s} \sigma^2_{\Delta s} \tag{13.56}$$

where $\bar{X} = \bar{X}_a + \bar{X}_p$ and $S = \frac{1}{2}(S_a + S_p)$. Noting that $\overline{\Delta X} = g\frac{\bar{X}}{S}\overline{\Delta S}$, the relative noise intensity $(\sigma_{\Delta x}/\bar{X})^2$ is given by

$$\left(\frac{\sigma_{\Delta x}}{\overline{\overline{\Delta X}}}\right)^2 = \frac{1}{g\bar{X}} \left(\frac{S}{\overline{\overline{\Delta S}}}\right)^2 + \frac{\tau_s}{\tau + \tau_s} \left(\frac{\sigma_{\Delta s}}{\overline{\overline{\Delta S}}}\right)^2 \tag{13.57}$$

The accuracy of gradient sensing at the level of the second messenger is considered as the signal to noise ratio SNR $\overline{\Delta X}/\sigma_{\Delta X}$, which is obtained by the inverse of the square root of Eq. 13.57.

In Fig. 13.7a, the SNR $\overline{\Delta X}/\sigma_{\Delta X}$ is plotted as a function of the average chemoattractant L. To calculate the SNR, the parameter values for *Dictyostelium* cells has

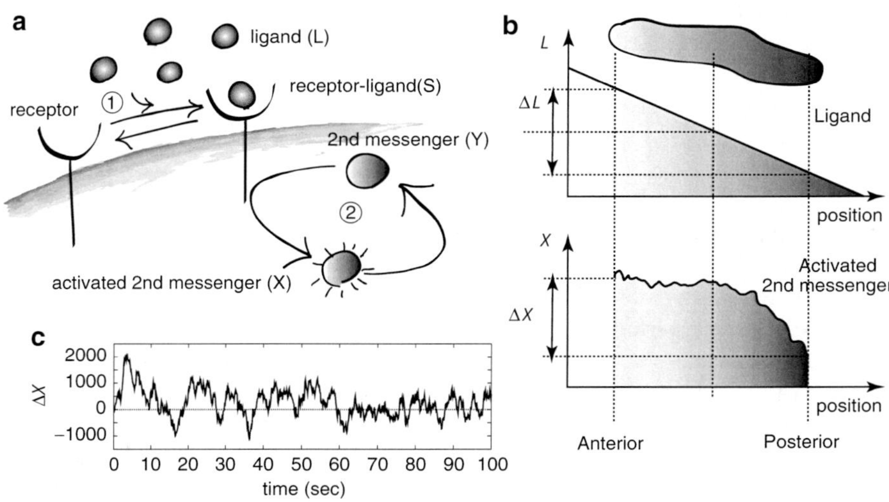

Fig. 13.6 Signal transduction system of eukaryotic chemotaxis. (**a**) Signal transduction reactions by chemoreceptors. The chemoattractant ligand L binds to the receptor forming the receptor–ligand-complex S. The active form of receptor S produces and activates of second messenger X from the inactive precursor Y. (**b**) The cell is on a chemoattractant gradient with the average concentration L and the difference between the anterior and posterior ends, ΔL. The anterior and posterior regions sense $\bar{L} + \overline{\Delta L}/4$ and $\bar{L} - \overline{\Delta L}/4$ on average, respectively. (**c**) The external chemoattractant gradient induces the difference in the activated receptor, ΔS, which then leads to the production of the spatial signal in the second messenger, ΔX, which is plotted as a function of time.

been used (see Ref. [31] in detail). We also performed stochastic numerical simulation showing agreement with our theory (Fig. 13.7a). The SNR of chemotactic signals attains a maximum at the ligand concentration between the affinity of the receptor, K_d, and the EC_{50} concentration where the G-protein activation reaches half-maximum. In Fig. 13.7b, the intrinsic and extrinsic noise contributions to the SNR are plotted. In the lower ligand concentration range, the SNR is determined mainly by the contribution of the extrinsic noise. This indicates that the fluctuations in active receptor dominantly affect the quality of the chemotactic signals. In the higher ligand concentration range, the SNR deteriorates with an increase in ligand concentration, because receptors are gradually saturated, making them unable to produce the large differences in second messenger concentration between the anterior and posterior halves of cells, leading to an increase in intrinsic noise.

The chemotactic accuracy of *Dictyostelium* cells has been measured experimentally by Fisher et al. [7, 28]. The dependence of chemotactic accuracy on ligand concentration exhibits a profile quite similar to our calculated SNR shown in Fig. 13.7 (see [31]). In the experiment, the accuracy of chemotaxis attained a maximum value at 25 nM of cAMP concentration. This optimal value is almost the same as the concentration at which the SNR reaches the maximum (Fig. 13.7a). The agreement between the SNR and chemotactic accuracy indicates that the ability of directional sensing is limited by the inherently generated stochastic noise during the transmembrane signaling of receptors. Note that Eq. 13.57 does not depend on a particular detail of the spatial sensing mechanism, and can be applied to other systems. In fact, similar dependence of chemotactic accuracy has been observed in mammalian leukocytes and neurons [24, 34].

When the chemoattractant concentration L is sufficiently small compared to the receptor's dissociation constant, the SNR changes in proportion to $SNR \propto \Delta L/\sqrt{L}$. If the cell requires a signal exceeding a threshold SNR to detect chemical gradients, exists a threshold gradient $\Delta L_{\mathrm{threshold}}$ for chemotaxis, which can be dependent on L. Suppose that such threshold SNR is independent of ligand concentration L. Then, we obtain the relation $\Delta L_{threshold} \propto \sqrt{L}$, which has been also obtained experimentally [33].

In conclusion, the gain-fluctuation relation can be successfully applied to the problem of accuracy of eukaryotic chemotaxis. The result indicates that stochastic properties of receptors at the most upstream stages of the signaling system determine the chemotactic accuracy of the cells. The noise generated at the receptor level limits the precision of the directional sensing, suggesting that receptor-G protein coupling and its modulation has an important role on chemotaxis efficiency in the cells.

13.8 Propagation of Noise in Linear Cascade Reaction

The output of these signal processing at the most upstream reactions induces the signal processing at the downstream reactions to induces cellular behaviors. Therefore, here we consider a simple linear cascade as shown in Fig. 13.8,

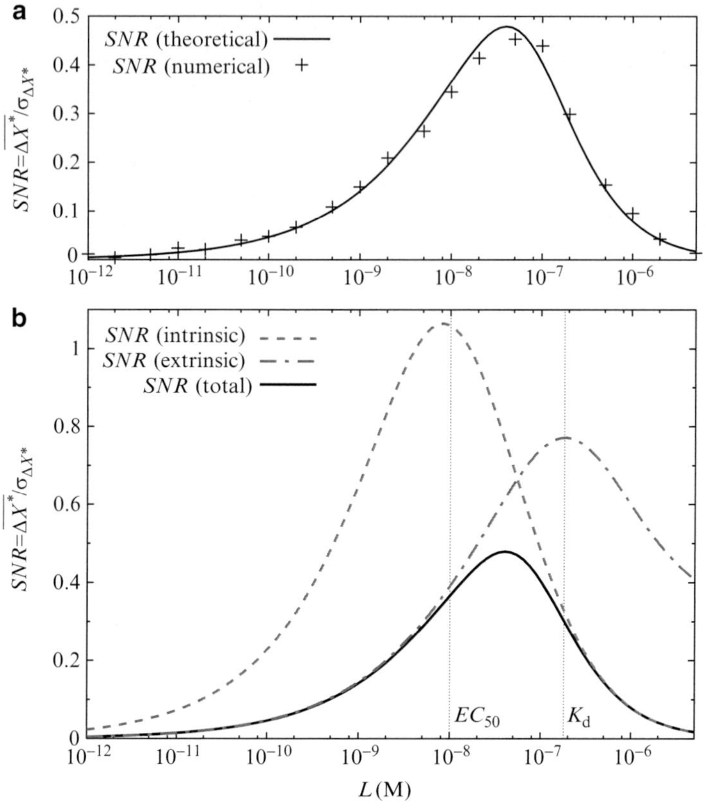

Fig. 13.7 The accuracy of eukaryotic chemotaxis. (**a**) The dependence of signal to noise ration (SNR) on the chemoattractant ligand concentration L obtained by theory and stochastic numerical simulation. (**b**) The contributions of extrinsic and intrinsic noise on the total SNR.

where the activated enzyme X_i of the ith reaction catalyzes the next step reaction ($(i+1)$th reaction) from inactive form Y_{i+1} to active form X_{i+1}. The most upstream reaction contains noise with the strength (standard deviation) σ_S. The downstream reactions are also stochastic processes, which produces intrinsic noises. These noises can affect the behavior of downstream reactions, and thus can be transmitted. For example, in mitogen-activated protein kinase (MAPK) cascade, the reactions of the MAPK kinase kinase activation produces intrinsic noise and the noise may affect the behavior of MAPK kinase. The total noise of each reaction step is sum of the intrinsic noise and the transmitted noise (called extrinsic noise), and the noise strength at the ith reaction step is denoted by σ_i.

For the reactions of the cascade, we consider an ultra-sensitive reaction with high gain. Here, we consider one of the typical signaling reaction, push–pull reaction scheme [11, 29]. The signaling molecule, that is an enzyme, switches its substrate

Fig. 13.8 Schematic diagram of a cascade of signal transduction reaction.

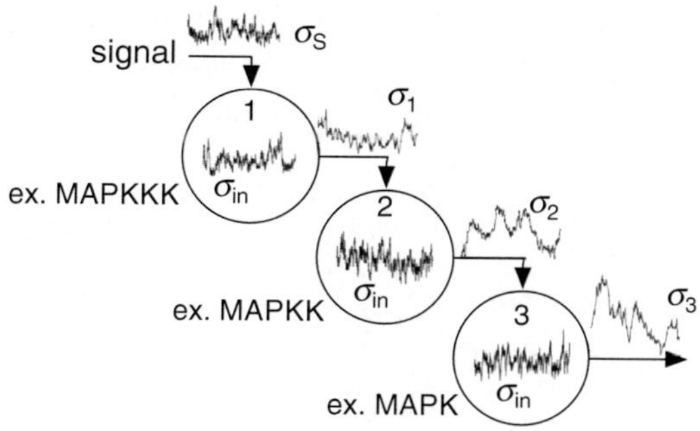

protein from an inactive state to an active state, whereas another enzyme switches the protein off. The reaction is described by a combination of Michaelis–Menten kinetics, given by

$$\begin{cases} \mathrm{Y}_i + \mathrm{X}_{i-1} \leftrightarrow \mathrm{Y}_i\ \mathrm{X}_{i-1} \rightarrow \mathrm{X}_i + \mathrm{X}_{i-1} \\ \mathrm{X}_i + \mathrm{E}_i \leftrightarrow \mathrm{X}_i\ \mathrm{E}_i \rightarrow \mathrm{Y}_i + \mathrm{E}_i \end{cases} \quad (i = 1, 2, \ldots, n) \tag{13.58}$$

where enzyme X_{i-1} switches inactive state Y_i to active state X_i of the ith reaction, and an enzyme E_i switches X_i off. For the first reaction, X_0 is the input signal S. If each Michaelis–Menten reaction works near saturation, sharp response is obtained (Fig. 13.3b).

As in the case of single push–pull reaction given in Eq. 13.28, the number of activated form, X_i, in the ith step can be described by the chemical Langevin equation

$$\frac{dX_i}{dt} = \Gamma_a(Y_i, X_{i-1})Y_i - \Gamma_d(X_i)X_i + \sigma_{\xi_i}\xi_i(t), \quad (i = 1, 2, \ldots, n) \tag{13.59}$$

where $\Gamma_a(Y_i, X_{i-1}) = k_a X_{i-1}\frac{K_a}{K_a+Y_i}$ and $\Gamma_d(X_i) = k_d E_d \frac{K_d}{K_d+X_i}$, are the activation and deactivation rates, respectively, which depend on the numbers X_{i-1}, X_i and Y_i, $\xi_i(t)$ is the white Gaussian noise, which is the source of the intrinsic noise of the ith reaction step, with $\langle \xi_i(t) \rangle = 0$, and $\langle \xi_i(t)\xi_j(t') \rangle = \delta_{i,j}\delta(t - t')$, and X_0 is the input signal intensity S. The noise strength $\sigma_{\xi_i}^2$ is given by $\sigma_{\xi_i}^2 = \Gamma_a(Y_i, X_{i-1})Y_i + \Gamma_d(X_i)X_i$ as in the case of a single push–pull reaction. The temporal evolution of the small deviation x_i from the average value $\overline{X}_i$ can be described by the linearized Langevin equation, obtained by linearizing Eq. 13.59, given by

$$\frac{dx_i}{dt} = \gamma_i x_{i-1} - \Gamma_i x_i + \sigma_{\xi_i}\xi_i(t) \tag{13.60}$$

with the regression coefficient $\Gamma_i = \frac{\partial}{\partial Y_i}\Gamma_a\overline{Y}_i + \frac{\partial}{\partial X_i}\Gamma_d\overline{X}_i$ and the response coefficient $\gamma_i = \frac{\partial}{\partial X_{i-1}}\Gamma_a\overline{Y}_i$ as in the case of single push–pull reaction given by Eqs. 13.32 and 13.33. With the linear noise approximation, the noise intensity $\sigma^2_{\xi_i}$ is given by

$$\sigma^2_{\xi_i} = \frac{2\Gamma_a\Gamma_d N_i}{\Gamma_a + \Gamma_d} \tag{13.61}$$

The parameters k_a, K_a, k_d, K_d and E_d can be also dependent on the step i. Thus, Γ_a and Γ_d are dependent on the step. Here the subscript i is omitted to avoid complications. The input signal intensity S and its small deviation $s(t)$ are given by X_0 and $x_0(t)$, respectively.

The power spectrum density of X_i, $I_i(f)$, is obtained by solving Eq. 13.60 as

$$\frac{I_i(f)}{\overline{X_i}^2} = \frac{\sigma^2_{\xi_i}}{\overline{X_i}^2}\frac{1}{(2\pi f)^2 + \Gamma_i^2} + g_i^2\frac{\Gamma_i^2}{(2\pi f)^2 + \Gamma_i^2}\frac{I_{i-1}(f)}{\overline{X}^2_{i-1}} \tag{13.62}$$

where g_i is the gain characterizes the amplification of signal at the ith step defined as the ratio between the fractional change in the output signal X_i, ΔX_i and the fractional change in the input signal X_{i-1}, ΔX_{i-1},

$$g_i = \frac{|\Delta X_i|/\bar{X}_i}{|\Delta X_{i-1}|/\bar{X}_{i-1}} \tag{13.63}$$

where $\bar{X}$ denotes the average of the concentration X. The power spectrum density $I_i(f)/\overline{X_i}^2$ gives the frequency-dependent total noise intensity of the ith reaction. The first and the second terms in Eq. 13.62 indicate the intrinsic and extrinsic noise, respectively.

Let us focus on the effect of noise of the most upstream signal S on the ith reaction step. Suppose that for simplicity the autocorrelation function of the stochastic modulation $s(t)$ decays exponentially with time constant τ_s. Then, the power spectrum density $I_s(f)$ of the input signal S is given by

$$I_s(f) = \frac{\sigma^2}{(2\pi f)^2 + \tau_s^{-2}} \tag{13.64}$$

The variance of the noise in signal, $\sigma_s{}^2$, is given by $\sigma_s^2 = \int_{-\infty}^{\infty} I_s(f)df = \frac{1}{2}\sigma^2\tau_s$. Then, the contribution of the noise in signal S to the power spectrum density of the nth reaction is given by

$$\begin{aligned}\frac{I_n(f)}{\bar{X}_n^2} &= \prod_{i=1}^{n} g_i^2\frac{\Gamma_i^2}{(2\pi f)^2 + \Gamma_i^2}\frac{I_s(f)}{\bar{S}^2} \\ &= G_n^2\frac{2\sigma_s^2}{\bar{S}^2}\frac{\tau_s^{-1}}{(2\pi f)^2 + \tau_s^{-2}}\prod_{i=1}^{i=n}\frac{\Gamma_i^2}{(2\pi f)^2 + \Gamma_i^2}\end{aligned} \tag{13.65}$$

where G_n is the total gain that quantifies the sensitivity of an entire cascade made up of n reaction steps, which is defined as the ratio between the fractional change in the output signal X_i and the fractional change in the input signal S [5], given by

$$G_n = \frac{|\Delta X_n|/\bar{X}_n}{\Delta S/\bar{S}} = \prod_{i=1}^{i=n} \frac{|\Delta X_i|/\bar{X}_i}{|\Delta X_{i-1}|/\bar{X}_{i-1}} = \prod_{i=1}^{i=n} g_j. \tag{13.66}$$

Here, in Eq. 13.65 the noise contribution of input signal S is only considered. The contribution of input noise σ_s to the noise of the nth reaction, σ_n, can be obtained by frequency integral of Eq. 13.65 as

$$\frac{\sigma_n^2}{\bar{X}_n^2} = \Lambda_n G_n^2 \frac{\sigma_s^2}{\bar{S}^2} \tag{13.67}$$

where Λ_n is given by

$$\Lambda_n = 2\int_{-\infty}^{\infty} \tau_s \prod_{i=0}^{n} \frac{1}{(2\pi f\tau_i)^2 + 1} df = \sum_{i=0}^{n} \frac{\tau_s}{\tau_i} \prod_{j=0, j\neq i}^{n} \frac{\tau_i}{\tau_i + \tau_j} \frac{\tau_i}{\tau_i - \tau_j} \tag{13.68}$$

with $\tau_i = \Gamma_i^{-1}$ and $\tau_0 = \tau_s$. The coefficient Λ_n describes the temporal averaging of noise by the reactions, and is determined only by the intrinsic time constants of reactions. The averaging effect along the cascade is easily seen, when we consider the case where the time constants are approximately the same value, and thus τ_i is given by a single parameter τ. In such a case, Λ_n is obtained as a simpler form by

$$\Lambda_n = \int_{-\infty}^{\infty} \frac{2\tau}{((2\pi f\tau)^2 + 1)^{n+1}} df = \frac{(2n)!}{2^{2n}(n!)^2} \tag{13.69}$$

As the number of steps n increases, Λ_n decreases in proportion to the inverse of square root of n, $\Lambda_n \propto 1/\sqrt{n}$. Hence, as the number of step increases, the effect of averaging gradually increases. From Eq. 13.67, the amplification rates of the input noise is given by

$$\frac{\sigma_n/\overline{X_n}}{\sigma_s/\overline{S}} = G_n\sqrt{\Lambda_n} \tag{13.70}$$

Therefore, the noise amplification rate is linearly proportional to the total gain G_n. Since G_n denotes the signal amplification rate of the cascade, a cascade that amplifies a signal also increases the noise amplitude.

When the extrinsic noise propagated from the most upstream reaction dominates the downstream reaction and another noise contributions including intrinsic noise are minor, it leads to the proportional relationship between the amplification rate and the total gain G_n, $\sigma_n/\bar{X}_n \propto G_n$. In this extrinsic noise dominant case, according to Eq. 13.70, the noise strength σ_n is proportional to the average concentration $\bar{X}_n$, if the total gain G_n is not so changed. These are the characteristic properties of extrinsic noise in a cascade reaction.

The intrinsic noise dominates the total noise, when the gain is small. In such an intrinsic noise dominant case, the variance σ_n^2 is proportional to the average value $\overline{X_n}$ [27]. Since the intrinsic noise is not affected by the upstream reactions, the intrinsic noise strength is not dependent on the total gain G_n. These are the characteristic properties of intrinsic noise in a cascade.

In the present cascade, each reaction works as a low-pass filter for the fluctuation of upstream components, in which higher frequency fluctuations in an input noise are damped out. Such property is clearly seen in the extrinsic noise term (second term) on the right hand side of Eq. 13.62. As the frequency f increases, the coefficient for the input noise decreases as $\Gamma_i^2/\left((2\pi f)^2 + \Gamma_i^2\right)$. When the frequency f is much higher than $\Gamma_i/2\pi$, i.e., $2\pi f \gg \Gamma_i$, the extrinsic noise term can be approximately neglected. When the frequency f is sufficiently lower than $\Gamma_i/2\pi$, i.e., $2\pi f \ll \Gamma_i$, the input noise intensity increases with the amplification rate g_i. Thus, in the higher frequency region, the intrinsic noise dominates, whereas in the lower frequency region, the extrinsic noise dominates if the total gain is sufficiently large.

Since the higher frequency component of extrinsic noise is damped out along the cascade, while the intensity of lower frequency component increases, the fluctuation at the downstream reaction becomes effectively slower along the cascade. In order to see this increase in the characteristic time scale more quantitatively, consider the noise transmitted from the top of cascade to the ith reaction. For the sake of simplicity, the time constant of each reaction is almost constant to be τ. The characteristic time of the stochastic fluctuation is given by the inverse of the frequency f^* with which the magnitude of the power spectrum density is half of the maximum value, i.e., $I(f^*) = I(0)/2$. Substituting this expression into Eq. 13.65 with $\tau_i = \tau$, the characteristic time T_i is given by

$$T_i = \frac{1}{\sqrt{2^{\frac{1}{i+1}} - 1}}\tau \tag{13.71}$$

From this expression, when $i \gg 1$, T_i is approximately given by $\sqrt{i}$. When the reactions are low-pass filter type, this result is generally valid for the cascade of reactions that exhibit a gain larger than unity.[1]

[1] When feed-back reaction is indispensable, reaction can have band-passed filter type property. The extrinsic noise is not slowing down along the cascade.

When the reactions in the cascade do not exhibit a gain larger than unity, the extrinsic noise contribution is minor in each reaction. Therefore, the characteristic time of the fluctuation is determined by its own intrinsic noise.

13.9 Concluding Remarks

As seen in the previous sections, *Dictyostelium* cells exhibit chemotaxis in a shallow chemoattractant gradient. The gain-fluctuation relation indicates that the high gain reactions may not necessarily be appropriate for the most upstream signaling reaction, because the small signal can be deteriorated by the large noise. The accuracy of chemotaxis has been shown to be determined by the noise at the most upstream reactions of the receptor and the subsequent signaling molecule directly activated by the receptor. However, the signal transduction network responsible for chemotaxis is not restricted to these reactions, but huge number of reactions works to produce chemotaxis. Based on the analysis of a signal transduction cascade, one possibility may be that the downstream reactions are devoted to amplification of the upstream signals. Under the absence of chemoattractant gradient, *Dictyostelium* cell also exhibits motility in random direction. How the processing of small signal and production of large fluctuations to induce the random motility are compatible remains as a future problem.

Acknowledgements I thank M. Ueda for providing the single molecule imaging data and discussions, and M. Nishikawa for discussions.

References

1. Alon U, Camarena L, Surette MG, Aguera y Arcas B, Liu Y, Leibler S, Stock JB (1998) Response regulator output in bacterial chemotaxis. Embo J 17(15):4238–4248
2. Berg HC, Brown DA (1972) Chemotaxis in escherichia coli analysed by three-dimensional tracking. Nature 239(5374):500–504
3. Berg OG, Paulsson J, Ehrenberg M (2000) Fluctuations and quality of control in biological cells: zero-order ultrasensitivity reinvestigated. Biophys J 79(3):1228–1236
4. Changeux JP, Edelstein SJ (1998) Allosteric receptors after 30 years. Neuron 21(5):959–980
5. Detwiler PB, Ramanathan S, Sengupta A, Shraiman BI (2000) Engineering aspects of enzymatic signal transduction: photoreceptors in the retina. Biophys J 79(6):2801–2817
6. Elowitz MB, Levine AJ, Siggia ED, Swain PS (2002) Stochastic gene expression in a single cell. Science 297(5584):1183–1186
7. Fisher PR, Merkl R, Gerisch G (1989) Quantitative analysis of cell motility and chemotaxis in dictyostelium discoideum by using an image processing system and a novel chemotaxis chamber providing stationary chemical gradients. J Cell Biol 108(3):973–984
8. Funatsu T, Harada Y, Tokunaga M, Saito K, Yanagida T (1995) Imaging of single fluorescent molecules and individual atp turnovers by single myosin molecules in aqueous solution. Nature 374(6522):555–559

9. Gillespie DT (1977) Exact stochastic simulation of coupled chemical-reactions. J Phys Chem 81(25):2340–2361
10. Gillespie DT (2000) The chemical langevin equation. J Chem Phys 113(1):297–306
11. Goldbeter Jr A, Koshland DE (1981) An amplified sensitivity arising from covalent modification in biological systems. Proc Natl Acad Sci USA 78(11):6840–6844
12. Huang Jr CY, Ferrell JE (1996) Ultrasensitivity in the mitogen-activated protein kinase cascade. Proc Natl Acad Sci USA 93(19):10078–10083
13. Janetopoulos C, Jin T, Devreotes P (2001) Receptor-mediated activation of heterotrimeric g-proteins in living cells. Science 291(5512):2408–2411
14. van Kampen NG (1992) Stochastic processes in physics and chemistry, revised and enlarged edition. North-Holland, Amsterdam
15. Korobkova E, Emonet T, Vilar JM, Shimizu TS, Cluzel P (2004) From molecular noise to behavioural variability in a single bacterium. Nature 428(6982):574–578
16. Koshland Jr DE (1998) The era of pathway quantification. Science 280(5365):852–853
17. Koshland Jr DE, Nemethy G, Filmer D (1966) Comparison of experimental binding data and theoretical models in proteins containing subunits. Biochemistry 5(1):365–385
18. Monod J, Wyman J, Changeux JP (1965) On the nature of allosteric transitions: a plausible model. J Mol Biol 12:88–118
19. Oosawa F (1975) The effect of field fluctuation on a macromolecular system. J Theor Biol 52 (1):175–186
20. Oosawa F (2001) Spontaneous signal generation in living cells. Bull Math Biol 63(4):643–654
21. Parent CA (2004) Making all the right moves: chemotaxis in neutrophils and dictyostelium. Curr Opin Cell Biol 16(1):4–13
22. Parent CA, Devreotes PN (1999) A cell's sense of direction. Science 284(5415):765–770
23. Paulsson J (2004) Summing up the noise in gene networks. Nature 427(6973):415–418
24. Rosoff WJ, Urbach JS, Esrick MA, McAllister RG, Richards LJ, Goodhill GJ (2004) A new chemotaxis assay shows the extreme sensitivity of axons to molecular gradients. Nat Neurosci 7(6):678–682
25. Sako Y, Minoguchi S, Yanagida T (2000) Single-molecule imaging of egfr signalling on the surface of living cells. Nat Cell Biol 2(3):168–172
26. Shibata T (2004) Amplification of noise in a cascade chemical reaction. Phys Rev E Stat Nonlin Soft Matter Phys 69(5 Pt 2):056218
27. Shibata T, Fujimoto K (2005) Noisy signal amplification in ultrasensitive signal transduction. Proc Natl Acad Sci USA 102(2):331–336
28. Song L, Nadkarni SM, Bodeker HU, Beta C, Bae A, Franck C, Rappel WJ, Loomis WF, Bodenschatz E (2006) Dictyostelium discoideum chemotaxis: threshold for directed motion. Eur J Cell Biol 85(9-10):981–989
29. Stadtman ER, Chock PB (1977) Superiority of interconvertible enzyme cascades in metabolic regulation: analysis of monocyclic systems. Proc Natl Acad Sci USA 74(7):2761–2765
30. Ueda M, Sako Y, Tanaka T, Devreotes P, Yanagida T (2001) Single-molecule analysis of chemotactic signaling in dictyostelium cells. Science 294(5543):864–867
31. Ueda M, Shibata T (2007) Stochastic signal processing and transduction in chemotactic response of eukaryotic cells. Biophys J 93(1):11–20
32. Van Haastert PJ, Devreotes PN (2004) Chemotaxis: signalling the way forward. Nat Rev Mol Cell Biol 5(8):626–634
33. Van Haastert PJ, Postma M (2007) Biased random walk by stochastic fluctuations of chemoattractant-receptor interactions at the lower limit of detection. Biophys J 93:17871796
34. Zigmond SH (1977) Ability of polymorphonuclear leukocytes to orient in gradients of chemotactic factors. J Cell Biol 75(2 Pt 1):606–616

Index

FSC
www.fsc.org
MIX
Papier | Fördert gute Waldnutzung
FSC® C083411

Kolibri 360 GmbH
Ferdinand-Jühlke-Straße 7
99095 Erfurt, Deutschland
produktsicherheit@kolibri360.de